T0205933

Biotechnology to Enhance Sugarcane Productivity and Stress Tolerance

Biotechnology to Enhance Sugarcane Productivity and Stress Tolerance

Edited by
Kalpana Sengar

CRC Press
Taylor & Francis Group
Boca Raton London New York

CRC Press is an imprint of the
Taylor & Francis Group, an **informa** business

CRC Press
Taylor & Francis Group
6000 Broken Sound Parkway NW, Suite 300
Boca Raton, FL 33487-2742

First issued in paperback 2021

© 2018 by Taylor & Francis Group, LLC
CRC Press is an imprint of Taylor & Francis Group, an Informa business

No claim to original U.S. Government works

ISBN-13: 978-1-03-209582-0 (pbk)
ISBN-13: 978-1-4987-5465-1 (hbk)

Contents

About the book

Sugarcane is the most important cash crop of the world. Consequent to change in the soil temperature, soil moisture and the composition of gases in the root zone, there are likely changes with respect to root growth, composition of root exudates, soil processes, nutrient dynamics, decomposition and so forth. The increase in CO_2 content will have a beneficial effect on the C_3 crop, and the dicot weeds may compete with the sugarcane crop. There could be a higher incidence of pests and disease under the altered temperature regime. If the ambient temperature remains within the favourable rang for pests, insect species will complete more generations, thereby leading to a larger population than normal. Climate change is likely to affect the pathogen, host or the host–pathogen interaction. Abiotic stresses, especially drought and salinity, are the most limiting abiotic factors for crop production, restricting useful areas for agriculture and yield under adverse conditions. Previous studies have shown that both stress types are intimately related, recruiting similar genes, especially in the initial steps after stress perception. The development of tolerant plants is therefore fundamental for the maintenance and increase of crop yields.

Sugarcane is the most important plant source for sugar and alcohol production, and it is cultivated in more than 80 countries (tropical and subtropical areas), being presently the most promising renewable biofuel source. However, environmental factors negatively influence its yield and jeopardize the prospect to meet the increasing demand for bioethanol.

The identification of genes that lead to stress tolerance is urgent and necessary. Sugarcane agricultural biotechnology encompasses a growing list of techniques that range from simple probes to determine whether an individual plant is carrying a specific gene to measurement of the activity of all an organism's genes (its genome) simultaneously. It is possible to select sugarcane genotypes under moderate water deficit conditions with higher productivity associated with higher stalk numbers, stalk height, and stalk weight. Therefore, these traits could be considered as useful tools during crop breeding procedures in order to make this process more rapid and less expensive. Development of stress-tolerant sugarcane germplasm is thus crucial for sustaining production in areas where water supply is limited. One technique common in agricultural/ biotechnology is the integration of a gene from one species into the genome of another. This could not generally be accomplished alone with standard breeding techniques, thus that knowledge has led to the development of some practical tools. Marker-assisted breeding uses conventional breeding techniques informed by specific genetic sequences, or 'markers', that segregate according to particular traits. Markers speed up breeding programs by allowing researchers to determine, early in the life of a progeny, whether the traits they hoped to combine from two organisms are present simply by checking for the presence of the markers. Tissue culture is used in clonal propagation of plants for which sexual breeding has proved inefficient. It has been important for reproducing crops used across the African continent, including oil palm, plantain, banana, date, eggplant, pineapple, rubber tree, cassava, maize, sweet potato, yam, and tomato. Cloning and *in vitro* fertilization allow the manipulation of germ cells for animal-breeding programs, genetic-resource conservation, and germplasm enhancement. Gene profiling or association mapping tracks the patterns of heritability of variations (alleles) of many genes. The quantitative trait loci (QTLs) collectively contribute to complex plant traits, such as drought tolerance and robust seed production, and an understanding of the groupings of QTLs provides insights into how genes work in concert to produce a particular characteristic. Metabolomics provides a snapshot of all the metabolites being produced in a plant cell at any given time under different environmental condition. This book provides an overview of some available techniques in agriculture and biotechnologies to improve crop productivity. The traits correlated with stress tolerance are suitable indicators for selection of stress-tolerant genotypes in breeding programs to reduce the impact of environmental stress on crop yields. For sugarcane, researchers have concentrated on agronomic and physiological traits, which could lead to adaptation to these conditions or be correlated to stress tolerance and could be used for the development of new varieties.

To optimize sugarcane improvement, it is necessary to know the impact a selected trait will have on the general physiology of the plant. However, this is not yet possible, as there are too many gaps in our knowledge of the unique development and physiology of sugarcane. Such gaps impair our ability to enhance desired agronomical traits, for example, selection for sugarcane varieties with increased photosynthetic capacity may be useless if sugar accumulation is constrained by temperature, water deficit or nutrient availability. It may prove difficult to consistently increase sucrose levels in the culm without first knowing the factors that affect sugarcane yield and carbon partitioning.

The chapters in this book present an overview of the differentially expressed genes in sugarcane under different biotic and abiotic stress, and the use of new biotechniques to develop tolerance in sugarcane against all stresses.

Preface

This book reports on the results of experiments to assess the effects of global climate change on sugarcane crop productivity. It covers issues such as temperature, salinity and stress on crop growth. Agriculture is a complex sector involving different driving parameters (environmental, economic and social). It is now well recognized that crop production is very sensitive to climate change with different effects according to region. The book underlines such concerns about the current status of our environment and agriculture.

This text analyses the global consequences to sugarcane crop yields, production and risk of hunger of linked socioeconomic and climate scenarios. Potential impacts of climate change are estimated to evaluate consequent changes in global sugarcane and cereal production, sugar and cereal prices and the number of people at risk from hunger. The crop yield results elucidate the complex regional patterns of projected climate variables, CO_2 effects and agricultural systems that contribute to the aggregation of global sugarcane crop production.

This book contains 12 chapters that discuss the impact of changing climate on agriculture and sugarcane production; environmental stress physiology; adaptation mechanisms; climate change data of recent years; and the impact of global warming and climate change on different crops like sugarcane, wheat, rice and medicinal plants. This book also attempts to underline various strategies for reducing agriculture's vulnerability to climate change and for adaptation to ongoing climate change.

This book will be useful for agriculturists, environmentalists, climate change specialists, policy makers and research scholars engaged in research on climate agriculture-related issues.

Editor

Kalpana Sengar, PhD, is a recipient of the Women Scientist Fellowship awarded by the Department of Science and Technology, Government of India. She has 10 years of research experience to her credit in biotechnology, environment and agriculture. She holds an MSc in biotechnology from Chaudhary Charan Singh (CCS) University and a PhD from Mahatma Jyotiba Phule (MJP) Rohilkhand University. Sengar is currently working as a scientist at the Sardar Vallabhbhai Patel University of Agriculture and Technology. She has published more than 25 papers in reputed national and international journals, and presented papers at several national and international conferences. In 2009, she received the Best Report Diploma Award from Bioinfobank Library Poland; a VIB PhD Scholarship to attend its International Symposium in Belgium 2010; and a scholarship to attend the 17th Annual Conference of the International Sustainable Development Research Society (ISDRS) hosted by the Earth Institute, Columbia University, in 2011. Sengar's research interests include agriculture, human nutrition, biotechnology and plant tissue culture.

Contributors

Anuradha Chelliah
National Research Centre
on Banana
Tirchy, India

Rachayya M. Devarumath
Molecular Biology and Genetic
Engineering Section
Vasantdada Sugar Institute
Pune, India

Pooja Dhansu
ICAR–Sugarcane Breeding Institute
Karnal, India

Paulo Cavalcanti Gomes Ferreira
Laboratório de Biologia Molecular
de Plantas
Instituto de Bioquímica Médica
Leopoldo de Meis
Universidade Federal do Rio de
Janeiro
Rio de Janeiro, Brazil

Sanjay Kumar Garg
Tissue Culture Lab
College of Biotechnology
Sardar Vallabhbhai Patel
University of Agriculture and
Technology
Meerut, India

Clícia Grativol
Laboratório de Química e Função
de Proteínas e Peptídeos
Centro de Biociências e
Biotecnologia
Universidade Estadual do Norte
Fluminense Darcy Ribeiro
Campos dos Goytacazes, Brazil

Adriana Silva Hemerly
Laboratório de Biologia Molecular
de Plantas
Instituto de Bioquímica Médica
Leopoldo de Meis
Universidade Federal do Rio de
Janeiro
Rio de Janeiro, Brazil

Tayyab Husnain
Center of Excellence in Molecular
Biology
University of the Punjab
Lahore, Pakistan

Lakshmi Kasirajan
Division of Crop Improvement
Sugarcane Breeding Institute
Coimbatore, India

Boomiraj Kovilpillai
Department of Environmental
    Sciences
Tamil Nadu Agricultural
    University
Coimbatore, India

Neeraj Kulshreshtha
ICAR–Sugarcane Breeding Institute
Karnal, India

Ashwani Kumar
ICAR–Central Soil Salinity
    Research Institute
Karnal, India

Rajinder Kumar
Punjab Agricultural University
Regional Research Station
Kapurthala, India

Ravinder Kumar
ICAR–Sugarcane Breeding Institute
Karnal, India

Anita Mann
ICAR–Central Soil Salinity
    Research Institute
Karnal, India

B.L. Meena
ICAR–Central Soil Salinity
    Research Institute
Karnal, India

Idrees Ahmad Nasir
Center of Excellence in Molecular
    Biology
University of the Punjab
Lahore, Pakistan

Gauri A. Nerkar
Molecular Biology and Genetic
    Engineering Section
Vasantdada Sugar Institute
Pune, India

B. Parameswari
ICAR–Sugarcane Breeding Institute
Karnal, India

Madhavi V. Purankar
Molecular Biology and Genetic
    Engineering Section
Vasantdada Sugar Institute
Pune, India

Gulzar S. Sanghera
Punjab Agricultural University
Regional Research Station
Kapurthala, India

Kalpana Sengar
Tissue Culture Lab
College of Biotechnology
Sardar Vallabhbhai Patel
    University of Agriculture and
    Technology
Meerut, India

R.S. Sengar
Tissue Culture Lab
College of Biotechnology
Sardar Vallabhbhai Patel University
    of Agriculture and Technology
Meerut, India

Vaishali Shami
Department of Biotechnology
Sardar Vallabhbhai Patel
    University of Agriculture and
    Technology
Meerut, India

Contributors

Manoj Kumar Sharma
Tissue Culture Lab
College of Biotechnology
Sardar Vallabhbhai Patel
　University of Agriculture and
　Technology
Meerut, India

Suman Sheelavantmath
Department of Biotechnology
Sinhgad College of Science
　Ambegaon (Bk)
Pune, India

Parvender Sheoran
ICAR–Central Soil Salinity
　Research Institute
Karnal, India

Ashu Singh
Tissue Culture Lab
College of Biotechnology
Sardar Vallabhbhai Patel
　University of Agriculture and
　Technology
Meerut, India

Naresh Pratap Singh
Department of Biotechnology
Sardar Vallabhbhai Patel
　University of Agriculture and
　Technology
Meerut, India

Yogendra Singh
Department of Plant Breeding and
　Genetics
Jawaharlal Nehru Krishi Vishwa
　Vidyalaya
Jabalpur, India

Bushra Tabassum
Center of Excellence in Molecular
　Biology
University of the Punjab
Lahore, Pakistan

Flávia Thiebaut
Laboratório de Biologia Molecular
　de Plantas
Instituto de Bioquímica Médica
　Leopoldo de Meis
Universidade Federal do Rio de
　Janeiro
Rio de Janeiro, Brazil

chapter one

Biotechnological approach

A new dimension for sugarcane improvement

Gauri A. Nerkar, Madhavi V. Purankar,
Suman Sheelavantmath and Rachayya M. Devarumath

Contents

Introduction

Sugarcane is the highest yielding crop worldwide and accounts for ~80% of the sugar (sucrose) production in the world (Nayak et al., 2014; Zhou et al., 2016). The genus falls in the tribe Andropogoneae in the grass family Poaceae. The tribe includes other tropical grasses such as *Sorghum* and *Zea* (maize). Very closely related to *Saccharum* are another four genera (*Erianthus* section Ripidium, *Miscanthus* section Diandra, *Narenga* and *Sclerostachya*) that readily interbreed, forming what is now commonly referred to as the *Saccharum* complex (Daniels and Roach, 1987). The polyploid-aneuploid (unbalanced number of chromosomes) nature of these genera creates a challenge for the taxonomist (Daniels and Roach, 1987; Sreenivasan et al., 1987). The genus *Saccharum* comprises six species: *S. barberi, S. edule, S. officinarum, S. robustum, S. sinense* and *S. spontaneum* (Daniels and Roach, 1987). Of these, *S. officinarum* (the domesticated sugar-producing species) and *S. spontaneum* (a vigorous wild species with many aneuploidy forms) are thought to be the ancestors of cultivated sugarcane. *S. officinarum* originally derived from *S. robustum*, while *S. barberi* and *S. sinense* are thought to have been derived by crossing *S. officinarum* and *S. spontaneum* (Asano et al., 2004; Sandhu et al., 2012). However, Irvine (1999) suggested that there are only two true species – *S. officinarum* and *S. spontaneum* – and therefore, current sugarcane commercial cultivars are thought to be hybrids with 80% to 90% of the genome from *S. officinarum* and 10% to 20% of the genome from *S. spontaneum* (Grivet et al., 1996; Hoarau et al., 2002). The chromosome number of these species ranges from 40 to 200 (Asano et al., 2004). Genetic improvement of sugarcane through conventional breeding programmes is difficult because of the polyploid and aneuploid nature of the genus *Saccharum*.

As a relatively recently domesticated species, sugarcane exhibits little of the available genetic diversity having been incorporated or actively analysed for introgression into domesticated varieties (Dillon et al., 2007; Sreenivasan et al., 1987), and breeding programs in the early 1900s focused on hybridization of *S. officinarum* clones, but quickly progressed to interspecific crosses incorporating *S. spontaneum*. This resulted in improved agronomic traits, such as tilling, stand and trashiness abilities, ratooning and disease resistance, but required a backcrossing program to *S. officinarum*, called 'nobilization,' to elevate the sucrose content (Dillon et al., 2007; Edmé et al., 2005). Since then, the majority of breeding programs have focused on intercrossing between the hybrids, though in recent decades the larger increases in genetic gains have been made by incorporating more diverse germplasm into the cultivated backgrounds (Edmé et al., 2005; Dillon et al., 2007) not only to increase sucrose production, but also to diversify into other alternative products to regain profitability.

Sugarcane is one of the most productive cultivated plants due to its C_4 carbohydrate metabolism and a perennial life cycle. The production and productivity of sugarcane is adversely affected due to various biotic and abiotic stresses. Damages caused by these stresses are responsible for enormous economic losses in sugarcane production. Traditional breeding technologies and proper management strategies continue to play a vital role in sugarcane improvement. The conventional breeding programmes are being employed to integrate favourable genes of interest from intercrossing genera and species into the crops to induce stress tolerance. However, the complex genome, narrow genetic base, poor fertility, and susceptibility to biotic and abiotic stresses of sugarcane also limit the traditional breeding (Suprasanna et al., 2011). Therefore, we need to deploy the biotechnological tools for addressing the critical problems of sugarcane improvement for sustainable agriculture. In this chapter, we review the role of biotechnology in increasing sugarcane productivity and tolerance to stress.

Sugarcane productivity and how it is affected by stress

Sugarcane productivity

Sugarcane grows under a wide array of agro-climatic conditions extending from the tropics to subtropics. It is grown between 30N and 30S and is cultivated in 20.42 million ha producing 1,333.2 million tonnes with an average cane productivity of 65.20 tonnes/ha (Shrivastava et al., 2015). The cane productivity in the tropical regions is comparatively higher due to even sunshine all through the year, well distributed rainfall and ideal conditions for the luxurious growth, whereas in the subtropics, the crop experiences pronounced winter, which affects sprouting and growth, erratic rainfall and drought and high temperatures during late season (Solomon, 2014). In India, the average yield of cane productivity hovers between 66 and 70 tons/ha/year (Solomon, 2014).

Stress and its impact on sugarcane productivity

Stress is a condition that hampers the expression of the full genetic potential of a living being. As sessile organisms, plants face many stressful conditions that require the evolutionary establishment of diverse developmental and physiological strategies to cope with or avoid the stress condition. However, these strategies usually require metabolic energy that should rather be directed to useful production. Biotic and abiotic stresses affect sugarcane productivity (Nair, 2010).

Impact of abiotic stress on sugarcane productivity

Sugarcane productivity is significantly affected by abiotic stresses such as salinity, alkalinity, drought and waterlogging. Drought is a major cause of decreased productivity because of the morpho-physiological effects that water deficiency causes to sugarcane (e.g., reduced photosynthesis and growth inhibition) (Andrade et al., 2015). Sugarcane, being a typical glycophyte, exhibits stunted growth or no growth under salinity, with its yield falling to 50% or even more as compared to its true potential (Akhtar et al., 2003; Wiedenfeld, 2008). Apart from this, in the flood-prone areas under sugarcane cultivation, waterlogging affects all stages of crop growth and can reduce germination, root establishment, tillering and growth resulting in yield reduction (Solomon, 2014).

Elucidating tolerance mechanisms would enable the development of cultivars more tolerant to drought and salinity, allowing cultivation in marginal areas, while ensuring the sustainability and viability of the industry in such drought-prone areas (Scortecci et al., 2012). Plant irrigation is a good option for agriculture, but it also increases salinity in the soil. Besides, it corresponds to 65% of global water demand, and considering the expansion of cultivation to areas without fresh water, tolerance to drought will become increasingly important. Drought tolerance will also contribute to reduce irrigation and water use (Rocha et al., 2007). Even though sugarcane can survive long dry periods, it demands a fair amount of water for optimal yield, leading to the use of irrigation in many areas.

Impact of biotic stress on sugarcane productivity

Genetic resistance to pests and diseases is an indispensable and essential condition in plant breeding. Pests and pathogens often conquer new territories and are well known to dynamically evolve towards breaking resistances, always posing new challenges. Indeed, biotic stresses are of special concern in sugarcane breeding programs, because they may cause great economical impact in plantations with susceptible cultivars.

Sugarcane diseases and pests are constraints to crop production throughout India, and losses due to disease are estimated to be about 10% to 15%. Among them red rot, smut, wilt and sett rot are the important fungal diseases. Bacterial diseases like leaf scald disease (LSD) and ratoon stunting disease (RSD) are found to cause considerable yield loss in certain regions. Among the viral diseases, mosaic is prevalent in all the states; however, its severity is felt in specific situations. Besides these, grassy shoot caused by phytoplasma is also a potential disease, which can cause considerable damage to sugarcane production. Emergence and spread of newly recorded yellow leaf disease (YLD) has become a major constraint in many locations. Borer pests, pyrilla, scale insect, white fly, termite and mealy bugs are present almost throughout the country.

Development of stress-tolerant plants through in vitro selection and mutation breeding

A tissue culture technique has emerged as a feasible and cost-effective alternative tool for developing stress-tolerant plants in recent years (Rai et al., 2011). This technique can operate under controlled conditions with limited space and time (Sakhanokho and Kelley, 2009), and has the potential for the selection of stress-tolerant variants using a low-cost laboratory setup. *In vitro* selection through enhanced expression of pathogenesis-related (PR) proteins, antifungal peptides or biosynthesis of phytoalexins is an important tool for desirable plant selection (Ganesan and Jayabalan, 2006; Kumar et al., 2008, as reviewed by Rai et al., 2011). This technology is easy and cost effective compared to the transgenic approach for the improved disease tolerance (Jayashankar et al., 2000). *In vitro* selection for resistance to a pathogen can be carried out using organogenic or embryogenic calli, shoots, somatic embryos or cell suspensions by exposing them to toxins produced by the pathogen, pathogen culture filtrate or to the pathogen itself (Kumar et al., 2008). *In vitro* selection in sugarcane has also been used for selection of eyespot resistant lines (Heinz et al., 1977; Larkin and Scowcroft, 1983; Prasad and Naik, 2000), red rot resistance (Mohanraj et al., 2003), salt stress (Gandonou et al., 2004) and drought stress (Errabii et al., 2006). Sengar et al. (2009) studied the phytotoxic effects of pathogen culture filtrate (PCF) on 5- to 8-month-old sugarcane calli and association between selection at callus and plant levels were studied for red rot resistance. Selection and development of a glyphosate-tolerant cellular line has been achieved by using cell suspensions derived from *Saccharum* spp. varieties (Zambrano et al., 2003).

Although studies of salt selection are available for diverse plant species, limited research has been conducted in sugarcane. Sugarcane embryogenic callus has been shown to be sensitive to sodium chloride (NaCl) (Gandanou et al., 2006) and gamma radiation (Patade et al., 2008). Saif-Ur-Rasheed et al. (2001) reported the isolation of salt-tolerant mutants from irradiated sugarcane callus. Although these studies have demonstrated the application of mutagenesis and *in vitro* techniques to study radio sensitivity or isolation of mutants in sugarcane, little information on characterization of salt-tolerant callus and progeny is available. Studies of the application of ionizing radiation for developing novel mutant germplasm in sugarcane will accordingly be beneficial for sugarcane improvement. Nikam et al. (2014a, 2014b) applied gamma ray-induced mutagenesis to isolate sugarcane mutants with improved tolerance to salinity, followed by morphological and agronomical characterization of selected mutants.

Patade et al. (2006) characterized the molecular diversity using the RAPD (randomly amplified polymorphic DNA) technique, among the putative salt and drought tolerant regenerants of sugarcane cv. CoC 671, induced by gamma ray mutagenesis *in vitro*, suggesting that the application of molecular marker technique will prove helpful to establish an efficient system for *in vitro* selection of the abiotic stress-tolerant clones through *in vitro* mutagenesis.

Marker-assisted selection for stress-tolerant varieties of sugarcane

Molecular markers such as RFLP (restriction fragment length polymorphism), RAPD (randomly amplified polymorphic DNA), ISSR (inter-simple sequence repeats) and SSR (simple sequence repeats) have shown excellent potentiality to assist selection of quantitative trait loci (QTLs) associated with these markers (Stuber, 1992). RAPD markers were detected for salt and drought stresses in sugarcane (Abdel-Tawab et al., 2003a, 2003b; Piperidies et al., 2004). Costa et al. (2011) determined the genetic diversity between sugarcane varieties with contrasting features under drought. They observed that drought-tolerant and susceptible varieties were clearly separated. This evaluation revealed important contrasting parental candidates regarding their drought response, very promising for future mapping approaches aiming for the identification of QTLs) associated with drought in sugarcane.

Attempts have been made to identify salt-tolerant sugarcane genotypes from available genetic resources using a multitude of biochemical, physiological and morphological indices (Chaum et al., 2012; Saxena et al., 2010). However, most of the genotypes identified as salt tolerant were based on field trials, which are affected by many environmental factors (Chaum et al., 2012). Virupakshi and Naik (2008) have used ISSR markers to characterize chloroplast and mitochondrial DNA for red rot disease in sugarcane.

SCAR (sequence-characterized amplified region) markers linked to the smut-resistance gene have been identified and utilized in the breeding program (Que et al., 2008; Srivastava et al., 2012). However, until now, the applicability and authenticity of SCAR markers have been found to be limited in the case of biotic stresses or for the identification of a particular trait corresponding to a single gene.

A study using AFLP (amplified fragment length polymorphism) markers by Selvi et al. (2006) revealed that Indian varieties grown under the subtropical belt facing extreme climatic conditions retained more of *Saccharum spontaneum* alleles than the tropical cultivars. Natarajan et al. (2001) found that an increase in *S. spontaneum* chromosome in the progenies increases the horizontal resistance component against red rot. A relatively higher amount of *S. spontaneum* genome in the subtropical varieties may have contributed to a higher level of resistance in subtropical clones, but the high selection pressure for red rot resistance in subtropical breeding programmes compared to that in tropical programmes also may be involved.

In India, systematic breeding of sugarcane has resulted in the development of a number of varieties with high productivity and stress tolerance by interspecific hybridization (ThuljaramRao, 1987). However, the genetic base of modern Indian sugarcane cultivars is considered narrow due to use of a limited number of parental species clones in cross-hybridization and repeated intercrossing of hybrids (Daniels and Roach, 1987). Understanding the extent of natural variation at the molecular level is essential to develop new strategies for sugarcane improvement.

Earlier, molecular markers such as RAPD, AFLP, and maize and sugarcane genomic microsatellites have been used for this purpose (Nair et al., 2002; Parida et al., 2009; Selvi et al., 2003, 2005, 2006). However, little effort has yet been made to understand the genetic diversity of Indian sugarcane cultivars based on functionally relevant genic regions of its complex genome. Parida et al. (2010) mined the available unigene sequences of sugarcane (*Saccharum* sp.) to understand the microsatellite structure and distribution in the expressed genic component of the genome, assess their functional significance *in silico*, design primers from the flanking regions of the identified microsatellites, assessed the efficiency of a set of fluorescent dye labeled primers in genotyping using an automated fragment analysis system, and determined functional diversity among different species, related genera and Indian varieties of sugarcane. Functional annotation of 364 unigenes carrying microsatellites in the functional domains of encoded proteins revealed that these unigenes corresponded maximally (47.7%) to the domains responsible for photosynthesis (cytochrome b/c and chlorophyll A/B binding domains) and carbohydrate metabolism (sucrose synthase and alpha amylase domains) followed by transcription factor associated basic leucine zipper, zinc finger, TATA box, Myb and WRKY DNA binding domains (22%), and minimally to the abiotic and biotic stress-related leucine-rich repeat, protein kinase and chitinase domains (5%).

Genetic engineering for developing biotic- and abiotic-resistant transgenic lines of sugarcane

Sugarcane is an ideal candidate for genetic engineering because of its complex polyploid nature, variable fertility, genotype versus environment interactions and availability of high frequency *in vitro* regeneration system (Suprasanna et al., 2011). The establishment of a suitable transformation system for sugarcane (Arencibia et al., 1998) has made possible the introduction of several genes (for disease/pest resistance, salt and drought tolerance, and sugar accumulation) targeted towards sugarcane improvement (Altpeter and Oraby, 2010; Hotta et al., 2011; Srikanth et al., 2011; Table 1.1).

Genes from bacteria such as *Bacillus thuringiensis* (Bt) and *Bacillus sphaericus*, protease inhibitors, plant lectins, ribosome inactivating proteins, secondary plant metabolites and small RNA viruses have been used alone or in combination with conventional host plant resistance to develop crop cultivars that suffer less damage from insect pests (Srikanth et al., 2011). Transgenic sugarcane for borer resistance was also reported using *Cry1Aa3* gene (Kalunke et al., 2009). Arvinth et al. (2010) determined the efficacy of native *Cry1Aa*, *Cry1Ab* and *Cry1Ac* against *C. infuscatellus* in *in vitro* bioassays through the diet-surface contamination method, and observed that

Table 1.1 Genetic engineering of sugarcane for different traits related to stress tolerance

Trait	Gene	Reference
Herbicide resistance		
Bialophos	*bar*	Gallo-Meagher and Irvine, 1996
Phosphinothricin	*bar*	Enriquez-Obregon et al., 1998
Glufosinate ammonium	*pat*	Leibbrandt and Snyman, 2003
Disease resistance		
SCMV	*SCMV-CP*	Joyce et al., 1998
Sugarcane leaf scald	*albD*	Zhang et al., 1999
SrMV	*SrMV-CP*	Ingelbrecht et al., 1999
Puccinia melanocephala	*Glucanase, chitanase, AP24*	Enriquez et al., 2000
Fiji leaf gall	*FDVS9 ORF 1*	McQualter et al., 2004
SCYLV	*SCYLV-CP*	Gilbert et al., 2005
SCYLV	*SCYLV-CP*	Zhu et al., 2011
Pest resistance		
Sugarcane stem borer	*cry1A*	Arencibia et al., 1999
Sugarcane stem borer	*cry1Ab*	Braga et al., 2003
Sugarcane stem borer	*cry1Ab*	Arvinth et al., 2010
Sugarcane borer	*cry1Ac*	Jing-Sheng et al., 2008
Sugarcane stem borer	*cry1Aa3*	Kalunke et al., 2009
Proceras venosatus	Modified *cry1Ac*	Weng et al., 2010
Sugarcane canegrub	*gna*	Legaspi and Mirkov, 2000
Mexican rice borer	*gna*	Sétamou et al., 2002
Ceratovacuna lanigera	*gna*	Zhangsun et al., 2007
Scirpophaga excerptalis	Aprotinin	Christy et al., 2009
Abiotic stress tolerance		
Osmotic stress tolerance	*Trehalose synthase*	Wang et al., 2005
Cold tolerance	*Isopentenyl transferase*	Belintani et al., 2012
Drought and salinity	*Arabidopsis vacuolar pyrrophosphatase*	Kumar et al., 2014
Salt tolerance	*D1-pyrroline-5-carboxylate synthetase*	Guerzoni et al., 2014
Cell membrane thermostability	*Pea DNA Helicase 45*	Augustine et al., 2015

the *Cry1Ab* as the most toxic among the three compounds. Christy et al. (2009) transferred aprotinin genes to sugarcane cultivars. The *in vivo* bioassay studies showed that larvae of top borer *Scirpophaga excerptalis* Walker (Lepidoptera: Pyralidae) fed on transgenics showed significant reduction in weight and impairment of larval development (Christy et al., 2009). Zhu et al. (2011) produced transgenic sugarcane with an untranslatable coat protein gene of SCYLV with the aim of improving resistance to this virus.

Controlling weeds is one of the most important tasks in sugarcane cultivation and management at the early growth stage. Unfortunately, owing to the lack of herbicide resistant genes in the gene pool, just like other crop species, sugarcane is sensitive to the herbicide and thus it needs exogenous genes to improve its herbicide resistance. Genetically modified sugarcane resistant to herbicide, usually with transformation of the genes such as *bar* and *epsps* into sugarcane genome, has been reported (Falco et al., 2000; Gallo-Meagher and Irvine, 1996). Transformation of *bar* gene conferring glufosinate ammonium tolerance has also been reported in sugarcane (Enriquez-Obregon et al., 1998; Falco et al., 2000; Gallo-Meagher and Irvine, 1996; Leibbrandt and Snyman, 2001).

Trehalose synthase gene (TSase) was transferred into sugarcane using the *Agrobacterium*-mediated method to improve sugarcane drought tolerance (Wang et al., 2005; Zhang et al., 2000).

Kumar et al. (2014) reported the transformation of sugarcane cultivar CP-77-400 with *AVP1* (*Arabidopsis Vacuolar Pyrophosphatase*) gene using apical meristem as the target tissue. This report suggested that transgenic plants expressing higher levels of *AVP1* transcripts in sugarcane are able to withstand salt and drought stress regimes probably due to profuse root system development in these plants.

In sugarcane, there are reports on proline accumulation in calli, plantlets and whole plants in field trials when exposed to salt stress (Gandonou et al., 2004; Patade et al., 2008; Wahid and Ghazanfar, 2006) and water deficit (Errabii et al., 2007). Several works report that higher proline accumulation in P5CS-transgenic plants confers increased tolerance to abiotic stress (Hong et al., 2000; Kumar et al., 2010; Molinari et al., 2007). Guerzoni et al. (2014) evaluated the response to salt stress of sugarcane plants transformed with the *Vigna aconitifolia P5CS* gene, which encodes Δ1-pyrroline-5-carboxylate synthetase, under the control of a stress-induced promoter AIPC (ABA-inducible promoter complex) and found that the transgenic lines accumulated up to 25% higher amounts of proline when compared with non-transformed control plants.

To improve the drought and salinity tolerance of sugarcane, a DEAD-box helicase gene isolated from pea (Pea DNA helicase 45, *PDH45*) with a constitutive promoter, Port *Ubi2.3*, was transformed into the commercial sugarcane variety Co86032 through *Agrobacterium*-mediated transformation, and the transgenics were screened for tolerance to soil moisture stress and salinity (Augustine et al., 2015). The V_1 transgenic events exhibited significantly higher cell membrane thermostability, transgene expression, relative water content, gas exchange parameters, chlorophyll content, and photosynthetic efficiency under soil moisture stress compared to wild type (WT). The overexpression of PDH45 transgenic sugarcane also led to the upregulation of DREB2-induced downstream stress-related genes. The transgenic events demonstrated higher germination ability and better chlorophyll retention than WT under salinity stress. These results suggest the possibility for development of increased abiotic stress-tolerant sugarcane cultivars through overexpression of *PDH45* gene.

In order to enhance cold tolerance in sugarcane, biolistic transformation into sugarcane (*Saccharum* spp.) cv. RB855536 with the gene encoding the enzyme isopentenyltransferase (*ipt*) under control of the cold inducible gene promoter AtCOR15a has been reported (Belintani et al., 2012). After being subjected to freezing temperatures, leaf total chlorophyll contents of transgenic plants were up to 31% higher than in wild-type plants. Also, lower malondialdehyde content and electrolyte leakage indicated less damage induced by cold intransgenic plants. Thus, the expression of *ipt* driven by the stress-inducible COR15a promoter did not affect plant growth while providing a greater tolerance to cold stress.

Genomics for stress tolerance in sugarcane

The lack of genetic and molecular information about drought-tolerance mechanisms and inheritance in sugarcane has limited the development of improved cultivars. There is a need to distinguish genes definitely associated with the response to water deficit, which hold an adaptive function to water deprivation and in stress environments. Genes associated with regulation of expression under water deficit or during the establishment of drought tolerance are potential candidates to evaluate differential expression between contrasting sugarcane genotypes. Studies conducted with rice, *Arabidopsis* and sugarcane have used microarray analyses (Rabbani et al., 2003; Rocha et al., 2007), with subsequent validation through quantitative amplification of reversed transcripts (RT-qPCR). This is to further investigate differentially expressed genes at distinct moments during drought. Some studies have used these differentially expressed genes to obtain transformed plants more tolerant to water deficit (Kirch et al., 2005; Zheng et al., 2004). The identification of genes encoding structural proteins directly related with the establishment of drought tolerance could be useful to develop genetic markers to select tolerant and/or sensitive genotypes. This helps to obtain improved cultivars by direct manipulation (transgenic) or classical breeding.

Genomics for biotic stress tolerance

Sugarcane is constantly challenged by herbivorous insects, nematodes, fungi and bacteria viruses. Plant defence responses to such perturbations are largely mediated by phytohormones through triggering conserved defence mechanisms, each with an intricate signaling pathway leading to plant protection. It has been shown that both the ethylene and jasmonic acid signaling pathways act synergistically in plant defence.

Sugarcane smut is a worldwide disease and can cause considerable losses in sugarcane yield (Hoy et al., 1986; Padmanaban et al., 1998; Que et al., 2012). Su et al. (2013) quantitatively determined the differences of b-1,3-glucanase enzyme activity between Yacheng 05-179 (resistant) and Liucheng 03-182 (susceptible) inoculated with S. scitamineum. Two β-1,3-glucanase genes from sugarcane were cloned and characterized. They were allocated in apoplast and involved in different expression patterns in biotic and abiotic stress (Su et al., 2013).

Sathyabhama et al. (2016) used suppression subtractive hybridization technique to identify differentially expressed genes in sugarcane in response to *Colletotrichum falcatum*, the fungal pathogen that causes red rot in sugarcane. At the end of subtraction, cloning and sequencing, 136 EST sequences were assembled into 10 clusters/contigs. Based on TIGR (The Institute of Genomic Research) homology search, the clusters were found to be involved in reactive oxygen species signaling, defence and the secretory pathway of plant innate immunity associated with hypersensitive response-mediated programmed cell death.

In a study to identify red-rot-related genes, Gupta et al. (2010) used an enriched subtractive cDNA library prepared from the *C. falcatum* challenged stem of sugarcane variety (Co 1148) and reported at least 85 red-rot-specific clusters, unique and not reported in the database previously.

In an attempt towards studying the host–pathogen interaction and deciphering the molecular basis of virulence of SCGS (Sugarcane Grassy Shoot) disease, Kawar et al. (2010a) isolated the partial genome of the first Asiatic strain of phytoplasma (SCGS) by genomic suppression subtractive hybridization (SSH). The library yielded 83 SCGS-specific fragments representing approximately 42% of the chromosome of sugarcane grassy shoot phytoplasma, comprising approximately 85 predicted partial phytoplasmal CDS. Further, a species-specific detection method was developed for early detection of SCGS infection (Kawar et al., 2010b).

Genomics for abiotic stress tolerance

Gene discovery and genomics are essential tools for the future of sugarcane improvement (Hotta et al., 2011). In recent years, much attention has been focused on the transcription factors (AP2/EREBP, bZIP, WRKY,

MYB, and zinc finger proteins), which play an essential role in stress responses by regulating the physiological and biochemical function of the organism (Ambawat et al., 2013; Du et al., 2012, 2013; Grotewold, 2008). Consequently, isolation and development of transgenic plants over-expressing some of these genes enhances tolerance to various stresses and, in particular, increases their water-use efficiency.

The mechanisms of plant stress responses are highly important to agriculture because of their direct link to production systems. Proteins such as heat-shock proteins, peroxidases and water transport proteins are involved in plant protection mechanisms under conditions of low water availability (Borges et al., 2001; Casu et al., 2005; Wang et al., 2004). In addition, differential expression of some sugarcane genes associated with tolerance to water stress has been reported. Rodrigues et al. (2011) analysed 3575 ESTs in a drought-tolerant sugarcane cultivar and found 165 differentially expressed genes, indicating a large number of genes associated with drought tolerance. A data mining analysis for sugarcane done using the Sugarcane Expressed Sequence Tags (SUCEST) database revealed an abundant expression of genes encoding chaperones, co-chaperones and other proteins linked to protection against stress in sug-arcane (Borges et al., 2001). The genes most commonly encountered by Borges et al. (2001) were those responsible for the synthesis of chaperone HSP70 (heat-shock protein) and its co-factors such as HSP40, in addition to encoders of the proteins HSP90, HSP100 and small HSP chaperones (Wang et al., 2004). The chaperone activity of small HSPs has been associ-ated with the heat stress response in sugarcane (Tiroli and Ramos, 2007). Gene expression profiles analysed by microarrays in sugarcane leaves identified 165 genes in response to water stress (Rodrigues et al., 2011).

Andrade et al. (2015) investigated the expression profiles of 12 genes in the leaves of a drought-tolerant genotype (RB72910) of sugarcane and compared the results with those of other studies. They observed that the patterns of gene expression vary in different genotypes classified as drought tolerant. This variability suggests a high degree of complexity in the response of sugarcane to water stress.

Gupta et al. (2010) studied various tissue-specific EST libraries' sequence data of Indian subtropical sugarcane variety (CoS 767) and observed 25 water-deficit stress-related clusters that showed greater than two-fold relative expression during 9 h dehydration stress. Prabu et al. (2010), based on sqRT-PCR (semi-quantitative reverse transcription polymerase chain reaction) analysis, showed higher transcript expression of WRKY, a 22-kDa drought-induced protein, MIPS and ornithine-oxo-acid amino transferase at initial stages of stress induction with a gradual decrease in advanced stages. They also identified differentially expressed transcripts in response to water deficiency stress in sugarcane cv. Co740 using PCR-based cDNA suppression subtractive hybridization technique. Annotation of these differentially ESTs indicated their possible function in cellular organization, protein metabolism, signal transduction and transcription.

Vantini et al. (2015) conducted a comparative study to identify gene expression profiles under water stress in tolerant sugarcane roots. Two different cultivars, one drought tolerant (RB867515) and one drought susceptible (SP86-155), were evaluated at four sampling time points (1, 3, 5 and 10 days) using the cDNA-amplified fragment length polymorphism technique. A total of 173 fragments were found to be differentially expressed in response to water stress in the tolerant cultivar. Seventy of these were cloned, sequenced and categorized. Similarity analysis using BLAST revealed that 64% of the fragments differentially expressed code proteins classified as no hits (23%), hypothetical (21%) or involved in stress response (20%), with others that were involved in communication pathways and signal transduction, bioenergetics, secondary metabolism, and growth and development.

Patade et al. (2011c) identified salinity-induced shaggy-like kinase (designated as sugarcane shaggy-like protein kinase [SuSk]). The expression was induced by salt as well as PEG (polyethylene glycol) stress indicating that the induction of this gene probably occurred in response to the osmotic component of salt stress rather than the ionic component.

Transcriptomic study of short-term (up to 24 h) salt (NaCl, 200 mM) or iso-osmotic PEG 8000 (20% w/v) stress has revealed altered expression of representative stress responsive genes in sugarcane leaves (Patade et al., 2011b). Efficient sequestration of Na to vacuole, which reduces the cytosolic Na concentration, is an important aspect of tissue tolerance to salinity. The authors reported downregulation of a sugarcane homologue of NHX belonging to the family of Na/H and K/H antiporters in response to the salt stress. On long-term exposure to salt or PEG stress, the steady-state levels of both P5CS and PDH gene expression increased (Patade, 2009), which also correlated to proline accumulation under these stress conditions (Patade et al., 2011a).

Transcriptome analysis of sugarcane hybrid CP72-1210 (cold susceptible) and *Saccharum spontaneum* TUS05-05 (cold tolerant) using sugarcane assembled sequences (SAS) from the SUCEST-FUN Database showed that a total of 35,340 and 34,698 SAS genes, respectively, were expressed before and after chilling stress (Park et al., 2015).

Involvement of sugarcane microRNA in stress responses

MicroRNAs (miRNAs) are small (20–24 nt), single-stranded, non-coding, naturally occurring, highly conserved families of transcripts. Several miRNAs are either upregulated or downregulated by abiotic stresses, suggesting that they may be involved in stress-responsive gene expression and stress adaptation (Shriram et al., 2016; Sunkar and Zhu, 2012).

The involvement of miRNAs in abiotic stress has been studied in plants in response to dehydration or NaCl by using expression analysis, suggesting stress-specific regulation of expression of miRNA (Patade and Suprasanna, 2010b) in sugarcane. In response to long-term (15 days) isoosmotic (−0.7 MPa) NaCl or PEG stress, no change in mature transcript level of miR159 over the control was detected. However, under the short-term (up to 24 h) salt stress, the transcript level of the mature miRNA increased to 112% of the control at 16 h treatment. The mature transcript level of miR159 was higher under all the PEG-induced osmotic stress treatments as compared to the control, and it progressively increased with the stress exposure period (1.3-fold at 8 h treatment). This indicated that expression of miR159 gene was more responsive to osmotic stress than ionic stress. The authors studied the expression of one of the predicted target MYB under the same stress (NaCl or PEG) conditions to study the changes in target gene expression in response to over- or underexpression of miR159. The results on the expression of specific miR159 and its targets could be useful in developing appropriate markers for selection of tolerant cultivars in sugarcane.

Lin et al. (2014) reported the expression profile of miRNA families synthesized by leaves of the drought-resistant cultivar ROC22 stressed by addition of PEG to the growth substrate. Twenty-three conserved miRNA families and 34 new miRNA families were identified, and 438 putative target genes of 44 miRNA families. Expression analysis revealed that 11 miRNA families were differentially expressed in control and drought-exposed plants. Of these, 9 families were upregulated and two down-regulated. The potential targets of the 11 miRNA families were genes associated with plant growth and stress resistance, specifically *SPBP, MYB*, the *AGO1-like gene, NCBP, BCP, CPI* and *LSG*. With the exceptions of *SPBP, NCBP* and *BCP*, the other genes were downregulated in response to drought stress.

There are several miRNAs that have been identified in a wide array of species, but only a few studies have been performed to identify the mature miRNA sequences and analyse their expression in response to drought stress in sugarcane (Ferreira et al., 2012; Gentile et al., 2013; Thiebaut et al., 2012). Thiebaut et al. (2012) separated eight sugarcane cultivars into two groups based on their tolerance to drought. Plants were grown in a greenhouse for 3 months and then drought stressed by holding irrigation for 24 h. Although the number of detected miRNAs was higher in the more tolerant cultivars, no miRNA was found to be induced by drought under these conditions. Similarly, in two sugarcane cultivars that differ in their tolerance to drought stress, RB867515 (higher tolerance) and RB855536 (lower tolerance), were both grown in a greenhouse for 3 months and then kept without water for 2 or 4 days (Ferreira et al., 2012). Some miRNAs were found only in plants that grew in the greenhouse (miR164, miR397 and miR399), while others were found only in field-grown plants (miR160, miR166, miR169, miR171 and miR172). This most likely reflects differences in the growth conditions, with field-grown plants giving a better picture of the real growth conditions that plants face in nature. However, some miRNAs had opposing expression profiles depending on the cultivar, such as miR164, miR399 and miR1432 (up-regulated by drought in one cultivar and down-regulated in the other) while others changed from down-regulated (miR399 and miR1432) to up-regulated depending on the cultivar. The timing of stress also affected the miRNA expression profile. For example, comparing greenhouse-grown plants that were stressed for 2 or 4 days, miR397 showed an altered expression profile, changing from induced to repressed, while miR399 remained invariantly downregulated. The other miRNAs found in the greenhouse-grown plants had variable expression profiles without a specific pattern. Gentile et al. (2013) identified 18 miRNA families comprising 30 mature miRNA sequences in two sugarcane cultivars differing in drought tolerance were grown in the field with and without irrigation (rain fed) for 7 months. Among these families, 13 mature miRNAs were differentially expressed in drought-stressed plants. Seven miRNAs were differentially expressed in both cultivars. After 7 months, five miRNAs from the rain-fed field-grown plants presented the same profile in both cultivars; two were induced (miR160 and miR399) and three were repressed (miR166, miR171 and miR396). Only two miRNAs (miR399 and miR1432) had opposite profiles among the different cultivars. Gentile et al. (2015) revealed the expression pattern of miRNAs was observed to be dependent on the species, type of stress tissue (seedlings, leaves, spikelets and root) and growth condition (field, greenhouse, hydroponic culture system). In most of the cases, miR396 and miR171 were differentially expressed. All of these data increase our understanding of the role of miRNAs in the complex regulation of drought stress in field-grown sugarcane, providing valuable tools to develop new sugarcane cultivars tolerant to drought stress.

An earlier study report of miRNA regulation has shown that it has a key role in cold-related molecular mechanisms in sugarcane (Thiebaut et al., 2012; Yang et al., 2016). In sugarcane cultivar SP90–1638 (sensitive cultivar) and SP83–2847 and SP83–5073 tolerant cultivars, miRNA was observed and the expression profiles of eight miRNAs were verified in leaf libraries in response to water depletion. Bioinformatics analyses identified 28 (leaf) and 36 (root) conserved miRNA families. MiRNAs are differentially expressed in leaves and roots upon treatment with water depletion. The expression profiles of eight miRNAs (miR156, miR159 (two isoforms), miR164, miR167, miR168, miR169 and miR397) differentially expressed in leaves. Khan et al. (2014) performed gene expression in sugarcane under waterlogging stress is still patchy, so the present study was undertaken to generate ESTs induced under waterlogging stress and to have a glimpse of genes expressed under such stress. In this study, they used a bioinformatics search based on sequence similarity or the secondary structure of precursors, and have identified seven novel miRNAs (miR 1 to miR 7) expressed in waterlogged sugarcane.

Vishwanthan et al. (2014) demonstrated that in miRNA encoded by sugarcane streak mosaic virus (SCSMV) infecting sugarcane further, SCSMV miR16 was found to target many cellular transcripts to establish a favourable environment for the virus life cycle and evade the host defence. These results provide a useful resource for further in-depth studies on streak mosaic disease in sugarcane, and targeting the SCSMV miR16 might be a new strategy for controlling SCSMV in sugarcane. Overall, a better understanding on the regulatory network from miRNAs and their targets under stresses has a great potential to contribute to sugarcane improvement, and possibilities for using miRNA-mediated gene regulation to enhance plant stress tolerance will become enormous.

Concluding remarks

Sugarcane is a source of food and fuel. For sustainable sugarcane production, an improved productivity, tolerance to biotic and abiotic stresses, and improved sugar recovery are the major challenges. Stress is one of the major factors that affect the productivity of sugarcane. Elucidating mechanisms for stress tolerance in sugarcane using biotechnological tools shall facilitate development of stress-tolerant varieties of sugarcane. The *in vitro* selection and mutation breeding have enabled rapid propagation and generation of novel germplasm with desirable traits. Molecular markers have enabled the identification of stress-tolerant clones which may be used as parents in crossing programmes towards rapid varietal improvement for cane and sugar yield as well as to enhance the resistance against different biotic and abiotic stresses for increased sugarcane productivity. Although there has been considerable progress in sugarcane transformation, the successful application of genetic transformation will require reliable, high levels of transgene expression and their stability over generations. Functional genomics may aid in understanding responses to biotic and abiotic stresses and their management. Studying gene expression profiles of sugarcane cultivars under conditions that affect crop yield can aid in target gene selection. Some of the important challenges include gene discovery, transgenics and controlled transgene expression, sucrose metabolism, and photosynthesis. Studying stress responsive miRNAs and their target expression can help to explore the possibilities for enhancing plant stress tolerance. All these advances in sugarcane biotechnology could become remarkable in the coming years, both in terms of improving productivity as well as substantially increasing the value and utility of this crop.

References

Abdel-Tawab FM, Fahmy EM, Allam AI, El-Rashidy HA, Shoaib RM (2003a) Development of RAPD and SSR marker associated with stress tolerance and some technological traits and transient transformation of sugarcane (*Saccharum* spp.). *Proceedings of the International Conference 'The Arab Region and Africa in the World Sugar Context,'* 10–12 March, Aswan, Egypt. pp. 1–23.

Abdel-Tawab FM, Fahmy EM, Allam AL, El Rashidy HA, Shoaib RM (2003b) Marker assisted selection for abiotic stress tolerance in sugarcane (*Saccharum* spp). *Egyptian Journal of Agricultural Research* 81: 635–646.

Abdel-Tawab FM, Rashed MA, Dhindsa RS, Bahieldin A, AboDoma A (1998) Molecular markers for salt tolerance in *Sorghum bicolor*. International Congress on Molecular Genetics, 21–26 February, Cairo, Egypt.

Akhtar S, Wahid A, Rasul E (2003) Emergence, growth and nutrient composition of sugarcane sprouts under NaCl salinity. *Biologia Plantarum* 46: 113.

Altpeter F, Oraby H (2010) Sugarcane. In: *Genetic modification of plants* (Biotechnology in agriculture and forestry, vol. 64), F Kempken C Jung, eds., 453–472. New York: Springer.

Ambawat S, Sharma P, Yadav NR, Yadav RC (2013) MYB transcription factor genes as regulators for plant responses: An overview. *Physiology and Molecular Biology of Plants* 19: 307–321.

Andrade JCF, Terto H, Silva JV, Almeida C (2015) Expression profiles of sugarcane under drought conditions: Variation in gene regulation. *Genetics and Molecular Biology* 38(4): 465–469.

Arencibia A, Carmona E, Cornide MT, Castiglione S, O'Relly J, Cinea A, Oramas P, Sala F (1999) Somaclonal variation in insect resistant transgenic sugarcane (*Saccharum* hybrid) plants produced by cell electroporation. *Transgenic Research* 8: 349–360.

Arvinth S, Arun S, Selvakesavan RK, Srikanth J, Mukunthan N, Ananda Kumar P, Premachandran MN, Subramonian N (2010) Genetic transformation and pyramiding of aprotinin-expressing sugarcane with cry1Ab for shoot borer (Chiloinfuscatellus) resistance. *Plant Cell Reports* 29(4): 383–395.

Asano T, Tsudzuki T, Takahashi A, Shimada H, Kadowaki K (2004) Complete nucleotide sequence of the sugarcane (*Saccharum officinarum*) chloroplast genome: A comparative analysis of four monocot chloroplast genomes. *DNA Research* 11: 93–99.

Augustine SM, Ashwin Narayan J, Syamaladevi DP, Appunu C, Chakravarthi M, Ravichandran V, Tuteja N, Subramonian N (2015) Introduction of pea DNA helicase 45 into sugarcane (*Saccharum* spp. hybrid) enhances cell membrane thermostability and upregulation of stress-responsive genes leads to abiotic stress tolerance. *Molecular Biotechnology* 57: 475–488. doi:10.1007/s12033-015-9841-x.

Belintani NG, Guerzoni JTS, Moreira RMP, Vieira LGE (2012) Improving low-temperature tolerance in sugarcane by expressing the ipt gene under a cold inducible promoter. *Biologia Plantarum* 56(1): 71–77.

Borges JC, Peroto MC, Ramos CHI (2001) Molecular chaperone genes in the sugarcane expressed sequence database (SUCEST). *Genetics and Molecular Biology* 24: 85–92.

Braga DPV, Arrigoni EDB, Filho MCS, Ulian EC (2003) Expression of the Cry1Ab protein in genetically modified sugarcane for the control of *Diatraea sacch-aralis* (Lepidoptera: Crambidae). *Journal of New Seeds* 5(2/3): 209–221.

Casu RE, Manners JM, Bonnett GD, Jackson PA, McIntyre CL, Dunne R, Chapman SC, Rae AL, Grof CPL (2005) Genomics approaches for the identification of genes determining important traits in sugarcane. *Field Crops Research* 92: 137–147.

Chaum S, Chuencharoen S, Mongkolsiriwatana C, Ashraf M, Kirdmanee C (2012) Screening sugarcane (*Saccharum* spp.) genotypes for salt tolerance using multivariate cluster analysis. *Plant Cell, Tissue and Organ Culture* 110(1): 23–33.

Christy LA, Aravinth S, Saravanakumar M, Kanchana M, Mukunthan N, Srikanth J, Thomas G, Subramonian N (2009) Engineering sugarcane culti-vars with bovine pancreatic trypsin inhibitor (aprotinin) gene for protec-tion against top borer (*Scirpophaga excerptalis* Walker). *Plant Cell Reports* 28: 175–184.

Costa M, Amorim L, Onofre A, Melo L, Oliveira M, Carvalho R, Benko-Iseppon A (2011) Assessment of genetic diversity in contrasting sugarcane varieties using inter-simple sequence repeat (ISSR) markers. *American Journal of Plant Sciences* 2: 425–432.

Daniels J, Roach BT (1987) Taxonomy and evolution. In: *Sugarcane improvement through breeding*, vol. 11, DJ Heinz, ed., 7–84. Amsterdam: Elsevier.

Dillon SL, Shapter FM, Henry RJ, Cordeiro G, Izquierdo L, Lee LS (2007) Domestication to crop improvement: Genetic resources for *Sorghum* and *Saccharum* (andropogoneae). *Annals of Botany* 100: 975–989.

Du H, Feng BR, Yang SS, Huang YB, Tang YX (2012) The R2R3-MYB transcription factor gene family in maize. *PLoS ONE* 7: 1–12.

Du H, Wang YB, Xie Y, Liang Z, Jiang SJ, Zhang SS, Huang YB, Tang YX (2013) Genome-wide identification and evolutionary and expression analyses of MYB-related genes in land plants. *DNA Research* 20(5): 437–448.

Enriquez GA, Trujillo LA, Menndez C, Vazquez RI, Tiel K, Dafhnis F, Arrieta J, Selman G, Hernandez L (2000) Sugarcane (*Saccharum* hybrid) genetic trans-formation mediated by *Agrobacterium tumefaciens*: Production of trans-genic plants expressing proteins with agronomic and industrial value. *Developments in Plant Genetics and Breeding* 5: 76–81.

Enriquez-Obregon GA, Vazquez PRI, Prieto SDL, Riva-Gustavo ADL, Selman HG (1998) Herbicide resistant sugarcane (*Saccharum officinarum* L.) plants by *Agrobacterium*-mediated transformation. *Planta* 206: 20–27.

Errabii T, Gandonou C, Essalmani H, Abrini J, Idaomar M, Senhaji N (2006) Growth, proline and ion accumulation in sugarcane callus cultures under drought-induced osmotic stress and its subsequent relief. *African Journal of Biotechnology* 4: 1250–1255.

Falco MC, Neto AT, Ulian EC (2000) Transformation and expression of a gene for herbicide resistance in a Brazilian sugarcane. *Plant Cell Reports* 19: 1188–1194.

Ferreira TH, Gentile A, Vilela RD, Costa GGL, Dias LI, Endres L, Menossi M (2012) microRNAs associated with drought response in the bioenergy crop sugar-cane (*Saccharum* spp.). *PLOS One* 7: e46703.

Gallo-Meagher M, Irvine JE (1996) Herbicide resistant transgenic sugarcane plants containing the bar gene. *Crop Science* 36: 1367–1374.

Gandonou C, Errabii J, Abrini M, Idaomar M, Senhaji N (2004) Selection of callus cultures of sugarcane (*Saccharum* spp.) tolerant to NaCl and their response to salt stress. *Plant Cell, Tissue and Organ Culture* 87: 9–16.

Ganesan M, Jayabalan N (2006) Isolation of disease-tolerant cotton (*Gossypium hirsutum* L. cv. SVPR 2) plants by screening somatic embryos with fungal culture filtrate. *Plant Cell, Tissue and Organ Culture* 87: 273–284.

Gentile A, Dias LI, Mattos RS, Ferreira TH, Menossi M (2015) MicroRNAs and drought responses in sugarcane. *Frontiers in Plant Science* 6: 58.

Gentile A, Ferreira TH, Mattos RS, Dias LI, Hoshino AA, Carneiro MS, Souza GM, Calsa T Jr, Nogueira (2013) Effects of drought on the microtranscriptome of field-grown sugarcane plants. *Planta* 237: 783–798.

Gilbert RA, Gallo-Meagher M, Comstock JC, Miller JD, Jain M, Abouzid A (2005) Agronomic evaluation of sugarcane lines transformed for resistance to sugarcane mosaic virus strain E. *Crop Science* 45: 2060–2067.

Grotewold E (2008) Transcription factors for predictive plant metabolic engineering: Are we there yet? *Current Opinion in Biotechnology* 19: 138–144.

Guerzoni JTS, Belintani NG, Moreira RMP, Hoshino AA, Domingues DS, Filho JCB, Vieira LGE (2014) Stress-induced Δ1-pyrroline-5-carboxylate synthetase (P5CS) gene confers tolerance to salt stress in transgenic sugarcane. *Acta Physiologiae Plantarum* 36: 2309–2319.

Gupta V, Raghuvanshi S, Gupta A, Saini N, Gaur A, Khan MS et al. (2010) The water-deficit stress and red-rot-related genes in sugarcane. *Functional & Integrative Genomics* 10: 207–214.

Heinz DJ, Krishnamurthi M, Nickel LG, Maretzki A (1977) Applied and functional aspects of plant cell, tissue and organ culture. In: *Cell tissue and organ culture in sugarcane improvement*, J Reinert, YPS Bajaj, eds., 3–17. New York: Springer-Verlag.

Hong Z, Lakkineni K, Zhang Z, Verma DPS (2000) Removal of feedback inhibition of D1-pyrroline-5-carboxylate synthetase results in increased proline accumulation and protection of plants from osmotic stress. *Plant Physiology* 122: 1129–1136.

Hotta CT, Lembke CG, Domingues DS, Ochoa EA, Cruz GMQ, Melotto-Passarin DM et al. (2011) The biotechnology roadmap for sugar cane improvement. *Tropical Plant Biology* 3: 75–87.

Hoy JW, Hollier CA, Fontenot DB, Grelen LB (1986) Incidence of sugarcane smut in Louisiana and its effects on yield. *Plant Disease* 70: 59–60.

Ingelbrecht IL, Irvine JE, Mirkov TE (1999) Posttranscriptional gene silencing in transgenic sugarcane. Dissection of homology-dependent virus resistance in a monocot that has a complex polyploid genome. *Plant Physiology* 119: 1187–1197.

Jayashankar S, Li Z, Gray DJ (2000) *In vitro* selection of *Vitis vinifera* 'Chardonnay' with Elsinoeampelina culture filtrate is accompanied by fungal resistance and enhanced secretion of chitinase. *Planta* 211: 200–208.

Jing-Sheng X, Shiwu G, Liping X, Rukai C (2008) Construction of expression vector of CryIA(c) gene and its transformation in sugarcane. *Sugar Tech* 10(3): 269–273.

Joyce PA, McQualter RB, Handley JA, Dale JL, Harding RM, Smith GR (1998) Transgenic sugarcane resistant to sugarcane mosaic virus. In *Proceedings of the Australia Society of Sugarcane Technologists* 20: 204–210.

Kalunke RM, Kolge AM, BabuKH, Prasad DT (2009) *Agrobacterium*-mediated transformation of sugarcane for borer resistance using Cry 1Aa3 gene and one-step regeneration of transgenic plants. *Sugar Tech* 11(4): 355–359.

Kawar PG, Pagariya MC, Dixit GB, Theertha Prasad D (2010a) Identification and isolation of SCGS phytoplasma-specific fragments by riboprofiling and development of specific diagnostic tool. *Journal of Plant Biochemistry and Biotechnology* 19: 185–194.

Kawar PG, Pagariya MC, Patel SR, Dixit GB, Theertha Prasad D (2010b) An overview of partial genome sequence of first Asiatic phytoplasma strain (SCGS)-Indian isolate. *Asian Journal of Plant Pathology* 4(1): 16–19.

Khan MS, Khraiwesh B, Pugalenthi G, Gupta RS, Singh J, Duttamajumder SK, Kapur R (2014) Subtractive hybridization-mediated analysis of genes and in silico prediction of associated microRNAs under waterlogged conditions in sugarcane (*Saccharum* spp.) *FEBS Open Bio* 4: 533–541.

Kumar T, Uzma, Khan MR, Abbas Z, Ali GM (2014) Genetic improvement of sugarcane for drought and salinity stress tolerance using *Arabidopsis vacuolar* pyrophosphatase (*AVP1*) gene. *Molecular Biotechnology* 56: 199–209.

Kumar V, Shriram V, Kavi-Kishor PB, Jawali N, Shitole MG (2010) Enhanced proline accumulation and salt stress tolerance of transgenic indica rice by over-expressing P5CSF129A gene. *Plant Biotechnology Reports* 4: 37–48.

Kumar JV, Ranjitha Kumari BD, Sujatha G, Castano E (2008) Production of plants resistant to *Alternaria carthami* via organogenesis and somatic embryogenesis of safflower cv. NARI-6 treated with fungal culture filtrates. *Plant Cell, Tissue and Organ Culture* 93: 85–96.

Larkin PJ, Scowcroft WR (1983) Somaclonal variation and crop improvement. In: *Genetic engineering of plants: An agricultural perspective*, T Kosuge, CP Mereditch, A Hollaender, eds., 289–314. New York: Plenum Press.

Legaspi JC, Mirkov TE (2000) Evaluation of transgenic sugarcane against stalk borers. *Proceedings of the International Society of Sugar Cane Technologists, Sugar Cane Entomology Workshop* 4: 68–71.

Leibbrandt N, Snyman S (2001) Initial field testing of transgenic glufosinate ammonium-resistant sugarcane. *Proceedings of the South African Sugar Technologists Association*, 108–111.

Leibbrandt NB, Snyman SJ (2003) Stability of gene expression and agronomic performance of a transgenic herbicide-resistant sugarcane line in South Afr. *Crop Science* 43: 671–678.

Lin S, Chen T, Qin X, Wu H, Khan MA, Lin W (2014) Identification of microRNA families expressed in sugarcane leaves subjected to drought stress and the targets thereof. *Pakistan Journal of Agricultural Sciences* 51(4): 925–934.

McQualter RB, Dale JL, Harding RH, McMahon JA, Smith GR (2004) Production and evaluation of transgenic sugarcane containing a Fiji disease virus (FDV) genome segment S9-derived synthetic resistance gene. *Australian Journal of Agricultural Research* 55: 139–145.

Mohanraj D, Padmanabhan P, Karunakaran M (2003) Effect of phytotoxin of *Colletotrichum falcatum* Went (Physalosporatucumanensis) on sugarcane in tissue culture. *Acta Phytopathologica et Entomologica Hungarica* 38: 21–28.

Molinari HBC, Marur CJ, Daros E, Campos MKF, Carvalho JFRP, Bespalhok JCF, Pereira LFP, Vieira LGE (2007) Evaluation of the stress-inducible production of proline in transgenic sugarcane (*Saccharum* spp.): Osmotic adjustment, chlorophyll fluorescence and oxidative stress. *Physiologia Plantarum* 130: 218–229.

Nair NV, Selvi A, Srinivasan TV, Pushpalatha KN (2002) Molecular diversity in Indian sugarcane cultivars as revealed by randomly amplified DNA poly- morphisms. *Euphytica* 127: 219–225.

Nair NV (2010) The challenges and opportunities in sugarcane agriculture. Souvenir STAI, 117–135.

Natarajan US, Balasundaram N, RamanaRao TC, Padmanabhan P, Mohanraj D (2001) Role of *Saccharum spontaneum* in imparting stable resistance against sugarcane red rot. *Sugarcane International* 10: 17–20.

Nayak SN, Song J, Villa A, Pathak B, Ayala-Silva T, Yang X et al. (2014) Promoting utilization of *Saccharum* spp. Genetic resources through genetic diversity analysis and core collection construction. *PLOS ONE* 9: e110856. doi:10.1371 /journal.pone.0110856.

Nikam AA, Devarumath RM, Ahuja A, Babu H, Shitole MG, Suprasanna P (2014a) Radiation-induced *in vitro* mutagenesis system for salt tolerance and other agronomic characters in sugarcane (*Saccharum officinarum* L.). *The Crop Journal* 3: 46–56.

Nikam AA, Devarumath RM, Shitole MG, Ghole VS, Tawar PN, Suprasanna P (2014b) Gamma radiation, *in vitro* selection for salt (NaCl) tolerance and characterization of mutants in sugarcane (*Saccharum officinarum* L.). *In Vitro Cellular and Developmental Biology – Plant* 50(6): 766–776.

Padmanaban P, Alexander KC, Shanmugan N (1998) Effect of smut on growth and yield parameters of sugarcane. *Indian Phytopathology* 41: 367–369.

Parida SK, Pandit A, Gaikwad K, Sharma TR, Srivastava PS, Singh NK, Mohapatra T (2010) Functionally relevant microsatellites in sugarcane unigenes. *BMC Plant Biology* 10: 251.

Parida SK, Kalia SK, Kaul S, Dalal V, Hemaprabha G, Selvi A et al. (2009) Informative genomic microsatellite markers for efficient genotyping appli- cations in sugarcane. *Theoretical and Applied Genetics* 118: 327–338.

Park JW, Benatti TR, Marconi T, Yu Q, Solis-Gracia N, Mora V et al. (2015) Cold responsive gene expression profiling of sugarcane and *Saccharum spon- taneum* with functional analysis of a cold inducible *Saccharum* homolog of NOD26-like intrinsic protein to salt and water stress. *PLOS ONE* 10(5): e0125810. doi:10.1371/journal.pone.0125810.

Patade VY (2009) Studies on salt stress responses of sugarcane (*Saccharum offi- cinarum* L.) using physiological and molecular approaches. PhD thesis, University of Pune, Pune, Maharashtra, India.

Patade VY, Bhargava S, Suprasanna P (2011a) Salt and drought tolerance of sugar- cane under iso-osmotic salt and water stress: Growth, osmolytes accumula- tion and antioxidant defense. *Journal of Plant Interactions*. doi:10.1080/174291 45.2011.557513.

Patade VY, Bhargava S, Suprasanna P (2011b) Transcript expression profiling of stress responsive genes in response to short-term salt or PEG stress in sug- arcane leaves. *Molecular Biology Reports*. doi:10.1007/s11033-011-1100-z.

Patade VY, Rai AN, Suprasanna P (2011c) Expression analysis of sugarcane shaggy-like kinase (SuSK) gene identified through cDNA subtractive hybridization in sugarcane (*Saccharum officinarum* L.). *Protoplasma* 248(3): 613–621.

Patade VY, Suprasanna P, Bapat VA (2008) Gamma irradiation of embryogenic cal- lus cultures and *in vitro* selection for salt tolerance in sugarcane (*Saccharum officinarum* L.). *Agricultural Sciences in China* 7: 101–105.

Patade VY, Suprasanna P, Kulkarni UG, Bapat VA (2006) Molecular profiling using RAPD technique of salt and drought tolerant regenerants of sugarcane. *Sugar Tech* 8: 63–68.

Piperidis GA, Rattey R, Taylor GO, Cox MC, Hogarth DM (2004) DNA markers: A tool for identifying sugarcane varieties. Conference of Australian Society of Sugarcane Technologists, Brisbane, Queensland, Australia, 4–7 May.

Prabu G, Kawar PG, Pagariya MC, Prasad DT (2010) Identification of water deficit stress upregulated genes in sugarcane. *Plant Molecular Biology Reporter* 29: 291. doi:10.1007/s11105-010-0230-0.

Prasad V, Naik GR (2000) *In vitro* strategies for selection of eye-spot resistant sug- arcane lines using toxins of Helmintho sporium sacchari. *Indian Journal of Experimental Biology* 38: 69–73.

Que YX, Xu LP, Lin JW, Chen RK, Grisham MP (2012) Molecular variation of *Sporisorium scitamineum* in Mainland China revealed by RAPD and SRAP markers. *Plant Disease* 96: 1519–1525.

Que YX, Chen TS, Lin JW, Han XY (2008) Development of one SCAR marker for detection of sugarcane smut resistance. *Proceedings of China Association for Science and Technology* 4: 147–150.

Rabbani MA, Maruyama K, Abe H, Khan MA, Katsura K, Ito Y, Yoshiwara K, Seki M, Shinozaki K, Yamaguchi-Shinozaki K (2003) Monitoring expression profiles of rice genes under cold, drought, and high-salinity stresses and abscisic acid application using cDNA microarray and RNA gel-blot analy- ses. *Plant Physiology* 133: 1755–1767.

Rai MK, Kalia RK, Singh R, Gangola MP, Dhawan AK (2011) Developing stress tol- erant plants through *in vitro* selection: An overview of the recent progress. *Environmental and Experimental Botany* 71: 89–98.

Rocha FR, Papini-Terzi FS, Nishiyama MY Jr, Vêncio RZ, Vicentini R, Duarte RD et al. (2007) Signal transduction related responses to phytohormones and environmental challenges in sugarcane. *BMC Genomics* 8: 71.

Saif-Ur-Rasheed M, Asad S, Zafar Y, Waheed RA (2001) Use of radiation and *in vitro* techniques for development of salt tolerant mutants in sugarcane and potato. *IAEA TECDOC* 1227: 51–60.

Sakhanokho HF, Kelley RY (2009) Influence of salicylic acid on *in vitro* propagation and salt tolerance in *Hibiscus acetosella* and *Hibiscus moscheutos* (cv. 'Luna Red'). *African Journal Biotechnology* 8: 1474–1481.

Sandhu SK, Thind KS, Singh P (2012) Variability trends for brix content in general crosscombinations of sugarcane (*Saccharum* spp.) complex. *World Journal of Agricultural Scienes* 8: 113–117.

Sathyabhama M, Viswanathan R, Malathi P, Ramesh Sundar A (2016) Identification of differentially expressed genes in sugarcane during pathogenesis of *Colletotrichum falcatum* by suppression subtractive hybridization (SSH). *Sugar Tech* 18(2): 176–183.

Saxena P, Srivastava RP, Sharma ML (2010) Studies on salinity stress tolerance in sugarcane varieties. *Sugar Tech* 12(1): 59–63.

Scortecci KC, Creste S, CalsaJr T, Xavier MA, Landell MGA, Figueira A, Benedito VA (2012) Challenges, opportunities and recent advances in sugarcane breeding. In: *Plant breeding*, I Abdurakhmonov, ed. InTech. doi:10.5772/28606.

Selvi A, Nair NV, Balasundaram N, Mohapatra T (2003) Evaluation of maize mic- rosatellite markers for genetic diversity analysis and fingerprinting in sug- arcane. *Genome* 46: 394–403.

Selvi A, Nair NV, Noyer JL, Singh NK, Balasundaram N, Bansal KC, Koundal KR, Mohapatra T (2005) Genomic constitution and genetic relationship among the tropical and subtropical Indian sugarcane cultivars revealed by AFLP. *Crop Science* 45: 1750–1757.

Selvi A, Nair NV, Noyer JL, Singh NK, Balasundaram N, Bansal KC, Koundal KR, Mohapatra T (2006) AFLP analysis of the phenetic organization and genetic diversity in the sugarcane complex, Saccharum and Erianthus. *Genetic Resources and Crop Evolution* 53: 831–842.

Sengar AS, Thind KS, Bipen Kumar, Mittal Pallavi, Gosal SS (2009) *In vitro* selection at cellular level for red rot resistance in sugarcane (*Saccharum* sp.). *Plant Growth Regulation* 58: 201–209.

Sétamou M, Bernal JS, Legaspi JC, Mirkov TE, Legaspi BC (2002) Evaluation of lectin-expressing transgenic sugarcane against stalkborers (Lepidoptera: Pyralidae): Effects on life history parameters. *Journal of Economic Entomology* 95: 469–477.

Shriram V, Kumar V, Devarumath RM, Khare TS, Wani SH (2016) MicroRNAs as potential targets for abiotic stress tolerance in plants. *Frontiers in Plant Science* 7: 817.

Shrivastava AK, Solomon S, Rai RK, Singh P, Chandra A, Jain R, Shukla SP (2015) Physiological interventions for enhancing sugarcane and sugar productivity. *Sugar Tech* 17(3): 215–226.

Solomon S (2014) Sugarcane agriculture and sugar industry in India: At a glance. *Sugar Tech* 16(2): 113–124.

Sreenivasan TV, Ahloowalia BS, Heinz DJ (1987) Cytogenetics. In: *Sugarcane improvement through breeding*, Heinz DJ, ed. New York: Elsevier, pp. 211–253.

Srikanth J, Subramonian N, Premachandran MN (2011) Advances in transgenic research for insect resistance in sugarcane. *Tropical Plant Biology* 4: 52–61.

Srivastava MK, Li C-N, Li Y-R (2012) Development of sequence characterized amplified region (SCAR) marker for identifying drought tolerant sugarcane genotypes. *Australian Journal of Crop Science* 6(4): 763–767.

Stuber CW (1992) Biochemical and molecular markers in plant breeding. *Plant Breeding Reviews* 9: 37–61.

Su Y, Xu L, Xue B, Wu Q, Guo J, Wu L, Que Y (2013) Molecular cloning and characterization of two pathogenesis related β-1,3-glucanase genes ScGluA1 and ScGluD1 from sugarcane infected by Sporisorium scitamineum. *Plant Cell Reports* 32: 1503–1519.

Sunkar, R, Li YF, Jagadeeswaran G (2012) Functions of microRNAs in plant stress responses. *Trends in Plant Science* 17: 196–203.

Suprasanna P, Patade VY, Desai NS, Devarumath RM, Kawar PG, Pagariya MC, Ganapathi A, Manickavasagam M, Babu KH (2011) Biotechnological developments in sugarcane improvement: An overview. *Sugar Tech* 13(4): 322–335.

Thiebaut F, Grativol C, Carnavale-Bottino M, Rojas CA, Tanurdzic M, Farinelli L, Martienssen RA, Hemerly AS, Ferreira PC (2012) Computational identification and analysis of novel sugarcane microRNAs. *BMC Genomics* 13: 290.

Thiebaut F, Grativol C, Tanurdzic M, Carnavale-Bottino M, Vieira T, Motta MR et al. (2014) Differential sRNA regulation in leaves and roots of sugarcane under water depletion. *PLOS ONE* 9(4): e93822.

ThuljaramRao J (1987) Sugarcane origin, taxonomy, breeding and varieties. In: *Sugarcane varietal improvement*, M Naidu, X Screenivasan, M Premachandran, eds. Coimbatore, India: Sugarcane Breeding Institute.

Tiroli AO, Ramos CH (2007) Biochemical and biophysical characterization of small heat shock proteins from sugarcane: Involvement of a specific region located at the N-terminus with substrate specificity. *International Journal of Biochemistry and Cell Biology* 39(4): 818–831.

Vantini JS, Dedemo GC, Jovino Gimenez DF, Fonseca LFS, Tezza RID, Mutton MA, Ferro JA, Ferro MIT (2015) Differential gene expression in drought-tolerant sugarcane roots. *Genetics and Molecular Research* 14(2): 7196–7207.

Virupakshi S, Naik GR (2008) ISSR analysis of chloroplast and mitochondrial genome can indicate the diversity in sugarcane genotypes for red rot resistance. *Sugar Tech* 10: 65–70.

Viswanathan C, Anburaj J, Prabu G (2014) Identification and validation of sugarcane streak mosaic virus encoded microRNAs and their targets in sugarcane. *Plant Cell Reports*. doi:10.1007/s00299-013-1527-x.

Wahid A, Ghazanfar A (2006) Possible involvement of somesecondary metabolites in salt tolerance of sugarcane. *Journal of Plant Physiology* 163: 723–730.

Wang W, Vinocur B, Shoseyov O, Altman A (2004) Role of plant heat-shock proteins and molecular chaperones in the abiotic stress response. *Trends in Plant Science* 9(5): 244–252.

Wang Z-Z, Zhang S-Z, Yang B-P, Li Y-R (2005) Trehalose synthase gene transfer mediated by *Agrobacterium tumefaciens* enhances resistance to osmotic stress in sugarcane. *Sugar Tech* 7(1): 49–54.

Weng LX, Deng HH, Xu JL, Li Q, Zhang YQ, Jiang ZD, Li QW, Chen JW, Zhang LH (2010) Transgenic sugarcane plants expressing high levels of modified cry1Ac provide effective control against stem borers in field trials. *Transgenic Research*. doi:10.1007/s11248-010-9456-8.

Wiedenfeld B (2008) Effects of irrigation water salinity and electrostatic water treatment for sugarcane production. *Agricultural Water Management* 95: 85–88.

Yang Y, Zhang X, Chen Y, Guo J, Ling H, Gao S, Su Y, Que Y, Xu L (2016) Selection of reference genes for normalization of microRNA expression by RT-qPCR in sugarcane buds under cold stress. *Frontiers in Plant Science* 7: 86.

Zambrano AY, Demey JR, Gonzalez V (2003) *In vitro* selection of a glyphosate-tolerant sugarcane cellular line. *Plant Molecular Biology Reporter* 21: 365–373.

Zhang L, Xu J, Birch RG (1999) Engineered detoxification confers resistance against a pathogenic bacterium. *Nature Biotechnology* 17: 1021–1024.

Zhang SZ, Zheng XQ, Lin JF, Guo LQ, Zan, LM (2000) Cloning of trehalose synthase gene and transformation into sugarcane. *Journal of Agricultural Biotechnology* 8(4): 385–388.

Zhangsun DT, Luo SL, Chen RK, Tang KX (2007) Improved Agrobacterium-mediated genetic transformation of GNA transgenic sugarcane. *Biologia* 62(4): 386–393.

Zhou D, Wang C, Li Z, Chen Y, Gao S, Guo J, Lu W, Su Y, Xu L, Que Y (2016) Detection of bar transgenic sugarcane with a rapid and visual loop-mediated isothermal amplification assay. *Frontiers in Plant Science* 7: 279.

Zhu YZ, McCafferty H, Osterman G, Lim S, Agbayani R, Lehrer A, Schenck S, Komor E (2011) Genetic transformation with untranslatable coat protein gene of sugarcane yellow leaf virus reduces virus titers in sugarcane. *Transgenic Research* 20: 503–512.

chapter two

Molecular mapping techniques
Application for the development of stress-tolerant sugarcane

Kalpana Sengar

Contents

Genetic diversity and mapping in plants

Conventional plant breeding schemes have typically used phenotypic data processed by complex statistical analysis to select new cultivars in breeding populations. This approach is still valid and used, although there are limitations in its efficiency. For instance, some agronomically useful traits are either very difficult to select on the basis of phenotype or cannot be selected on this basis alone.

On the other hand, there are traits showing continuous phenotypic variation because they are controlled by several genes, the individual effects of which are relatively small (Yano and Sasaki, 1997), making the selection base on these traits a hard task.

The use of molecular markers and genetic maps could overcome many of the limitations of conventional breeding. Molecular markers are the most modern and powerful tools that can be used in the old art of selection and they are used in a variety of approaches to characterize and conserve plant genetic resources.

A wide range of molecular markers and maps are now available in a great number of crops, revolutionizing breeding strategies through the genetic diversity studies, taxonomical identification of species and cultivars, the identification of the quantitative trait loci (QTL), and the integration of molecular maps with linkage maps based on observable phenotypes (Figure 2.1).

The major uses of molecular markers in agricultural research are related to estimating the genetic relationships between populations within species, establishing and managing gene banks, studying gene flow from domesticated populations to wild relatives, and estimating and monitoring the effective population size (FAO, 2005). At present, many studies can be found about the use of molecular markers in genetic diversity in plants, including those reported in the Brazilian upland rice gene bank by Ferreira (2005); sorghum (Grenier et al., 2000); sandalwood (Shashidhara et al., 2003), and wheat (Börner et al., 2000).

In Cuba, studies about the use of molecular markers for genetic diversity research were applied to sugarcane breeding. A group of 35 wild *Saccharum* complex clones (ECL clones) and representatives of the main germplasm diversity regularly used in introgression were studied for their and cytoplasmic genetic diversity by restriction fragment length polymorphism (RFLP). Results obtained suggested the occurrence

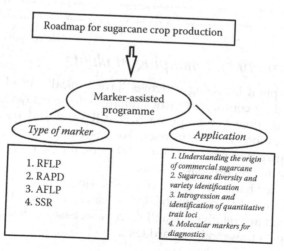

Figure 2.1 Roadmap for marker-assisted sugarcane production.

of genetic recombination among them. The diversity revealed contrasts with what is usually observed and points to an active population evolution (Coto et al., 2002). On the other hand, 89 *Saccharum* complex clones, including the aforementioned collection and chosen for their outstanding vigour and resistance to major diseases, were studied. Their genetic diversity revealed agronomical components, as their molecular groups were employed to recommend sources to increase the genetic base of the Cuban introgression program in sugarcane (Coto et al., 2005). Important results also obtained in Cuba using molecular markers in genetic diversity studies were reported in tropical fruits like avocado (Ramirez et al., 2005) and guava (Rodríguez et al., 2004).

Another use of molecular markers in the study of germplasm collections is related to their use for an accurate identification of plant materials. Traditionally, varieties are identified employing morphological and agronomical markers; these are sometimes complemented by physiological, biochemical, and cytogenetic markers. New varieties must be distinct from those already registered by the expression of at least one characteristic and they must achieve standards of uniformity and stability. In this sense, molecular markers are being successfully used for the taxonomical identification of wild cultivars collected for widening the genetic base in germplasm.

Frequently, wild cultivars are so similar that is almost impossible to detect morphological traits useful for the identification procedure. Consequently, misclassification problems of very similar varieties arise. For instance, two wild species of *Saccharum* spp. collected in Laos by an expedition of Cuban breeders were misclassified based on morphological characteristics, a mistake that was overcome using RFLP and specific primers for *Erianthus* spp. and 5S rDNA (Coto, 2001).

The International Union for Protection of New Varieties of Plants, commonly known as UPOV, established the requirements of distinctness, uniformity, and stability, referred to as DUS. A Test Guidelines developed by this organization was used for a first time using public maize inbreeds in India (Mauria, 2000).

An approach based on the development of molecular descriptors was established and employed to assist traditional methods in the selection of genetic resources for the Cuban sugarcane introgression programme. Based on this approach, a group of RFLPs reported elsewhere as characteristics for nuclear and cytoplasmic DNA and/or different taxa were employed to separate groups of clones in the collection assayed and to provide information for breeding and taxonomic purposes using two types of markers: a group marker (**G**) present in all members of the group (taxonomic, geographic, diversity, etc.) and an individual marker (**I**) that differentiates between accessions of the same group or among all the genotypes considered (Cornide et al., 2000). Other markers (amplified fragment length polymorphism [AFLP] and random amplified micro-satellite polymorphism [RAMP]) were also employed to characterize trans-genic sugarcane plants obtained previously by Arencibia et al. (1999).

Another extremely useful application of molecular markers technology is marker-assisted selection (MAS). MAS is a complementary technology, used in conjunction with more established conventional methods of genetic selection in plant improvement, and different approaches were investigated since the initial report made by Tanksley (1983). The MAS approach has been widely studied and many molecular maps in plants are available. Novel approaches are being applied including the use of marker-assisted backcross (MABC) selection in maize (Ribaut and Ragot, 2006).

Although in the last two decades a great deal of research has been dedicated to improving the selection process of several plant species employing the available molecular markers, some fundamental issues remain to be resolved in the MAS approach, particularly regarding complex traits, and including the development of high throughput precision phenotyping systems for QTL mapping, improved understanding of genotype by environment interaction and epistasis, and development of publicly available computational tools tailored to the needs of molecular breeding programs (Xu and Crouch, 2008).

Although the current impact of MAS on products useful for farmers in developing countries is still limited, their future possibilities and potential impacts are considerable. However, several obstacles should be solved, for instance, the high costs of the technology, a limited infrastructure, the absence of conventional breeding programmes, and problems related with intellectual property rights (FAO, 2004).

Molecular markers used in sugarcane

Molecular biology has begun to play an important role in agriculture due to its ability to modify microorganisms, plants, animals, and agricultural processes. Particularly, it can aid conventional plant breeding programmes using molecular markers. Several types of DNA markers and molecular breeding strategies are now available to plant breeders and geneticists to help overcome many problems faced in conventional breeding (Caetano-Anollés et al., 1991; Weising et al., 1998).

Sequences (and restriction site analyses) are the only molecular markers that contain a comprehensible record of their own history. Hence, appropriate analysis based on sequence data (or restriction site data) can provide hypotheses on the relationships between the different genotypic categories (or species) that they class together (Karp et al., 1996).

DNA markers have the potential to enable individual clones to be differentiated reliably and unambiguously. Such markers are widely used in human forensics and paternity testing; in animal paternity testing; and in fingerprinting a wide range of plants, animals, and microorganisms. To assess germplasm diversity within the genus *Saccharum*, a diversity of methods has been employed. One of the initial molecular methods for varietal identification of sugarcane cultivars has been in the development of isozymes and RFLP. The development of polymerase chain reaction (PCR)–based molecular markers such as random amplified polymorphic DNA (RAPD), amplified fragment length polymorphism (AFLP), and, more recently, microsatellites or simple sequence repeats (SSRs) has allowed for fine-scale genetic characterization of germplasm collections previously thought impossible. Of these techniques, SSRs appear promising in relation to reproducibility and interchangeability of results, which is required for cultivar identification.

Other methods tested for determining general relationships between the different *Saccharum* species include low-copy nuclear sequences (Lu et al., 1994a, 1994b), polymorphisms of nuclear ribosomal DNA (Glaszmann et al., 1990), and polymorphisms of cytoplasmic DNA (D'Hont et al., 1994). Studies relating to the characterization of the double genome structure and the genome organization of modern sugarcane cultivars (Irvine, 1999) using genomic in situ hybridization have also been published (D'Hont et al., 1994, 1996; Grivet et al., 1996).

Early research on sugarcane genotyping had indicated that isozymes could be used to differentiate sugarcane clones (Fautret and Glaszmann, 1988; Heinz, 1969; Thom and Maretzki, 1970; Waldron and Glasziou, 1971). This, however, had been limited to detecting the identity of varietal material and the verification of hybrid origins of offspring after a cross. Glaszmann et al. (1989) took this a step further and used isozymes to differentiate between wild and noble canes. Their research showed strong differentiation between *S. spontaneum* and *S. officinarum*, revealing that most of the diversity among sugarcane varieties is related to the presence or absence of *S. spontaneum* genes. Their research also indicated that multiple bands of unequal intensities were often present and this was consistent with the high polyploidy present in the various groups of sugarcane. This brings about practical difficulties in characterizing clones due to the high number of bands that may migrate at similar distances.

Gallacher et al. (1995) found that some isozyme markers are not polymorphic, may be weak and unreliable, and may produce different results in different laboratories (Glaszmann et al., 1989; Ortiz, 1983). Genetic interpretation of isozyme data in the highly polyploid *Saccharum* spp. is often complex. The inability to interpret much of the observed variation limits the practical application of isozyme visualization (Gallacher et al., 1995) and no set of isozymes can differentiate among all clones. They also give insufficient information to indicate confidently genetic similarity between clones. In addition, due to the requirement that samples used are freshly collected, their wider application is limited by the requirement for local cultivation of the plants for varietal identification (Eksomtramage et al., 1992).

Restriction fragment length polymorphisms (RFLPs)

RFLPs (Burnquist et al., 1992; D'Hont et al., 1995; Jannoo et al., 1999; Lu et al., 1994a, 1994) are the original plant DNA marker. They are robust, reliable, dominant, and are simply inherited, naturally occurring Mendelian characteristics. These markers, first used in the construction of genetic maps (Botstein et al., 1980), have now been successfully used in the assessment of genetic variability and in the elucidation of phylogenetic relationships among plant populations (Debener et al., 1990; Miller and Tanksley, 1990; Song et al., 1988, 1990; Wang et al., 1992).

Several research groups (Burnquist et al., 1992; Grivet et al., 1996; Lu et al., 1994a, 1994b) have tested this marker technique in sugarcane. It has been used to show a strong molecular differentiation between *S. officinarum* and *S. spontaneum* and that the major part of the diversity among sugarcane cultivars is due to the *S. spontaneum* fraction of the genome. However, only a limited variability within *S. officinarum* was identified, although only a small number of clones have been assessed. A more recent and larger study (Jannoo et al., 1999) used RFLP probes on 53 clones of *S. officinarum* and 109 cultivars from different sugarcane-growing regions of the world. It revealed a larger degree of RFLP polymorphism with the surveyed material, thereby allowing an analysis of the organization of genetic diversity within *S. officinarum* and the studied cultivars. Their results contrasted with those of Lu et al. (1994a, 1994b) indicating that a considerable diversity does exist within *S. officinarum*.

Random amplified polymorphic DNA (RAPDs)

RAPDs were the original PCR-based DNA marker, and are relatively easy to use, rapid, moderately reliable, and dominant (Welsh and McClelland, 1990; Williams et al., 1990), and suited for efficient nonradioactive DNA fingerprinting of genotypes (dos Santos et al., 1994; Thormann et al., 1994). In sugarcane, RAPDs have been used to detect polymorphisms in a quick and reproducible manner (Oropeza and Degarcia, 1997); to determine genetic diversity successfully in 20 commercial sugarcane hybrids (Harvey and Botha, 1996), as well as between members of the *Saccharum* complex; and in resolving taxonomical groups in cluster analyses (Harvey and Botha, 1996; Nair et al., 1999). RAPDs have also been shown to be efficient in detecting gross genetic changes in sugarcane cultivars subject to prolonged periods in tissue culture, although sensitivity was lacking when the technique was applied to detecting minor genetic changes, such as that which occurs during sugarcane genetic transformation (Taylor et al., 1995).

The use of RAPD markers in determining relationships between distantly related species or genera has, however, been questioned (Pan et al., 1997) because markers are nonlocus specific (Karp et al., 1996; Kesseli et al., 1994). Difficulties with reproducibility and amplification (Karp et al., 1997) have also been recorded.

Amplified fragment length polymorphisms (AFLPs)

This method rapidly generates hundreds of reproducible markers from DNA of any organism allowing the high-resolution genotyping of fingerprinting quality. The key feature of this technique is its capacity for the simultaneous screening of many different regions distributed randomly throughout the genome (Mueller and Wolfenbarger, 1999). As a means of determining genetic diversity, AFLPs have been found to be of value in many crop species (Loh et al., 2000; Lu et al., 1996; Muluvi et al., 1999; Sensi et al., 1996; Tohme et al., 1996) including sugarcane (Besse et al., 1998).

These markers have also been highly successful as a means of producing high-density linkage maps (Boivin et al., 1999; Jin et al., 2000; Schondelmaier et al., 1996; Van Eck et al., 1995) or as a means of saturating a map region containing a trait of interest (Haanstra et al., 1999; Hartl et al., 1999; Thomas et al., 1995).

The use of AFLP markers for determining genetic diversity and mapping in sugarcane has been assessed by Besse et al. (1998) and Xu et al. (1999). Besse et al. (1998) used these markers as a means of providing a preliminary characterization of the levels of diversity in wild collections of members of the *Saccharum* complex with the ultimate aim of using the technique to assist in the selection of diverse accessions as possible parents in introgression programmes. AFLP markers were able to reveal the major *Saccharum* complex groups corresponding to the genera under study, and the results were in accordance with data obtained using RFLP markers (Besse et al., 1997; Lu et al., 1994b). These results indicate that these markers have the potential as genetic markers in sugarcane to analyse relationships within and between species within the same genus, with the additional advantage of rapid data generation.

The method has been promoted as being highly reproducible allowing for high-resolution genotyping of fingerprinting quality. However, despite the encouraging results of Besse et al. (1998), there are questions regarding their reproducibility in sugarcane.

Simple sequence repeats (SSRs)

SSRs are the most frequent DNA markers used for fingerprinting and have been developed for use in sugarcane (Cordeiro et al., 1999, 2000). The principal reason for the increasing success of SSRs as a molecular tool is that they provide a higher incidence of detectable polymorphisms than other techniques such as RFLPs and RAPDs (Powell et al., 1996a).

An example of the success of this technique in fingerprinting a vegetatively propagated crop can be found in the grape. With over 6000 varieties identified worldwide based on morphological criteria, there has been a need for molecular markers for fingerprinting. SSRs have now been studied intensively in this crop. The use of these markers for fingerprinting (Cipriani, 1994; Lamboy and Alpha, 1998; Sefc et al., 1998) has led to the successful reconstruction of the complex genetic relationship among a number of European grapevine cultivars (Sefc et al., 1998). SSRs have also been used for germplasm management (Lopes et al., 1999) in the Portuguese grapevine gene pool. Studies in grape varietal identification have shown that SSRs (as compared with RFLPs and AFLPs) are the most promising molecular method for reproducible and interchangeable cultivar identification (Sanchez-Escribano et al., 1999).

The ability of SSRs to reveal high allelic diversity is particularly useful in distinguishing between genotypes. The success of using these markers in other crop species like barley (*Hordeum vulgare*) (Russell et al., 1997; Saghai Maroof et al., 1994), rice (*Oryza sativa*) (Wu and Tanksley, 1993), wheat (*Triticum aestivum*) (Röder et al., 1995), apple (*Malus domestica*) (Szewc-McFadden et al., 1996) and avocado (*Persea americana*) (Lavi et al., 1994) has encouraged the testing of SSRs in sugarcane. Where characterization and identification of germplasm for purposes of research, product development, conservation, measurement, and monitoring of genetic diversity in agriculture and for support of intellectual property are concerned, microsatellite repeats exceed the capabilities of RFLPs (Smith et al., 1997).

Much of the early characterization of SSRs has relied on database searches of published sequences or on the construction of genomic libraries. The recent development of new microsatellite enrichment techniques (Cordeiro et al., 1999; Edwards et al., 1996) has, however, increased the efficiency of microsatellite characterization in species for which little or no previous sequence knowledge is available. However, in contrast to methods such as RFLPs that do not require previous sequence knowledge, the development of SSRs requires an initial high cost and labour-intensive development.

Because of the high cost of SSR development, a consortium of sugarcane biotechnologists developed a collaborative effort to isolate approximately 200 microsatellite markers through the creation of an enriched

SSR DNA library (Cordeiro et al., 1999; Edwards et al., 1996). Testing of these markers on a small sample population of five sugarcane genotypes revealed each marker to have between 3 and 12 alleles, with an average of 8. Markers showing polymorphisms had a polymorphism information content (PIC) (Weir, 1996) value of between 0.48 and 0.8, with a mean value of 0.72 (Cordeiro et al., 2000). This PIC value, a determination of the value of a marker in detecting polymorphism, indicates that sugarcane microsatellite markers will be suitable for use in genotypic identification.

Application of molecular markers

Molecular markers have great potential for improving the efficiency of the breeding process not only by targeting traits to be selected in one generation but also by the precision and efficiency with which the genotypes can be selected. Information obtained with these markers has contributed to a better understanding of origin, evaluation, genetics, and QTL. Molecular markers are having an impact in several areas of sugarcane improvement and disease management. For each area, their impact to date and likely contribution to the future are discussed in turn next.

Understanding the origin of commercial sugarcane

Sugarcane belongs to the informal taxonomical group the "Saccharum complex," which contains species from the genera *Saccharum, Erianthus, Sclerostachya, Miscanthus,* and *Narenga* (Daniels et al., 1975). This gave rise to the suggestion that sugarcane may have emerged from hybridization involving different genera including *Saccharum, Erianthus,* and *Miscanthus* (Daniels and Roach, 1987). Various molecular marker systems including isozymes (Glaszmann et al., 1989), RFLPs (Lu et al., 1994), ribosomal RNA (Glaszmann et al., 1990), mitochondria and chloroplast genes (D'Hont et al., 1993), RAPDs (Nair et al., 1999), and SSRs (Cordeiro et al., 2003; Selvi et al., 2003) have shown that the three genera have highly contrasting patterns and can be easily differentiated. Further evidence of the distinction of the genera is provided by the occurrence of repeated elements specific to *Erianthus* and *Miscanthus,* which are not detected in the *Saccharum* genus (Alix et al., 1999). Relatively recently, *in situ* hybridization analysis of two ribosomal RNA gene families determined that *S. officinarum* has a basic chromosome number of $x = 10$, meaning that these plants are octoploid (D'Hont et al., 1998). Using the same method it was shown that *S. spontaneum* has a basic chromosome number of $x = 8$ and that the ploidy level of this species varies between 5 and 16 (D'Hont et al., 1998; He et al., 1999). These studies established the coexistence of two distinct chromosome organizations in modern sugarcane cultivars. Using genomic *in situ* hybridization (GISH), D'Hont et al. (1996) and Cuadrado et al. (2004)

have demonstrated that modern cultivars contain about 15% to 20% *S. spontaneum* chromosomes and less than 5% are recombinant or translocated chromosomes. Their high ploidy and complex genome structure creates challenges for both transgene expression and the development of molecular markers.

Sugarcane diversity and variety identification

Investigation of diversity within sugarcane cultivars is highly heterozygous with many distinct alleles at a locus (Jannoo et al., 1999b). This has also been demonstrated with SSRs (Selvi et al., 2003). The majority of the diversity is due to the more polymorphic and smaller contribution of the *S. spontaneum* portion of the genome (Jannoo et al., 1999b). Recently, AFLP markers have shown that they can be used to determine genetic similarity among sugarcane cultivars (Lima et al., 2002). In the development of modern cultivars, only a small number of meiotic divisions have levels of linkage disequilibrium expected among modern cultivars. A study by Jannoo et al. (1999a) on Mauritian cultivars confirmed this, where chromosome haplotypes were conserved over regions as long as 10 cM. This could have major implications on detection and location of genes involved in traits of interest using association-mapping techniques (Butterfield et al., 2003).

One of the major ways sugarcane industries have already benefited from molecular markers is the use of SSRs for cultivar identification. As sugarcane is highly heterozygous, it is vegetatively propagated and goes through a number of stages of selection, propagation, and replanting to produce a new cultivar. Consequently, there are many opportunities for mistakes to occur, varieties to be mislabelled, or propagated unintentionally. SSR markers have been used to fingerprint 180 sugarcane varieties and the data stored in a database (Piperidis et al., 2004). This provides a source of information to identify varieties of unknown or disputed origin. SSR profiles of sugarcane varieties can also be used as additional information in plant breeding rights application and for quality assurance for delivery of new cultivars to the industry (Piperidis et al., 2004).

Introgression and identification of quantitative trait loci (QTL)

As sugarcane cultivars have a narrow genetic base, sugarcane breeders have been interested in incorporating new germplasm into the breeding pool. *Erianthus arundinaceous* is one species that has been of interest to sugarcane breeders because of its good ratooning performance and tolerance to environmental stresses. Hybrid progeny have been produced from crossing *E. arundinaceous* and *S. officinarum*, and molecular markers and GISH have been used to verify these crosses (D'Hont et al., 1995; Piperidis et al., 2000). These hybrids were not themselves fertile. However,

Deng et al. (2002) have generated fertile hybrids between *E. arundinaceous* and *S. officinarum*, which have been confirmed by analysis of species-specific SSRs (Cai et al., 2005). This significant development provides a new genetic stock for exploitation in commercial sugarcanes. Introgression of genes from these types of hybrids would be greatly facilitated by identification of molecular markers linked to genes of interest. Molecular genetic mapping of *Saccharum* species has been limited due to its polyploid nature and the resulting mix of single- and multi-dose alleles. But using single-dose markers (Wu et al., 1992), linkage maps have been constructed with RAPD (Da Silva et al., 1995; Guimaraeas et al., 1997a; Ming et al., 1998, 2002b) markers for *S. spontaneum*, *S. officinarum*, and *S. robustum*. Single-dose linkage maps have also been constructed in cultivars using RFLP (Grivet et al., 1996), AFLP (Hoarau et al., 2001), and AFLP and SSR markers (Aitken et al., 2005; Rossi et al., 2003). Aitken et al. (2005) put over 1000 SSR and AFLP markers cluster 123 linkage groups into the eight homology groups, which correspond to the basic chromosome number of *S. spontaneum*. For two of these groups, two sets of linkage groups align to one larger linkage group giving 10 homology groups corresponding to the basic chromosome number of *S. officinarum*. There was large variation in map coverage for the different homology groups indicating that genome coverage was not complete. Chromosome pairing in sugarcane is still not fully understood. It does not appear to have either complete disomy or complete polysomy, but some disomic-like pairing behaviour has been observed (Aitken et al., 2005; Hoarau et al., 2001). Due to the size of the sugarcane genome (2C = 7440 Mbp), a large number of markers are required. Even the most complete of these linkage maps, which have over 1000 markers, still only cover about two-thirds of the genome. Development of new high-throughput marker systems like single nucleotide polymorphisms (SNPs) (Grivet et al., 2003) and diversity array technology (DArT) markers (Wenzl et al., 2004) for sugarcane are expected to have a major impact on this area in the future. New theoretical models are also being developed to improve accuracy and extract more information from mapping experiments in complex polyploids (Qu and Hancock, 2001).

Mapping of single gene traits are relatively easy and thus far two single major gene traits, eyespot susceptibility (Mudge et al., 1996) and rust resistance (Daugrois et al., 1996), have been mapped in sugarcane. Apparently, due to the polyploidy nature of the sugarcane genome, the majority of traits of interest quantitatively inherited. The initial QTL studies were done on small populations with sparse map coverage (Guimaraes et al., 1997a; Sills et al., 1995). More recent studies carried out on two interspecific *S. officinarum* × *S. spontaneum* crosses were used to identify QTL for sucrose content and related traits (Ming et al., 2001, 2002c). In total, 102 significant associations were detected, of these 61 could be assigned map locations and 50 clustered in 12 genomic regions on 7 sugarcane homologous groups.

The corresponding location of the QTL suggested the presence of several alleles at a locus that contributed to variation of a trait. Using map comparisons to a number of grasses, some of the alleles were given a candidate identity. Another similar study on the self-crossed progeny of a cultivar again identified numerous QTL of small effect for yield traits (Hoarau et al., 2002). The size of the effects was not consistent across crop cycles and total explained from 30% to 55% of the phenotypic variation. Comparative analysis of QTL affecting plant height and flowering between sorghum and sugarcane identified QTL clusters for both these traits in sugarcane, which corresponded closely to QTL previously mapped in sorghum (Ming et al., 2002a). The large number of alleles with small effect detected in all these studies is largely due to high ploidy and high levels of heterozygosity in sugarcane, which results in numerous alleles at a locus. Although theory is being developed for QTL detection in polyploids (Doerge and Craig, 2000), there is still a need for improved biometrical methods and tools to extract the maximum amount of information from QTL studies.

The conservation of chromosome organization across the grasses also has the potential to impact on location and identification of genes of interest in sugarcane. The colinearity between sugarcane and sorghum chromosomes (Guimaraes et al., 1997b; Ming et al., 1998) means that genes identified in this diploid relative could be of use in locating the same genes in sugarcane. Although markers linked to QTL for a number of agronomically important traits have been identified in sugarcane, the implementation of these markers has yet to commence. Work is under way to use marker-assisted selection introgress QTL for high sucrose identified in *S. officinarum* into commercial sugarcane (Aitken et al., 2002). Silva and Bressiani (2005) described the development of an expressed sequence tag (EST)-derived RFLP marker for sugarcane elite genotypes, which can be QTL tagging for sugar content. It is likely that cultivars produced with molecular marker input will arise through introgression studies initially. Use of markers in routine breeding activities will occur only when high-throughput marker detection and more robust statistical methods are developed.

Molecular markers for diagnostics

Development of new and improved molecular assays to detect various sugarcane pathogens is progressing in different laboratories worldwide (Braithwaite and Smith, 2001). Molecular diagnostic tests, based mostly on nucleic acids, are highly sensitive and relatively easy to use compared with the traditional detection method such as histology, electron microscopy, sap transmission onto indicator plants, or the isolation and culture of the causal agents. Due to their high sensitivity, molecular diagnostic methods are capable of detecting pathogens even in asymptomatic plants with an extremely low pathogen titer. Molecular tests for diseases such as ratoon stunting (Fegan et al., 1998; Pan et al., 1998), Fiji disease (Smith and van de Velde, 1994; Smith et al., 1994), mosaic (Smith and van de Velde, 1994), striate mosaic (Thompson et al., 1998), yellow leaf syndrome (Chatenet et al., 2001; Irey et al., 1997), smut (Albert and Sehenek, 1996), sorghum mosaic (Yang and Mirkow, 1997), and SCBV (Braithwaite et al., 1995) have been developed and are being applied to screen quarantine germplasm. With advances in genome sequencing and the continued refinement of various techniques in recombinant DNA technology, development of more reliable, faster, and cost-effective molecular diagnostic tests for all the important sugarcane pathogens can be expected in the near future. Molecular markers have been used to assess genetic diversity in sugarcane germplasm (Jannoo et al., 2001; Hemaprabhha and Sree Rangaswamy, 2001; Nair et al., 2002; Selvi et al., 2003).

Conclusion

The rapid development of molecular marker types, novel biotechnological tools, and their multiple applications in plant breeding is a permanent challenge for plant breeders in developed and developing countries. The present review is an effort to update plant breeders about the main biotechnological advances in the aim to contribute to the development of well-qualified human resources. RFLPs and SSRs require an initial investment in terms of probe or sequence information. SSRs, in particular, are very costly and time-consuming to develop owing to this requirement. However, both these techniques allow probes of primer pairs that are nonpolymorphic in the germplasm set or population to be excluded from future consideration. AFLPs and RAPDs, however, do not require any prior sequence information, but do allow empirical determination of which primer pairs and restriction enzymes are best for a given germplasm. Their dominant nature of inheritance, however, results in a lack of comparative information at each assayed locus thus precluding an accurate assessment of true genetic relationships. SSRs and AFLPs are, however, much simpler to apply, more sensitive, and have the advantage that they can be automated. Due to their multi-allelism and co-dominance, SSR markers appear suited for the analysis of outcrossing heterozygous individuals, which makes them ideal for sugarcane. AFLPs, unfortunately, have problems with reproducibility when used in sugarcane. Hence, the level of polymorphism, dominant or co-dominant inheritance of the marker, convenience, technical difficulty, availability of species-specific probes/primers, reproducibility, quantity of DNA required, and the ease of exchange of data between laboratories are all factors contributing to the choice of marker. It would be difficult to find a marker that meets all criteria, but a marker system can be identified that would fulfil a number of the desired qualities. In the near future, molecular markers can provide simultaneous and sequential selection of agronomically important genes in sugarcane breeding programs allowing screening for several agronomically important traits at early stages and effectively replace time-consuming bioassays in early generation screens. Although, the potential of biotechnology has often been exaggerated, a high level of optimism is clearly justified for its use in the improvement of sugarcane. Undoubtedly, functional genomics, as it is now termed, will revolutionize the way in which plant breeding is undertaken in the future. Basic research is leading to an improved understanding of the genetic mechanisms operating within a plant in response to the diverse stresses that it is exposed to, as well as the overall production of biomass and grain. The challenge for developing countries is to tap as much of this emerging technology as possible.

References

Arencibia A., Carmona E., Cornide M.T., Castiglioni S., O'Relly J., Chinea A., Oramas P. and Sala F. (1999). Somaclonal variation in insect-resistant transgenic sugarcane (*Saccharum* hybrid) plants produced by cell electroporation. *Transgenic Research* 2180, 1–12.

Besse P., Taylor G., Carroll B., Berding N., Burner D. and McIntyre C.L. (1998). Assessing genetic diversity in a sugarcane germplasm collection using an auto-mated AFLP analysis. *Genetica* 104, 143–153.

Boivin K., Deu M., Rami J.F., Trouche G. and Hamon P. (1999). Towards a saturated sorghum map using RFLP and AFLP markers. *Theoretical and Applied Genetics* 98, 320–328.

Börner A., Chebotar S. and Korzun V. (2000). Molecular characterization of the genetic integrity of wheat (*Triticum aestivum* L.) germplasm after long-term maintenance. *Theoretical and Applied Genetic* 100, 494–497.

Burnquist W.L., Sorrells M.E. and Tanksley, S. (1992). Characterization of genetic variability in *Saccharum* germplasm by means of restriction fragment length polymorphism (RFLP) analysis. *XXI Proceedings of the International Society of Sugarcane Technologists* 2, 355–365.

Caetano-Anollés G., Bassam, B.J. and Gresshoff, P.M. (1991). DNA-fingerprinting: a strategy for genome analysis. *Plant Molecular Biology Reporter* 9, 294–307.

Cornide M.T., Coto O., Calvo D., Canales E., de Prada F. and Pérez G. (2000). Molecular markers for the identification and assisted management of genetic resources for sugarcane breeding. *Plant Varieties and Seeds* 13, 113–123.

Coto O. (2001). Caracterización e identificación molecular de clones del complejo *Saccharum* para la ampliación de la base genética del mejoramiento cañero en Cuba. PhD thesis.

Coto O., Cornide M.T., Calvo D., Canales E., D'Hont A. and de Prada F. (2002). Genetic diversity among wild sugarcane germplasm from Laos revealed with markers. *Euphytica* 123, 121–130.

Coto O., Cornide M.T., Rodríguez M., Hernández I., Canales E., de Prada F. and Pérez G. (2005). Diversidad genética basada en marcadores moleculares de una colección de clones del complejo *Saccharum* utilizados en el programa de introgresión de la caña de azúcar en Cuba. *Cultivos Tropicales* 26(1), 41–48.

D'Hont A., Grivet L., Feldmann P., Rao S., Berding N. and Glaszmann J.C. (1996). Characterisation of the double genome structure of modern sugarcane cultivars (*Saccharum* spp.) by molecular cytogenetics. *Molecular and General Genetics* 250, 405–413.

D'Hont A., Lu Y.H., Gonzalez de Leon D., Grivet L., Feldmann P., Lanaud C. and Glaszmann J.C. (1994). A molecular approach to unravelling the genetics of sugarcane, a complex polyploid of the Andropogoneae tribe. *Genome* 37, 222–230.

Eksomtramage T.F., Laulet F., Noyer J.L., Feldmann P. and Glaszmann J.C. (1992). Utility of isozymes in sugarcane breeding. *Sugar Cane* 3, 14–21.

FAO. (2004). Molecular marker assisted selection as a potential tool for genetic improvement of crops, forest trees, livestock and fish in developing countries. FAO Biotechnology Forum, Summary Document, Conference 10. Available at http://www.fao.org/biotech/logs/C10/summary.htm

FAO. (2005). The role of biotechnology for the characterization and conservation of crop, forest, animal and fishery genetic resources in developing countries. Electronic Forum on Biotechnology in Food and Agriculture: Conference 13. http://www.fao.org/biotech/Conf13.htm

Fautret A. and Glaszmann J.C. (1988). Isozyme electrophoresis for sugarcane clonal identification at IRAT CIRAD. International Society of Sugarcane Technologists Cane Disease Committee, Pathology Workshop.

Ferreira M.E. (2005). Molecular analysis of genebanks for sustainable conservation and increased use of crop genetic diversity. *The Role of Biotechnology*, Villa Gualino, Turin, Italy, 5–7 March, pp. 77–82.

Gallacher D.J., Lee D.J. and Berding N. (1995). Use of isozyme phenotypes for rapid dis- crimination among sugarcane clones. *Australian Journal of Agricultural Research* 46, 601–609.

Glaszmann J.C., Fautret A., Noyer J.L., Feldmann P. and Lanaud C. (1989). Biochemical genetic markers in sugarcane. *Theoretical and Applied Genetics* 78, 537–543.

Glaszmann J.C., Lu Y.H. and Lanaud C. (1990). Variation of nuclear ribosomal DNA in sugarcane. *Journal of Genetics and Breeding* 44, 191–198.

Grenier C., Deu M., Kresovich S., Bramel-Cox P.J. and Hamon P. (2000). Assessment of genetic diversity in three subsets constituted from the ICRISAT *Sorghum* collection using random and non-random sampling procedures B. Using molecular markers. *Theoretical and Applied Genetics* 101, 197–202.

Grivet L., D'Hont A., Roques D., Feldmann P., Lanaud C. and Glaszmann J.C. (1996). RFLP mapping in cultivated sugarcane (*Saccharum* spp.)—Genome organization in a highly polyploid and aneuploid interspecific hybrid. *Genetics* 142, 987–1000.

Heinz D.J. (1969). Isozyme prints for variety identification. *International Society of Sugarcane Technologists Cane Breeders' Newsletter* 24, 8.

Jannoo N., Grivet L., Seguin M., Paulet F., Domaingue R., Rao P.S., Dookun A., D'Hont A. and Glaszmann J.C. (1999). Molecular investigation of the genetic base of sugarcane cultivars. *Theoretical and Applied Genetics* 99, 171–184.

Karp A., Seberg O. and Buiatti M. (1996). Molecular techniques in the assessment of botanical diversity. *Annals of Botany* 78, 143–149.

Lu Y.H., D'Hont A., Paulet F., Grivet L., Arnaud M. and Glaszmann J.C. (1994a). Molecular diversity and genome structure in modern sugarcane varieties. *Euphytica* 78, 217–226.

Lu Y.H., D'Hont A., Walker D.I.T., Rao P.S., Feldmann P. and Glaszmann J.C. (1994b). Relationships among ancestral species of sugarcane revealed with RFLP using single copy maize nuclear probes. *Euphytica* 78, 7–18.

Mauria S. (2000). DUS testing of crop varieties-a synthesis on the subject for new PVP-opting countries. *Plant Varieties and Seeds* 13, 69–90.

Ortiz Q.R. (1983). Culture de tissus somatiques chez *Saccharum*. Analyse de la variabilité enzymatique des plantes régénérées. These de Docteur-Ingénieur, Développmentet Amelioration des Végétauz, Université Paris-Sud, Orsay, France.

Ramírez I.M., Fuentes J.L., Rodríguez N.N., Coto O., Cueto J., Becker D. and Rhode W. (2005). Diversity analysis of Cuban avocado varieties based on agro-morphological traits and DNA polymorphisms. *Journal Genet Breeding* 59, 241–252.

Ribaut J.M. and Ragot M. (2006). Marker-assisted selection to improve drought adaptation in maize: the backcross approach, perspectives, limitations, and alternatives. *Journal of Experimental Botany* 58(2), 351–360.

Rodríguez N.N., Valdés-Infante J., Becker D., Velázquez B., Coto O., Ritter E. and Rohde W. (2004). Morphological, agronomical and molecular characterization of guava accessions (*Psidium guajava* L.). *Journal of Genet and Breeding* 58, 79–90.

Shashidhara G., Hema M.V., Koshy B. and Farooqi A.A. (2003). Assessment of genetic diversity and identification of core collection in sandalwood germplasm using RAPDs. *Journal of Horticultural Science and Biotechnology* 78, 528–536.

Song K.M., Osborn T.C. and Williams P.H. (1988). *Brassica* taxonomy based on nuclear restriction fragment length polymorphisms (RFLPs). 2. Premininary analysis of subspecies within *B. rapa* (syn. *campestris*) and *B. oleracea*. *Theoretical and Applied Genetics* 76, 593–600.

Song K.M., Osborn T.C. and Williams P.H. (1990). *Brassica* taxonomy based on nuclear restriction fragment length polymorphisms (RFLPs). 3. Genome relationships in *Brassica* and related genera and the origin of *B. oleracea* and *B. rapa* (syn. *campestris*). *Theoretical and Applied Genetics* 79, 497–506.

Szewc-McFadden A.K., Lamboy W.F., Hokanson S.C. and McFerson J.R. (1996). Utilization of identified simple sequence repeats (SSRs) in *Malus domestica* (apple) for germplasm characterization. *Horticultural Science* 31, 619.

Tanksley S.D. (1983). Molecular markers in plant breeding. *Plant Molecular Biology Reporter* 1, 1–3.

Thom M. and Maretzki A. (1970). Peroxidase and esterase isozymes in Hawaiian sugarcane. *Hawaiian Planters' Record* 58, 81–94.

Waldron J.C. and Glasziou K.T. (1971). Isozymes as a method of varietal identification in sugarcane. *Proceedings of the International Society of Sugarcane Technologists* 14, 249–256.

Weising K., Winter P., Huttel B. and Kshl G. (1998). Microsatellite markers for molecular breeding. *Journal of Crop Production* 1, 113–142.

Xu Y. and Crouch J.H. (2008). Marker-assisted selection in plant breeding: from publications to practice. *Crop Science* 48, 391–407.

Yano M. and Sasaki T. (1997). Genetic and molecular dissection of quantitative traits in rice. *Plant Molecular Biology* 35, 145–153.

chapter three

Coding the non-coding small RNA elements in sugarcane environmental responses

Clícia Grativol, Flávia Thiebaut, Adriana Silva Hemerly and Paulo Cavalcanti Gomes Ferreira

Contents

Introduction

Plants are sessile organisms that are constantly exposed to abiotic stress such as drought, salinity and nutrient deficiencies in the soil, and to biotic influences, such as interaction with beneficial and pathogenic microorganisms. Multiple mechanisms of gene regulation are activated to establish the cellular homeostasis during adverse conditions (Fujita et al., 2006). The identification of genes responsible for desirable agronomic features and the manipulation of these by molecular biology techniques can be used in plant breeding programmes. Gene regulation guided by endogenous small non-coding RNA has been described as essential to normal growth and plant development (Ding et al., 2013; Kidner, 2010; Vazquez, Gasciolli, Crété, and Vaucheret, 2004; Wu, 2013), as well as for adaptation to stress conditions (Ding et al., 2009; Lu, Sun, and Chiang, 2008). Plant small RNA (sRNA) can be divided into two categories: microRNA (miRNA) and small interfering RNA (siRNA) (Guleria, Mahajan, Bhardwaj, and Yadav, 2011). Small RNA act as repressors of their targets, at the transcriptional level – DNA

41

or histone modification – or the post-transcriptional level by target cleavage or inhibition of translation (Ramachandran and Chen, 2009). Both classes of sRNA are 20 to 24 nucleotides in length and the precursors are cleaved by RNAse II-like and DCLs (Dicer-like enzymes) (Llave, Kasschau, Rector, and Carrington, 2002; Reinhart, Weinstein, Rhoades, Bartel, and Bartel, 2002). However, the type of precursor, enzymes involved in the biogenesis and the mechanisms of gene silencing differentiate these two classes of plant sRNA (Axtell, 2013).

The first stage of the miRNA biogenesis process involves transcription of the gene *MIR* by action of RNA polymerase II (Pol II), NOT2 (At-Negative on TATA less 2) and Mediator, forming the primary miRNA (pri-miRNA) (Kim et al., 2011; Wang et al., 2013). Pri-miRNA have a cap-7-methylguanosine at the 5′ end and a poly A tail at the 3′ end, similar to mRNA structure (Lee et al., 2004). DCL1 interacts with a nuclear protein, HYPONASTIC LEAVES 1 (HYL1), which binds to double-stranded RNA to cleave the pri-miRNA (Kurihara, Takashi, and Watanabe, 2006). DCL1 is a canonical enzyme involved in the biogenesis of miRNAs (Liu, Axtell, and Fedoroff, 2012), but in *Arabidopsis* another three DCLs proteins were reported (Liu, Feng, and Zhu, 2009). Another protein that interacts with HYL1 is SERRATA (SE), whose function is necessary for pri-miR processing (Lobbes, Rallapalli, Schmidt, Martin, and Clarke, 2006). The pre-miRNA, a result of pri-miRNA cleavage, is also processed by DCL1 leading to a formation of a double-stranded miRNA/miRNA* with approximately 21 nucleotides in length. Methyl is inserted in the nucleotide of 3′ end in both strands by action of a nuclear methyltransferase (HEN1) to prevent degradation of miRNA (Yu et al., 2005). The double strand is then transported to the cytoplasm by the HASTY enzyme (Park, Wu, Gonzalez-Sulser, Vaucheret, and Poethig, 2005). In the cytoplasm, mature miRNA exerts its function in the regulation of the expression of its target gene (Wu, 2013). Regulation of gene expression via miRNA occurs in three different ways: cleavage of mRNA target, inhibition of target translation, or epigenetic modifications, consisting of target gene methylation (Chen, 2004; Kidner and Martienssen, 2005; Wu et al., 2010). In plants, the main mechanism of plant miRNA regulation is the cleavage of the mRNA target (Kidner and Martienssen, 2005). In this process, miRNA on the complex ribonucleoprotein RISC recognizes the mRNA target by a perfect or close-to-perfect Watson and Crick pairing (Lu and Huang, 2008). This RISC complex contains many proteins, but the best characterized is Argonaute (AGO). In *Arabidopsis* there are 10 AGO proteins, including the AGO1 that has the cleavage activity of the target mRNA (Baumberger and Baulcombe, 2005). After cleavage, the target mRNA fragments are released and RISC can cleave another target mRNA (Bartel, Lee, and Feinbaum, 2004).

In 2002, the first plant miRNA was identified in *Arabidopsis* by computational analysis and cloning (Llave et al., 2002). Due to the importance of miRNA in the regulation of gene expression, extensive investigation aiming to uncover new roles for miRNA in plant development and response to stresses has been carried out. The miRBase (http://microrna.sanger.ac.uk/) is a database where sequences of identified mature miRNAs are deposited, with information of their discovery, precursor structure and function (Kozomara and Griffiths-Jones, 2011). Currently, 28,645 sequences of miRNA precursors are deposited in the miRBase (release 21.0 of June 2014); comprising 35,828 mature miRNA sequences in 223 species, among them, 1,616 precursors were found in monocot plants. MiRNAs are grouped into families based on similarity of their sequences (Sunkar and Zhu, 2004). MiRNA precursors that generate identical or very similar mature miRNAs are grouped in the same family, and the number of members of each family vary (Jones-Rhoades, Bartel, and Bartel, 2006). Isoforms at the same family receive the name of miRNA family, followed by a letter to differ the members, such as miR408a and miR408b. In order to avoid inaccurate annotation of new microRNAs, the scientific community established a list of criteria to classify a miRNA: (1) they should have between 20 and 25 nucleotides in length; (2) miRNA must have single-strand precursors with a self-complementarity region, and lower free energy than transporter RNA (tRNA) or ribosomal RNA (rRNA); (3) miRNA must be contained in the self-complementary region; and (4) be conserved among plants (Zhang et al., 2006).

The first sugarcane miRNAs were identified by searching Expressed Sequence Tag (EST) databases against previously described *Arabidopsis* miRNAs (Zhang et al., 2005). In this study, 25 known and 2 potential miRNAs were identified from EST databases of *S. officinarum* and *Saccharum* spp., respectively. However, only nine of the sugarcane miRNAs identified – miR156, miR159a, miR159c, miR167b, miR168a, miR396, miR408a, miR408b and miR408d – were deposited at the miRBase database. These miRNAs have no more than two different nucleotides in comparison with *Arabidopsis* miRNAs, indicating that miRNAs are highly conserved between plant species. In another study, also based on computational analysis of sugarcane EST databases, seven more mature miRNAs – miR159b, miR159d, miR159c, miR167a, miR168b, miR408c and miR408d – were added to the miRBase (Dezulian, Palatnik, Huson, and Weigel, 2005). Another two studies also contributed to increase the number of sugarcane miRNA precursors available in the miRBase (Ferreira et al., 2012; Zanca et al., 2010). Nowadays, a sequence of 16 miRNAs from *S. officinarum* and 19 from *Saccharum* spp. are available at the miRBase (release 21.0 of June 2014).

Despite the high conservation of plant miRNA, some new miRNAs can be gained and old ones can be lost from the genomes (Ma, Coruh, and Axtell, 2010). Progress in sequencing and bioinformatics of sRNAs libraries have been increasing the number of identified miRNAs, allowing the identification of species-exclusive miRNAs (Ortiz-Morea et al., 2013). To prevent wrong miRNA annotation, additional criteria were established: (1) mature miRNA should have a complementary miRNA* sequence identified in a database; (2) up to four mismatches are allowed between miRNA and miRNA*; and (3) the presence of non-aligned nucleotide bulge formed in the alignment between the miRNA and miRNA* have to be minimal in size (one or two bases) and frequency (Meyers et al., 2008). Based on these criteria, 44 new sugarcane miRNAs were identified with high confidence, using data of high-throughput sugarcane sRNA sequencing mapped to the sorghum genome (JGI, v1.0) (Thiebaut, Grativol et al., 2012). Novel sugarcane miRNAs were also identified in sRNA libraries constructed with plants under pathogen infection, submitted to salt and drought stresses, using sorghum genome as reference. Sorghum was used as a reference genome because it is the species most phylogenetically related to sugarcane that has a genome completely sequenced (Paterson et al., 2009), while the whole sugarcane genome sequence is not available yet. Although these miRNA sequences had been identified in sugarcane sRNA libraries, they were deposited in the miRBase as sorghum miRNA.

Drought stress is one of the major abiotic stresses affecting crop productivity worldwide, including sugarcane. A group of sugarcane cultivars were classified as tolerant or sensitive to drought measurement of physiological parameters in plants under drought stress (Silva, Jifon, Da Silva, and Sharma, 1997). Like in other plants, studies have highlighted the importance of sugarcane miRNA in the response and adaptation to water availability in plants (Ferreira et al., 2012; Gentile et al., 2013; Thiebaut et al., 2014). Interestingly, several genes associated with plant responses to stress are targets for regulation by miRNAs, and many of these encode transcription factors (Golldack, Lüking, and Yang, 2011; Jones-Rhoades et al., 2006). More recently, studies involved the manipulation of plant small RNA (sRNA) and their target expressions have been intensively performed and these genetically modified plants have shown enhanced ability to cope abiotic stress, as reviewed by Zhou and Luo (2013). Thus, understanding the responses of sugarcane to environmental stresses regulated by sRNA is central for the development of biotechnological applications for stress adaptation.

Although most studies with sugarcane sRNA described the role of miRNAs, two reports have also identified sugarcane siRNAs (Ortiz-Morea et al., 2013; Thiebaut et al., 2014). This class is formed by a complex pool of small RNAs derived from double-stranded RNA precursors synthesized by RNA polymerase enzymes dependent of RNA (RDR) (Xie and Qi, 2008). In *Arabidopsis*, there are six members of the family RDR (Wassenegger and Krczal, 2006), but only RDR2 and RDR6 were shown to be involved in siRNA biogenesis. RDRs are responsible for the conversion of RNA into dsRNA, the substrate of DCL enzyme (Willmann, Endres,

Cook, and Gregory, 2011). SiRNAs can be subdivided into three groups: (1) secondary siRNAs, known as 'trans-acting' siRNA (ta-siRNAs, when the target regulated by ta-siRNA is a different precursor) or phasiRNA; (2) siRNA natural antisense (nat-siRNAs); and (3) siRNAs of heterochromatin (hc-siRNAs) (Axtell, 2013; Xie and Qi, 2008). The last group is derived from repetitive sequences in the genome, such as transposons, retro-elements, rDNAs and repeats in the centromeric region (Kasschau et al., 2007). The targets of siRNAs are the same loci from which they were derived, the exception being siRNA resulting from TAS (Pais, Moxon, Dalmay, and Moulton, 2011). Seven potential sugarcane ta-siRNAs, including the conserved TAS3, were identified and the data suggest that they could be regulating auxiliary bud development (Ortiz-Morea et al., 2013). In addition, sRNAs that matched with sugarcane ESTs classified as MITEs were found. Importantly, this study also identified 26 conserved miRNA families and 2 putative novel miRNAs in sRNA libraries from sugarcane auxiliary buds, being the first report of sugarcane miRNAs modulated in response to developmental processes. Sugarcane ta-siRNAs were also identified in sRNA libraries from sugarcane submitted to drought stress (Thiebaut et al., 2014). These ta-siRNAs were classified based on the criteria that miRNAs with 22-nt in length are capable of slicing their target gene to produce siRNA (Chen et al., 2010). Interestingly, this siRNA was induced in plants after stopping irrigation, suggesting its involvement in response to this stress. Moreover, the complex siRNA population was also analysed through the distribution of siRNA matches to repeats and genes.

Because miRNAs are master regulators of plant growth and development, and that their expression profiles are significantly altered during stress (Sunkar, Li, and Jagadeeswaran, 2012), we next summarize the potential roles of sRNA regulation in sugarcane submitted to environmental stresses. The knowledge of miRNA-derived regulatory mechanisms of stress response in sugarcane is an important step for the application of sRNA-target based molecular tools in sugarcane breeding to result in an improvement of sugarcane tolerance to biotic and abiotic stresses.

Computational analyses for small RNA discovery in sugarcane

Since the discovery of first plant miRNA in 2002, the approaches for identification of miRNA precursors and mature miRNAs, and for exploiting their function have been established. Historically, the researchers seek to discover patterns of miRNA expression in different developing and environmental conditions in order to elucidate the regulatory effects of miRNA on their targets and to modulate the miRNA activity for plant improvement. The strategies for miRNA characterization rely on two main methods: experimental approaches and computational screening (Ming Chen, Meng, Mao, Chen, and Wu, 2010). The experimental approaches are most applied to a fine-scale detection of a miRNA through Northern Blot (Xie, Kasschau, and Carrington, 2003), Stem-loop qRT-PCR (Varkonyi-Gasic, Wu, Wood, Walton, and Hellens, 2007), *in situ* hybridization (Chen, 2004) and GUS (β-glucuronidase) or GFP (green fluorescent protein) reporter genes (Wang et al., 2005); to validate the regulation of a target gene by 5′ RACE (Llave et al., 2002) and qRT-PCR; and to modulate the miRNA activity through overexpression with 35S promoter or inhibition of target cleavage using artificial target mimicry or a miRNA-resistant target (Palatnik et al., 2003; Wang et al., 2005). On the other hand, computer-based approaches are required for high-throughput characterization of sRNA classes including, miRNA, siRNA and novel miRNA.

In the last few years, remarkable improvements in the sequencing technology, from Sanger to next-generation sequencing (NGS), have allowed the large-scale characterization of sRNA in diverse plant species (Studholme, 2012). The most powerful sequencing technique applied to sRNA search is the Illumina (Meyers, Souret, Lu, and Green, 2006). The computational methods developed to deal with the output of sRNA sequencing and to characterize miRNA precursors *in silico* are based in two main strategies: comparative and non-comparative (Zhang et al., 2006). The comparative strategy is based on the conservation of mature miRNA and hairpin structure over the plant kingdom. A common application of the comparative analysis is the homology search for known mature miRNA in the sRNA libraries, genome or cDNA data of different plants. For precursor identification, the homology search of miRNA in the genome is followed by the inspection of the hairpin structure coming from the 5′ and 3′ flanking regions of the mature miRNA alignment site. The classification of a hairpin as pre-miRNA-like is based on rules proposed by researchers from the MFEI (minimum free energy index) value and the shape of stem-loop secondary structure observed in known precursors (Zhang et al., 2006). In studies aiming the prediction of sugarcane miRNA precursors, the Blastn and Mfold were the most used algorithms for homology search and hairpin structure inspection, respectively (Figure 3.1) (Khan et al., 2014; Sunkar and Jagadeeswaran, 2008; Zanca et al., 2010; Zhang et al., 2005). Unlike the aforementioned studies, the MIRcheck pipeline was applied to predict precursors in genomic data instead of EST data (Grativol et al., 2014). This program extracts sequences with similarity to know mature miRNAs from the genome and evaluates the robustness of the folding properties of the resulting stem-loops (http://bartellab.wi.mit.edu/software .html). More than 14,000 precursors sequences were identified in sugarcane methyl-filtered genome data (Grativol et al., 2014).

As mentioned before, the comparative analysis has been also very useful for identification of mature miRNAs in high-throughput sequencing data. The pipeline of miRNA detection in sRNA libraries consists of a first

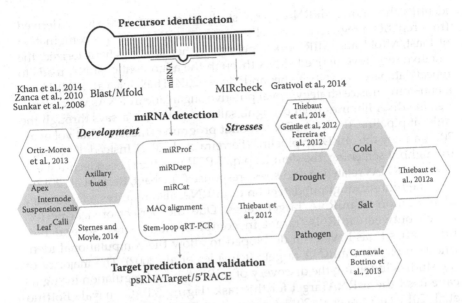

Figure 3.1 Summary of small discovery techniques applied to sugarcane. Starting with the precursor identification inside the sugarcane ESTs available at Genbank in 2005, followed works aimed the characterization of miRNA precursor sequences through a combination of Blast search and validation of hairpin with Mfold. Only one study performed the miRNA precursor identification using genome data and MIRcheck pipeline. The miRNA-based regulation related to development and stresses was mostly performed by Illumina sequencing coupled with the computational prediction of putative targets. Some experimental approaches as Stem-loop qRT-PCR and 5′ RACE were also employed to miRNA investigation in sugarcane.

step of Illumina 3′ adaptor sequence and length (18-28 nt) trimming and a second step of t/rRNA filtering. For helping the researchers in these tasks, some programs have been developed such as the UEA sRNA Toolkit pipeline (Studholme, 2012). The next step is the classification of sRNA sequences as known miRNA and calculation of expression by measuring the number of reads homologs to a specific miRNA. In sugarcane, the miRProf, part of the UEA sRNA toolkit, and miRDeep (Yang and Li, 2011) pipelines were applied to achieve these results (Carnavale Bottino et al., 2013; Ferreira et al., 2012; Gentile et al., 2013; Sternes and Moyle, 2014; Thiebaut et al., 2014). Instead of the miRProf pipeline, which the final output is the expression profile of known miRNAs, the miRDeep generates the list of both known and putative novel miRNA together with the related precursor's sequences. An additional homology search against the miRBase database is required to separate these two groups of sRNAs. The same procedure have been used in the study of Ortiz-Morea and co-workers (Ortiz-Morea et al., 2013), which used the MAQ program to align sRNA reads against several databases to identify the known miRNAs. This study also recognized sRNA derived from repetitive regions and novel sugarcane miRNAs using a combination of Blast/Mfold and MIRcheck pipelines. Other studies characterized the putative repetitive-derived sRNA through the alignment of sRNA reads to repeat databases as well (Sternes and Moyle, 2014; Thiebaut et al., 2014). Using a non-comparative strategy, novel putative sugarcane miRNAs were identified in sRNA libraries from drought, salt and pathogen assays through the miRCat pipeline, the UEA sRNA Toolkit program (Thiebaut, Grativol et al., 2012), and by the miRDeep pipeline (Ferreira et al., 2012). Instead of the high-throughput screening, the stem-loop qRT-PCR was the chosen technique to detect miRNAs related to cold stress response (Thiebaut, Rojas et al., 2012).

The high-throughput detection of miRNA also requires a large-scale identification of miRNA target genes. Due the perfect or near-perfect match between miRNA/target in complementary sites, several bioinformatics tools have been developed to allow the computational identification of plant miRNA targets (Rhoades et al., 2002). The majority of the studies involving the discovery of miRNA-based regulation in sugarcane used the psRNATarget for this task (Figure 3.1) (Carnavale Bottino et al., 2013; Ferreira et al., 2012; Gentile et al., 2013; Ortiz-Morea et al., 2013; Sternes and Moyle, 2014; Thiebaut et al., 2014). Although in some plants, high-throughput methods for miRNA-target validation (PARE, Parallel Analysis of RNA Ends) have been applied (Zhai, Arikit, Simon, Kingham, and Meyers, 2013), the 5′ RACE was the chosen technique to confirm the cleavage of a specific target identified *in silico* by sugarcane researchers.

Conserved microtranscriptome changes in stressed sugarcane

One important characteristic of miRNA is their conservation between plants, mainly in plants that are phylogenetically close (Cuperus, Fahlgren, and Carrington, 2011; Dezulian et al., 2005). Based on this, known miRNA were used to identify sugarcane miRNAs (Carnavale Bottino et al., 2013; Ferreira et al., 2012; Gentile et al. 2013; Thiebaut et al., 2014; Thiebaut, Grativol et al., 2012; Thiebaut, Rojas et al., 2012). Analysis of these miRNA expressions revealed that many of them have an important role in response to biotic and abiotic stresses. A total of 274 conserved mature miRNAs were found in sugarcane, comprising 41 miRNA families (Table 3.1). Among these conserved miRNA families identified in sugarcane, miR159 is the family with the most members (27 isoforms), following by miR169 and miR156 (25 and 23 isoforms, respectively). In contrast, 14 miRNA families had only one isoform (Table 3.1).

Data on sRNA regulation can assist the development of tools to enhance sugarcane resistance/tolerance against a specific stress. In the same family, isoforms can be regulated differentially in response to

Table 3.1 Sugarcane miRNAs identified in response to different stress

miRNA family	Total isoforms	Drought stress[a]				Salt stress[b]			Cold stress[c]	Biotic stress[d]
		Tolerant shoot	Tolerant root	Sensitive shoot	Sensitive root	1 h	6 h	24 h		
miR1118	1	–	–	–	0/1	–	–	–	–	–
miR1310	1	–	1/0	–	0/1	–	–	–	–	–
miR1432	4	0/2	2/1	1/1	0/3	–	–	–	–	–
miR1439	2	–	0/1	2/0	0/1	3/8	0/11	9/2	–	–
miR156	23	3/1	13/8	0/4	22/0	7/1	5/3	2/6	–	–
miR159	27	6/9	4/17	8/7	7/16	0/2	1/1	1/1	–	–
miR160	4	1/1	1/1	2/0	2/0	0/1	0/1	0/1	–	–
miR162	1	–	1/0	–	0/1	0/1	0/1	1/1	–	–
miR164	14	2/1	7/5	0/3	5/8	0/7	1/6	1/6	–	–
miR166	16	0/1	1/6	1/2	12/2	0/5	1/4	3/3	–	–
miR167	15	3/2	9/3	3/4	8/3	0/4	0/4	1/4	–	–
miR168	16	0/1	1/11	0/2	8/7	2/6	6/4	6/3	–	1/1
miR169	25	1/1	10/3	1/4	15/4	1/3	1/3	2/2	–	–
miR171	18	1/1	13/5	0/1	10/8	0/1	0/1	0/1	1/0	–
miR172	3	1/1	1/0	1/1	0/1	–	–	–	–	–
miR1878	1	1/0	1/0	1/0	0/1	–	–	–	–	–

(Continued)

Table 3.1 (Continued) Sugarcane miRNAs identified in response to different stress

miRNA family	Total isoforms	Drought stress[a]				Salt stress[b]			Cold stress[c]	Biotic stress[d]
		Tolerance shoot	Tolerance root	Sensitive shoot	Sensitive root	1 h	6 h	24 h		
miR2118	1	–	–	–	–	–	–	–	–	0/1
miR2916	1	0/1	–	–	–	–	–	–	–	–
miR319	11	0/1	–	–	–	–	–	–	–	–
miR3633	1	–	4/6	0/1	2/10	0/3	0/3	0/3	1/0	–
miR390	7	–	0/1	–	–	–	–	–	–	–
miR393	5	1/3	0/5	1/0	2/5	0/1	0/1	0/1	–	–
miR394	1	0/3	4/0	1/2	4/0	0/3	0/3	0/3	–	–
miR395	11	1/2	–	1/2	–	1/0	1/0	1/0	–	–
miR396	19	0/1	2/2	1/1	5/2	1/4	3/1	0/4	–	–
miR397	7	2/1	5/1	0/4	6/5	0/12	1/10	0/12	–	–
miR3979	3	–	5/1	3/1	1/4	1/1	0/2	0/2	–	–
miR398	6	0/1	0/1	–	3/0	–	–	–	–	–
miR399	6	4/2	0/1	0/2	1/0	4/2	2/2	1/3	–	–
miR408	4	–	1/1	4/3	2/0	0/1	0/1	0/1	–	–
miR437	2	–	1/0	–	–	2/1	2/1	0/3	1/0	0/1
miR444	4	2/0	1/2	2/1	2/1	2/0	2/0	2/0	–	–

(Continued)

Table 3.1 (Continued) Sugarcane miRNAs identified in response to different stress

miRNA family	Total isoforms	Drought stress[a]				Salt stress[b]			Cold stress[c]	Biotic stress[d]
		Tolerance shoot	Tolerance root	Sensitive shoot	Sensitive root	1 h	6 h	24 h		
miR5013	1	0/1	–	–	–	–	–	–	–	–
miR5021	1	–	–	1/0	–	–	–	–	–	–
miR5054	1	–	–	–	0/1	–	–	–	–	–
miR5072	3	–	0/3	–	0/3	2/0	1/1	2/0	–	–
miR5082	2	–	1/1	–	2/0	–	–	1/0	–	–
miR5139	1	–	–	–	–	1/1	0/2	0/2	–	–
miR528	3	3/1	1/0	2/2	1/0	1/0	1/0	1/0	–	–
miR529	1	0/1	1/0	0/1	1/0	1/0	1/0	0/1	–	–
miR827	1	–	1/0	–	–	–	–	–	–	–

Note: Number of mature miRNA differentially expressed in each situation (up/down). Drought stress was observed in four different experiment times: 24 h, 2 days, 4 days and 7 months. tol., drought tolerant cultivars; sens., drought-sensitive cultivars. Salt stress was performed using shoot of SP70-1143 harvested after 1, 6 and 24 h in 170 mM NaCl of hydroponic solution. SP70-1143 was also used in experiments of cold and biotic stress. For cold stress, data from qPCR using whole plants were submitted at 4°C for 24 h. For biotic stress, whole plants were infected with *Acidovorax avenae* for 7 days.

[a] Data from Ferreira et al., 2012, *PLoS ONE,* 7(10); Thiebaut, F. et al., 2012, *BMC Genomics,* 13, 290; Gentile, A. et al., 2013, *Planta,* 237(3), 783–798; and Thiebaut, F., 2014, *PLoS ONE,* 9(4).
[b] Data from Bottino, M. C. et al., 2013, *PLoS ONE,* 8(3).
[c] Data based on samples submitted to cold for 24 h. Data from Thiebaut, F., 2014, *PLoS ONE,* 9(4).
[d] Data from Thiebaut, F. et al., 2012, *BMC Genomics,* 13, 290.

determined stress. Based on this, Table 3.1 presents the number of sugarcane miRNA that showed upregulation or downregulation in each of the following treatments: drought, salt, cold and pathogenic infection. It is important to note that when the same miRNA was analysed in different studies, it has counted more than once. Among these stresses, drought stress was intensively studied, highlighting the importance of sugarcane miRNA in the response and adaptation to low water availability to plants (Ferreira et al., 2012; Gentile et al., 2013; Thiebaut et al., 2014). Drought stress is likely the abiotic stress that most affects agricultural productivity worldwide, including sugarcane yield (Azevedo, Carvalho, Cia, and Gratão, 2011). The primary response of the plant to drought stress is to decrease growth, resulting in inhibition of root development, reduced number of leaves and reduced stalk elongation, leaf rolling and reduction in water and other nutrients uptake (Inman-Bamber and Smith, 2005). Sugarcane cultivars can be classified as tolerant or sensitive to drought based on the values of a number of physiological parameters evaluated (Silva et al., 1997). Studies of the microtranscriptome changes in response to drought stress used cultivars previously classified as tolerant and sensitive (or less tolerant) to this stress. Tolerant cultivars used in these analyses were RB867515, CTC15, CTC6, SP83-2847 and SP83-5073, and cultivars classified as sensitive were RB855536, CTC9, CTC13, SP90-1638 and SP90-3414 (Ferreira et al., 2012; Gentile et al., 2013; Thiebaut et al., 2014). Analysis of miRNA expression revealed that many of them are expressed only in certain tissues and/or cell types (Breakfield et al., 2011; Mica et al., 2010), which explains the difference in miRNA regulation between root and shoot of sugarcane submitted to water deficit (Table 3.1). Moreover, miRNAs can be regulated at specific stages of development (Luo, Guo, and Li, 2013; Ortiz-Morea et al., 2013), or differentially regulated in response to the length of the exposure to the stress, as observed in experiment of salt stress (Table 3.1).

In contrast to salt and drought stress, which used sRNA libraries to find differentially expressed miRNA in response to these stresses, the study of sugarcane subjected to low temperature (4°C) analysed the expression profile of only 12 miRNAs. In Table 3.1, the information of miRNA regulated at 24 h of treatment is shown. Among these, miR319 was the most regulated miRNA with an approximately threefold increase in expression after 24 h of exposure to 4°C. The induction of this miRNA was observed in roots and shoots of sugarcane submitted to cold stress. Interestingly, miR319 targets, GAMYB and PCF6, were downregulated in sugarcane submitted to cold stress. The cleavage of miRNA-targets was confirmed by 5' RACE PCR assay. Furthermore, the comparison between two different cultivars, RB931011 and TAMBO FEPAGRO, classified as sensitive and tolerant to cold stress, respectively, showed that miR319 was induced later in the tolerant cultivar submitted to 4°C (Thiebaut, Rojas et al., 2012). Accordingly, the expression of miR319 mRNA target (GAMYB) decreased early in the sensitive cultivar.

Although infection by biotic agents cause severe losses on sugarcane yields (Rott, Bailey, Comstock, Croft, and Saumtally, 2000), much less is known about the regulation of microRNA in plants infected with pathogenic microbes. The pathogen *Acidovorax avenae* subsp. *avenae* causes red stripe disease, which leads to damages to the leaves and leaf sheaths, and affects crops worldwide (CMI, 1995; Fegan, 2006). In sugarcane plants infected with this pathogen, three conserved miRNAs – miR169, miR437 and miR2118 – were upregulated (Thiebaut, Grativol et al., 2012). Interestingly, the miR2118 is involved with the phasing cleavage of defence response nucleotide binding site–leucine rich repeat (NBS-LRR) genes leading to the production of ta-siRNA in legumes (Zhai et al., 2011). Sugarcane is plagued by a large and diverse group of pathogens, including other species of bacteria, fungus and virus. Thus, information of sRNA and the regulatory network of plant immune system are urgently needed and can be important for the development of tools to enhance plant resistance against pathogens.

Different strategies for developing miRNA-based genetically modified plants can be used to develop novel stress-resistant cultivars with better performance under certain environments (Zhou and Luo, 2013). For instance, under cold stress, rice miR319 is downregulated, and transgenic rice overexpressing miR319 results in enhanced cold tolerance (Yang et al., 2013). Yang and colleagues (2013) also showed that transgenic rice downregulating the expression of miR319-targeted genes leads to increased tolerance to cold stress. Thus, the manipulation of the expression of a miRNA or its target can result in plants with desirable agronomic traits. In order to understand the role of miRNA, it is necessary to identify their targets in sugarcane (Table 3.2). Many predicted miRNA sugarcane targets encode transcription factors, which is in agreement to the majority of canonical miRNA targets (Jones-Rhoades et al., 2006). These groups of targets have been considered as good candidates for genetic manipulation (Century, Reuber, and Ratcliffe, 2008). The targets of miR156 are squamosa-promoter binding protein-like (SPL), which are transcription factors with important functions in plant growth and development (Preston and Hileman, 2013). For instance, in switchgrass, manipulation of miR156 has been used for enhancing biomass production (Fu et al., 2012).

Genetic engineering based on miRNA-derived regulation can be used as a potential tool to improve environmental responses of crops, including sugarcane. A good candidate for this approach would be the miR169, because it was shown in a number of experiments to be responsive to all stresses to which sugarcane was submitted (Table 3.1). Modulation of the miR169 expression on the stresses was linked with the sugarcane genotype and tissue, and to exposure time to stress (Figure 3.2a). At 24 h of salt and cold stresses, sugarcane plants showed an miR169 expression increase of 90% and 27%, respectively, compared to control plants. For salt stress, a continuous increase of miR169 expression was observed during the whole

Table 3.2 Putative targets of sugarcane miRNA

miRNA family	Target	Sugarcane study
miR1118	–	–
miR1310	–	
miR1432	B-Zip transcription factor	Ferreira et al., 2012
miR1439	MADS-box transcription factor MADS2	Thiebaut et al., 2014
miR156	Squamosa promoter binding protein 1 (SPL1)	Ortiz-Morea et al., 2013
miR159	GAMYB	Ortiz-Morea et al., 2013
miR160	Auxin response factor - ARF	Ortiz-Morea et al., 2013
miR162	–	–
miR164	NAC trascription factor	Ferreira et al., 2012
miR166	Class III HD-Zip protein 4	Bottino et al., 2013
miR167	Auxin response factor	Zanca et al., 2010
miR168	ARGONAUTE1 (AGO1)-like protein	Zanca et al., 2010
miR169	Nuclear transcription factor Y subunit A-10	Gentile et al., 2013
miR171	SCARECROW-LIKE1 protein	Gentile et al., 2013
miR172	Floral homeotic protein APETALA 2-like	Gentile et al., 2013
miR1878	No target found	Thiebaut et al., 2014
miR2118	Protoporphyrin IX magnesium-chelatase	Thiebaut, Grativol et al., 2012
miR2916	Peroxisomal membrane marker - PEX14-like	Thiebaut et al., 2014
miR319	TCP/GAMYB	Thiebaut, Rojas et al., 2012
miR3633	–	
miR390	Protein of unknown function	–
miR393	Auxin-responsive factor TIR1 protein	Thiebaut et al., 2014 Gentile et al., 2013
miR394	Glyceraldehyde 3-phosphate dehydrogenase (GAPDH)	Ferreira et al., 2012
miR395	ATP sulfurylase	Thiebaut et al., 2014
miR396	Growth-regulating factor	Zanca et al., 2010
miR397	Laccase	Ferreira et al., 2012
miR3979	–	–
miR398	Selenium binding protein	Bottino et al., 2013
miR399	Inorganic pyrophosphatase 2-like	Ferreira et al., 2012

(Continued)

Table 3.2 (Continued) Putative targets of sugarcane miRNA

miRNA family	Target	Sugarcane study
miR408	Basic blue copper protein-like protein	Zanca et al., 2010
miR437	–	–
miR444	MADS-box transcription factor	Zanca et al., 2010
miR5013	Plastidic phosphate translocator - like protein 2	Thiebaut et al., 2014
miR5021	Rac-like GTP-binding protein 6 precursor	Thiebaut et al., 2014
miR5054	–	–
miR5072	–	–
miR5082	–	–
miR5139	–	–
miR528	Cu2+-binding domain containing protein	Zanca et al., 2010
miR529	SPL1	Ortiz-Morea et al., 2013
miR827	SPX domain-containing protein	Zanca et al., 2010

period that the plants were under stress. Tolerant and sensitive cultivars submitted to drought stress showed a contrasting regulation of miR169. In comparison to non-stressed plants, tolerant cultivars increased only 1% miRNA169 expression in roots, while in the sensitive ones the induction was 36%. On the other hand, miR169 was downregulated (−85%) in leaves of sensitive cultivars and upregulated (60%) in tolerant. It is important to note that these changes of miR169 expression were calculated considering the sum of normalized expression of all miR169 isoforms detected (Table 3.1). Curiously, studies of sugarcane submitted to 7 days of pathogen infection and 7 months of drought stress (Gentile et al., 2013; Thiebaut, Grativol et al., 2012) showed that the miRNA* sequences were downregulated after these stresses, instead of the mature miRNA. In the case of the 7 months of drought stress, a distinct change of miR169* expression in tolerant and sensitive cultivars was also observed.

Conservation of miR169 stress-modulated isoforms identified in the sugarcane studies (Carnavale Bottino et al., 2013; Gentile et al., 2013; Thiebaut et al., 2014; Thiebaut, Grativol et al., 2012; Thiebaut, Rojas et al., 2012) is highlighted in the box in Figure 3.2a. Three regions at the miR169 sequences were more preserved, including the region where the slicing of an mRNA target is performed (10–11 nt) (Shin et al., 2010). The psRNATarget prediction of mRNA target for miR169 sugarcane isoforms, including the miR169* (Gentile et al., 2013), showed the transcription factor nuclear factor-Y subunit A (NF-YA) as the most likely target. Also known as heme-activated protein (HAP) or CCAAT binding factor (CBF), this class of transcription

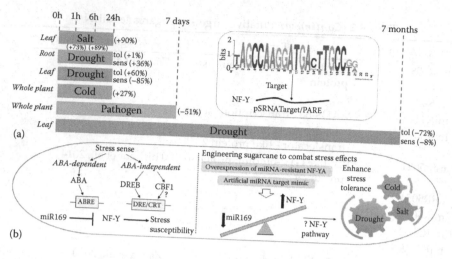

Figure 3.2 MiR169 is a good candidate to enhance sugarcane tolerance to drought, salt and cold. The expression of miR169 in different stress conditions and times is summarized in (a). The high conservation in the cleavage sites (10–12 nt) of mature miR169 sequences identified in the studies and in the validated target genes (NF-YA) is also shown (a). Stress pathways, which involve ABA-dependent and ABA-independent signaling, can induce the expression of the NF-Y transcription factors that have been shown to enhance stress tolerance (b). The increase in miR169 expression at 24 h of salt, cold and drought stresses affects the availability of NF-YA, which could lead to stress susceptibility. Engineering sugarcane with a miRNA-resistant NF-YA or artificial miRNA target mimicry could impact positively the expression of NF-YA, leading to enhanced stress tolerance. The tolerant sugarcane submitted to drought stress showed a more pronounced down-regulation of miR169 than the sensitive cultivar, corroborating the aforementioned idea. Although in pathogen and drought-stressed sRNA libraries the miRNA* sequences were identified as regulated, the NF-YA was reported as target in the drought stress assay. The numbers in parentheses represent the relative expression compared to control samples. -tol: tolerant cultivar; -sens: sensitive cultivar.

factors is composed by three subunits (NF-YA, NF-YB and NF-YC), which are all required for DNA binding (Mantovani, 1999). The role of NF-Y in the enhancement of stress tolerance have been emphasized by studies with both dicots and monocots plants. In *Arabidopsis* (Li et al., 2008), soybean (Ni et al., 2013), rice (Chen et al., 2015) and maize (Nelson et al., 2007), overexpression of different NF-Y subunits led to increased stress tolerance. NF-Y transcription factors can be activated by environmental stress through ABA-dependent and -independent pathways (Figure 3.2b). As shown in *Brassica napus* L. (Xu et al., 2014) and soybean (Ni et al., 2013), the ABA-responsive element (ABRE) and the dehydration responsive element/C-repeat (DRE/CRT) are *cis*-elements at NF-Y promoters. Although the C-repeat binding factors

(CBF) are important constituents of ABA-independent stress responses, there is no evidence of NF-Y activation by these factors at the DRE/CRT motif at NF-Y promoters (Figure 3.2b). Importantly, the NF-Y expression can be transcriptionally controlled by ABA-dependent and -independent pathways and posttranscriptionally by the miR169. The slice of NF-YA (HAP2) by miR169 at the ARGONAUTE protein complex can alter the availability of NF-YA (Figure 3.2b). The miR169-NF-YA regulation has been validated in soybean, cotton and rice using the 5' RACE experimental approach (Ni et al., 2013; Turner, Yu, and Subramanian, 2012; Zhao et al., 2009). Cleavage of this transcription factor by miR169 was also confirmed in PARE libraries of tomato and cotton (Karlova et al., 2013; Yang et al., 2013), and our group's PARE experiment with sugarcane also confirmed the regulation (unpublished data). Although the miRNA direct cleavage was only confirmed for NF-YA subunits, the slice of this subunit could complicate the functionality of the NF-Y complex since the three subunits are expressed at the same time (Li et al., 2008).

Upregulation of miR169 at 24 h of salt, drought and cold stresses would decrease the availability of NF-YA, which would increase the susceptibility of sugarcane plants in the first hours of stress. This profile can be corroborated in transgenic *Arabidopsis* where the overexpression of miR169 induce the sensitivity to drought stress compared to wild-type plants (Li et al., 2008). Interestingly, the increase of miR169 expression in rice plants submitted to high salinity is linked with the presence of ABRE and DRE motifs at miR169 promoters, which lead to dual activation of miRNA and its target by the same pathways (Zhao et al., 2009). This is in contrast with the downregulation of miR169 in *Arabidopsis* plants submitted to drought stress (Li et al., 2008). Thus, genetic engineering of sugarcane to cope with stress could use approaches that would increase the NF-YA availability. As illustrated in Figure 3.2b, two different strategies based on miRNA-target regulation can be used to genetically engineer sugarcane to cope with stresses, overexpression of a miRNA-resistant NF-YA transcription factor or the mimic of artificial miR169 target. Both methodologies were applied in *Arabidopsis* in order to analyse the root growth and branching (Sorin et al., 2014). In addition, as mentioned before, overexpression of NF-Y leaded to increased stress tolerance in different plants. However, in sugarcane whose miR169 is up-regulated during stresses, the strategy should be based on the engineering of plants overexpressing a NF-YA resistant to the miRNA. By using the technique of artificial target mimic (Franco-Zorrilla et al., 2007), the transgenic sugarcane could also inhibit the miRNA cleavage of NF-YA, resulting in plants with enhanced tolerance to stresses. Since miR169 were similarly induced in sugarcane at the early stages of drought, salt and cold stresses, genetic engineered plants using the aforementioned approaches could result in an enhanced tolerance to all three stresses (Figure 3.2b).

Novel miRNAs and siRNAs with potential roles in sugarcane stress responses

In the last years, miRNA discovery has been accelerated by high-throughput screening coupled with NGS (Meyers et al., 2006). In general, new miRNA are less conserved among plant species and are expressed only in certain tissues and/or cell types at low amounts (Breakfield et al., 2011; Lu et al., 2006; Mica et al., 2010; Wei, Yan, and Wang, 2011). In sugarcane, high-throughput sequencing allowed the discovery of 623 mature miRNAs in abiotic and biotic stress assays, which 37 of these are from 44 bona fide precursors (Thiebaut, Grativol et al., 2012). In addition, the work of Ortiz-Morea and colleagues (2013) identified a group of novel miRNAs regulated during sugarcane development. The work from Thiebaut, Grativol, and co-workers (2012) showed that a large fraction (37%) of the new miRNAs is only expressed upon stress induction, being not present in the control sample. They have also shown that new miRNAs could play an important role in both salt, drought and pathogen infection, since these adverse conditions increased the number of new miRNAs detected in sugarcane tissues submitted to them. Moreover, salt and pathogenic stresses induced an increase in the numbers of exclusive new miRNAs detected, 41% and 16%, respectively. Another miRNA subgroup was identified in this study: the new long miRNAs with their characteristic size of 24 nt. Although it was not possible to identify targets for most of the novel long miRNAs, the amounts of 24 nt species were increased in all stressed conditions, but not in tolerant cultivars under drought stress. Because subgroups of new miRNA seem to be involved on stress-activated pathways, they consist of a whole new group of candidates for enhanced sugarcane response to stresses.

The last sRNA class identified as responsive to an environmental stress in sugarcane was the siRNA. Two sub-classes have been detected in sugarcane submitted to drought stress: putative ta-siRNA and hc-siRNA (Thiebaut et al., 2014). The first one was identified based on the previous work of Chen and co-workers (2010), which showed that 22 nt miRNAs can cleave their targets, triggering siRNA production. They report a distinct ta-siRNA pathway where the slice of an MYB mRNA by a miR159 with 22 nt in length can trigger the production of ta-siRNA, which match to the same target gene as the original miRNA. Regardless of the sugarcane genotype – drought tolerant or sensitive – upregulation of this ta-siRNA was observed in drought stress, suggesting that the role of the ta-siRNA during water depletion could be the reinforcement of MYB silencing. However, further experimental confirmation needs to be performed to validate the role of this ta-siRNA.

The second siRNA sub-class identified in sugarcane plants under drought stress was the hc-siRNA (Thiebaut et al., 2014). The well-known epigenetic mechanism of transposon mobility suppression relies on the siRNA silencing pathway, of which Dicer-like (DCL) and RNA-dependent

RNA polymerase (RDR) take part (Ito, 2012; Law and Jacobsen, 2010). Under drought stress, the tolerant cultivars showed twice the amount of retrotransposon-derived siRNAs, overrepresented by LTR-gypsy-derived siRNAs, while in the sensitive cultivars these siRNAs remained mostly unchanged (Thiebaut et al., 2014). In addition, the rise of siRNA production from gene *loci* in both tolerant and sensitive sugarcane cultivars submitted to water depletion suggested a putative role for these siRNA on stress response. Although a larger siRNA increase was observed in tolerant cultivars, the siRNA-derived gene *loci* are related to genes of photosystem I and II, which may be linked with the already reported decrease of photosynthetic carbon reduction cycle enzymes activity during water depletion (Reddy, Chaitanya, and Vivekanandan, 2004). Therefore, the siRNA sub-classes constitute another important molecular tool to cope with drought stress in sugarcane.

Summary

The history of small RNA discovery in sugarcane started in 2005 when the first miRNA precursors were identified. Since then, the progress in the knowledge of sugarcane miRNA and siRNA-guided regulation during development and environmental stresses has been steady. In this chapter, we have summarized the regulation of the conserved microtranscriptome and the putative roles of novel miRNA and siRNA involved in sugarcane responses to stress. The high-throughput characterization of miRNA and siRNA associated with sugarcane responses to environmental changes was possible due to the remarkable improvement of sequencing technologies and advances in the computational analysis of plant sRNA. The analysis of miRNA expressions during drought, salt, cold and pathogenic infection using these approaches revealed that the miR169 is the top-ranked candidate for further biotechnological application in sugarcane stress adaptation. By engineering sugarcane to overexpress an NF-YA miR169-resistant version of this transcription factor, or artificial mimic of the NF-YA target, the transgenic sugarcane plants could block miRNA-mediated cleavage of NF-YA and enhance tolerance to the stresses. In addition, once novel miRNA and some siRNA classes were found to be involved in stress-activated pathways, they can comprise a whole new group of candidates to enhance sugarcane response to stresses. With the increased availability of sugarcane genomic data, the small RNA research field will continue to grow, adding more members to the miRNA-derived regulatory mechanisms of stress adaptation in sugarcane.

Acknowledgements

The authors thank INCT (Instituto Nacional de Ciência de Tecnologia) in Biological Nitrogen Fixation, FAPERJ (Fundação de Amparo à Pesquisa do Estado do Rio de Janeiro), CNPq (Conselho Nacional de Desenvolvimento

Científico e Tecnológico) and CAPES (Coordenação de Aperfeiçoamento de Pessoal de Nível Superior) for financial support.

References

Axtell, M. J. (2013). Classification and comparison of small RNAs from plants. *Annual Review of Plant Biology, 64*(January), 137–159. http://doi.org/10.1146 /annurev-arplant-050312-120043.

Azevedo, R. A., Carvalho, R. F., Cia, M. C., and Gratão, P. L. (2011). Sugarcane under pressure: An overview of biochemical and physiological studies of abiotic stress. *Tropical Plant Biology, 4*(1), 42–51. http://doi.org/10.1007 /s12042-011-9067-4.

Bartel, D. P., Lee, R., and Feinbaum, R. (2004). MicroRNAs: Genomics, biogenesis, mechanism, and function. *Genomics: The miRNA Genes, 116*, 281–297.

Baumberger, N., and Baulcombe, D. C. (2005). *Arabidopsis* ARGONAUTE1 is an RNA slicer that selectively recruits microRNAs and short interfering RNAs. *Proceedings of the National Academy of Sciences of the United States of America, 102*(33), 11928–11933. http://doi.org/10.1073/pnas.0505461102.

Bottino, M. C., Rosario, S., Grativol, C., Thiebaut, F., Rojas, C. A., Farrineli, L., … Ferreira, P. C. G. (2013). High-throughput sequencing of small RNA Transcriptome reveals salt stress regulated microRNAs in sugarcane. *PLoS ONE, 8*(3). http://doi.org/10.1371/journal.pone.0059423.

Breakfield, N. W., Corcoran, D. L., Petricka, J. J., Shen, J., Sae-Seaw, J., Rubio-Somoza, I., … Benfey, P. N. (2011). High-resolution experimental and computational profiling of tissue-specific known and novel miRNAs in *Arabidopsis*. *Genome Research*. http://doi.org/10.1101/gr.123547.111.

Century, K., Reuber, T. L., and Ratcliffe, O. J. (2008). Regulating the regulators: The future prospects for transcription-factor-based agricultural biotechnology products. *Plant Physiology, 147*(1), 20–29. http://doi.org/10.1104/pp.108.117887.

Chen, H., Chen, L., Patel, K., Li, Y., Baulcombe, D. C., and Wu, S. (2010). 22-nucleotide RNAs trigger secondary siRNA biogenesis in plants. *Proceedings of the National Academy of Sciences*, 1–6. http://doi.org/10.1073 /pnas.1001738107.

Chen, M., Meng, Y., Mao, C., Chen, D., and Wu, P. (2010). Methodological framework for functional characterization of plant microRNAs. *Journal of Experimental Botany, 61*(9), 2271–2280. http://doi.org/10.1093/jxb/erq087.

Chen, M., Zhao, Y., Zhuo, C., Lu, S., and Guo, Z. (2015). Overexpression of a NF-YC transcription factor from Bermuda grass confers tolerance to drought and salinity in transgenic rice. *Plant Biotechnology Journal, 13*, 482–491. http://doi .org/10.1111/pbi.12270.

Chen, X. (2004). A microRNA as a translational repressor of APETALA2 in *Arabidopsis* flower development. *Science (New York), 303*, 2022–2025. http:// doi.org/10.1126/science.1088060.

CMI. (1995). *Acidovorax avenae* subsp. *avenae*. (Distribution maps of plant diseases, no. 511). Wallingford, UK: CAB International.

Cuperus, J. T., Fahlgren, N., and Carrington, J. C. (2011). Evolution and functional diversification of MIRNA genes. *The Plant Cell Online*, 1–13. http://doi .org/10.1105/tpc.110.082784.

Dezulian, T., Palatnik, J. F., Huson, D., and Weigel, D. (2005). Conservation and divergence of microRNA families in plants. *Bioinformatics.*

Ding, D., Li, W., Han, M., Wang, Y., Fu, Z., Wang, B., and Tang, J. (2013). Identification and characterisation of maize microRNAs involved in developing ears. *Plant Biology (Stuttgart, Germany),* 1–7. http://doi.org/10.1111/plb.12013.

Ding, D., Zhang, L., Wang, H., Liu, Z., Zhang, Z., and Zheng, Y. (2009). Differential expression of miRNAs in response to salt stress in maize roots. *Annals of Botany, 103*(1), 29–38. http://doi.org/10.1093/aob/mcn205/.

Fegan, M. (2006). Plant pathogenic members of the genera *Acidovorax* and *Herbaspirillum.* In S. S. Gnanamanickam (Eds.), *Plant-associated bacteria,* 671–702. Dordrecht: Springer.

Ferreira, T. H., Gentile, A., Vilela, R. D., Costa, G. G. L., Dias, L. I., Endres, L., and Menossi, M. (2012). microRNAs associated with drought response in the bioenergy crop sugarcane (*Saccharum* spp.). *PLoS ONE, 7*(10). http://doi.org/10.1371/journal.pone.0046703.

Franco-Zorrilla, J. M., Valli, A., Todesco, M., Mateos, I., Puga, M. I., Rubio-Somoza, I., ... Paz-Ares, J. (2007). Target mimicry provides a new mechanism for regulation of microRNA activity. *Nature Genetics, 39*(8), 1033–1037. http://doi.org/10.1038/ng2079.

Fu, C., Sunkar, R., Zhou, C., Shen, H., Zhang, J.-Y., Matts, J., ... Wang, Z.-Y. (2012). Overexpression of miR156 in switchgrass (*Panicum virgatum* L.) results in various morphological alterations and leads to improved biomass production. *Plant Biotechnology Journal, 10,* 443–452.

Fujita, M., Fujita, Y., Noutoshi, Y., Takahashi, F., Narusaka, Y., Yamaguchi-Shinozaki, K., and Shinozaki, K. (2006). Crosstalk between abiotic and biotic stress responses: A current view from the points of convergence in the stress signaling networks. *Current Opinion in Plant Biology, 9*(4), 436–442. http://doi.org/10.1016/j.pbi.2006.05.014.

Gentile, A., Ferreira, T. H., Mattos, R. S., Dias, L. I., Hoshino, A. A., Carneiro, M. S., ... Menossi, M. (2013). Effects of drought on the microtranscriptome of field-grown sugarcane plants. *Planta, 237*(3), 783–798. http://doi.org/10.1007/s00425-012-1795-7.

Golldack, D., Lüking, I., and Yang, O. (2011). Plant tolerance to drought and salinity: Stress regulating transcription factors and their functional significance in the cellular transcriptional network. *Plant Cell Reports, 30*(8), 1383–1391. http://doi.org/10.1007/s00299-011-1068-0.

Grativol, C., Regulski, M., Bertalan, M., McCombie, W. R., Da Silva, F. R., Zerlotini Neto, A., ... Ferreira, P. C. G. (2014). Sugarcane genome sequencing by methylation filtration provides tools for genomic research in the genus *Saccharum. Plant Journal, 79*(1), 162–172. http://doi.org/10.1111/tpj.12539.

Guleria, P., Mahajan, M., Bhardwaj, J., and Yadav, S. K. (2011). Plant small RNAs: Biogenesis, mode of action and their roles in abiotic stresses. *Genomics, Proteomics and Bioinformatics/Beijing Genomics Institute, 9*(6), 183–199. http://doi.org/10.1016/S1672-0229(11)60022-3.

Inman-Bamber, N. G., and Smith, D. M. (2005). Water relations in sugarcane and response to water deficits. *Field Crops Research, 92*(2–3), 185–202. http://doi.org/10.1016/j.fcr.2005.01.023.

Ito, H. (2012). Small RNAs and transposon silencing in plants. *Development, Growth and Differentiation, 54*(1), 100–107. http://doi.org/10.1111/j.1440-169X.2011.01309.x.

Jones-Rhoades, M. W., Bartel, D. P., and Bartel, B. (2006). MicroRNAs and their regulatory roles in plants. *Annual Review of Plant Biology, 57*, 19–53. http:// doi.org/10.1146/annurev.arplant.57.032905.105218.

Karlova, R., Van Haarst, J. C., Maliepaard, C., Van De Geest, H., Bovy, A. G., Lammers, M., … De Maagd, R. A. (2013). Identification of microRNA targets in tomato fruit development using high-throughput sequencing and degradome analysis. *Journal of Experimental Botany, 64*(7), 1863–1878. http:// doi.org/10.1093/jxb/ert049.

Kasschau, K. D., Fahlgren, N., Chapman, E. J., Sullivan, C. M., Cumbie, J. S., Givan, S. A., and Carrington, J. C. (2007). Genome-wide profiling and analysis of *Arabidopsis* siRNAs. *PLoS Biology, 5*(3), e57. http://doi.org/10.1371/journal .pbio.0050057.

Khan, M. S., Khraiwesh, B., Pugalenthi, G., Gupta, R. S., Singh, J., Duttamajumder, S. K., and Kapur, R. (2014). Subtractive hybridization-mediated analysis of genes and in silico prediction of associated microRNAs under waterlogged conditions in sugarcane (*Saccharum* spp.). *FEBS Open Bio, 4*, 533–541. http:// doi.org/10.1016/j.fob.2014.05.007.

Kidner, C. A. (2010). The many roles of small RNAs in leaf development. *Journal of Genetics and Genomics = Yi Chuan Xue Bao, 37*(1), 13–21. http://doi.org/10.1016 /S1673-8527(09)60021-7.

Kidner, C. A., and Martienssen, R. A. (2005). The developmental role of microRNA in plants. *Current Opinion in Plant Biology, 8*(1), 38–44. http://doi.org/10.1016/j .pbi.2004.11.008.

Kim, Y. J., Zheng, B., Yu, Y. Y., Won, S. Y. S. Y., Mo, B., and Chen, X. (2011). The role of mediator in small and long noncoding RNA production in *Arabidopsis thaliana*. *EMBO Journal, 30*, 814e822. http://doi.org/10.1038/emboj.2011.3.

Kozomara, A., and Griffiths-Jones, S. (2011). miRBase: Integrating microRNA annotation and deep-sequencing data. *Nucleic Acids Research, 39*(Database issue), D152–D157. http://doi.org/10.1093/nar/gkq1027.

Kurihara, Y., Takashi, Y., and Watanabe, Y. (2006). The interaction between DCL1 and HYL1 is important for efficient and precise processing of pri-miRNA in plant microRNA biogenesis. *RNA (New York), 12*(2), 206–212. http://doi .org/10.1261/rna.2146906.

Law, J. A., and Jacobsen, S. E. (2010). Establishing, maintaining and modifying DNA methylation patterns in plants and animals. *Nature Reviews Genetics, 11*, 204–220. http://doi.org/10.1038/nrg2719.

Lee, Y., Kim, M., Han, J., Yeom, K.-H., Lee, S., Baek, S. H., and Kim, V. (2004). MicroRNA genes are transcribed by RNA polymerase II. *EMBO Journal, 23*, 4051–4060.

Li, W.-X., Oono, Y., Zhu, J., He, X.-J., Wu, J.-M., Iida, K., … Zhu, J.-K. (2008). The *Arabidopsis* NFYA5 transcription factor is regulated transcriptionally and posttranscriptionally to promote drought resistance. *The Plant Cell, 20*(8), 2238–2251. http://doi.org/10.1105/tpc.108.059444.

Liu, C., Axtell, M. J., and Fedoroff, N. V. (2012). The helicase and RNaseIIIa domains of *Arabidopsis* Dicer-like1 modulate catalytic parameters during microRNA biogenesis. *Plant Physiology, 159*, 748–758.

Liu, Q., Feng, Y., and Zhu, Z. (2009). Dicer-like (DCL) proteins in plants. *Functional and Integrative Genomics, 9*(3), 277–286. http://doi.org/10.1007 /s10142-009-0111-5.

Llave, C., Kasschau, K. D., Rector, M. A., and Carrington, J. C. (2002). Endogenous and silencing-associated small RNAs in plants. *Society*, 14(July), 1605–1619. http://doi.org/10.1105/tpc.003210.ruses.

Lobbes, D., Rallapalli, G., Schmidt, D. D., Martin, C., and Clarke, J. (2006). SERRATE: A new player on the plant microRNA scene. *EMBO Reports*, 7(10), 1052–1058. http://doi.org/10.1038/sj.embor.7400806.

Lu, C., Kulkarni, K., Souret, F. F., MuthuValliappan, R., Tej, S. S., Poethig, R. S., ... Meyers, B. C. (2006). MicroRNAs and other small RNAs enriched in the *Arabidopsis* RNA-dependent RNA polymerase-2 mutant. *Genome Research*, 16(10), 1276–1288. http://doi.org/10.1101/gr.5530106.

Lu, S., Sun, Y.-H., and Chiang, V. L. (2008). Stress-responsive microRNAs in Populus. *The Plant Journal: For Cell and Molecular Biology*, 55(1), 131–151. http://doi.org/10.1111/j.1365-313X.2008.03497.x.

Lu, X.-Y., and Huang, X.-L. (2008). Plant miRNAs and abiotic stress responses. *Biochemical and Biophysical Research Communications*, 368(3), 458–462. http://doi.org/10.1016/j.bbrc.2008.02.007.

Luo, Y., Guo, Z., and Li, L. (2013). Evolutionary conservation of microRNA regulatory programs in plant flower development. *Developmental Biology*, 380, 133–144.

Ma, Z., Coruh, C., and Axtell, M. J. (2010). *Arabidopsis lyrata* small RNAs: Transient MIRNA and small interfering RNA loci within the *Arabidopsis* genus. *The Plant Cell*, 22(4), 1090–1103. http://doi.org/10.1105/tpc.110.073882.

Mantovani, R. (1999). The molecular biology of the CCAAT-binding factor NF-Y. *Gene*, 239, 15–27.

Meyers, B. C., Axtell, M. J., Bartel, B., Bartel, D. P., Baulcombe, D., Bowman, J. L., ... Zhu, J.-K. (2008). Criteria for annotation of plant MicroRNAs. *The Plant Cell*, 20(12), 3186–3190. http://doi.org/10.1105/tpc.108.064311.

Meyers, B. C., Souret, F. F., Lu, C., and Green, P. J. (2006). Sweating the small stuff: microRNA discovery in plants. *Current Opinion in Biotechnology*, 17, 1–8. http://doi.org/10.1016/j.copbio.2006.01.008.

Mica, E., Piccolo, V., Delledonne, M., Ferrarini, A., Pezzotti, M., Casati, C., ... Horner, D. S. (2010). Correction: High throughput approaches reveal splicing of primary microRNA transcripts and tissue specific expression of mature microRNAs in *Vitis vinifera*. *BMC Genomics*, 11, 109. http://doi.org /10.1186/1471-2164-11-109.

Nelson, D. E., Repetti, P. P., Adams, T. R., Creelman, R. a, Wu, J., Warner, D. C., ... Heard, J. E. (2007). Plant nuclear factor Y (NF-Y) B subunits confer drought tolerance and lead to improved corn yields on water-limited acres. *Proceedings of the National Academy of Sciences of the United States of America*, 104(42), 16450–16455. http://doi.org/10.1073/pnas.0707193104.

Ni, Z., Hu, Z., Jiang, Q., and Zhang, H. (2013). GmNFYA3, a target gene of miR169, is a positive regulator of plant tolerance to drought stress. *Plant Molecular Biology*, 82(1–2), 113–129. http://doi.org/10.1007/s11103-013-0040-5.

Ortiz-Morea, F. A., Vicentini, R., Silva, G. F. F., Silva, E. M., Carrer, H., Rodrigues, A. P., and Nogueira, F. T. S. (2013). Global analysis of the sugarcane microtranscriptome reveals a unique composition of small RNAs associated with axillary bud outgrowth. *Journal of Experimental Botany*, 64(8), 2307–2320. http://doi.org/10.1093/jxb/ert089.

Pais, H., Moxon, S., Dalmay, T., and Moulton, V. (2011). Small RNA discovery and characterisation in Eukaryotes using high-throughput approaches. In L. J. Collins (Ed.), *RNA Infrastructure and Networks* (pp. 239–254). New York: Springer.

Palatnik, J. F., Allen, E., Wu, X., Schommer, C., Schwab, R., Carrington, J. C., and Weigel, D. (2003). Control of leaf morphogenesis by microRNAs. *Nature*, 425(6955), 257–263. http://doi.org/10.1038/nature01958.

Park, M. Y., Wu, G., Gonzalez-Sulser, A., Vaucheret, H., and Poethig, R. S. (2005). Nuclear processing and export of microRNAs in *Arabidopsis*. *Proceedings of the National Academy of Sciences of the United States of America*, 102(10), 3691–3696. http://doi.org/10.1073/pnas.0405570102.

Paterson, A. H., Bowers, J. E., Bruggmann, R., Dubchak, I., Grimwood, J., Gundlach, H., ... Rokhsar, D. S. (2009). The Sorghum bicolor genome and the diversification of grasses. *Nature*, 457(7229), 551–556. http://doi.org/10.1038/nature07723.

Preston, J. C., and Hileman, L. C. (2013). Functional evolution in the plant squamosa-promoter binding protein-like (SPL) gene family. *Frontiers in Plant Science*, 4(April), 80. http://doi.org/10.3389/fpls.2013.00080.

Ramachandran, V., and Chen, X. (2009). Small RNA metabolism in *Arabidopsis*. *Trends in Plant Science*, 13(7), 368–374. http://doi.org/10.1016/j.tplants.2008.03.008.Small.

Reddy, A. R., Chaitanya, K. V., and Vivekanandan, M. (2004). Drought-induced responses of photosynthesis and antioxidant metabolism in higher plants. *Journal of Plant Physiology*, 161(11), 1189–1202. http://doi.org/10.1016/j.jplph.2004.01.013.

Reinhart, B. J., Weinstein, E. G., Rhoades, M. W., Bartel, B., and Bartel, D. P. (2002). MicroRNAs in plants. *Genes and Development*, 16(13), 1616–1626. http://doi.org/10.1101/gad.1004402.

Rhoades, M. W., Reinhart, B. J., Lim, L. P., Burge, C. B., Bartel, B., and Bartel, D. P. (2002). Prediction of plant microRNA targets. *Cell*, 110(4), 513–520.

Rott, P., Bailey, A. R., Comstock, J. C., Croft, B. J., and Saumtally, A. S. (Eds.) (2000). Diseases caused by bacteria. In *A guide to Sugarcane Diseases* (pp. 21–67). Quae.

Shin, C., Nam, J.-W., Farh, K. K.-H., Chiang, H. R., Shkumatava, A., and Bartel, D. P. (2010). Expanding the microRNA targeting code: Functional sites with centered pairing. *Molecular Cell*, 38(6), 789–802. http://doi.org/10.1016/j.molcel.2010.06.005.

Silva, M., Jifon, J., Da Silva, J., and Sharma, V. (1997). Use of physiological parameters as fast tools to screen for drought tolerance in sugarcane. *Brazilian Journal of Plant Physiology*, 19(3), 193–201.

Sorin, C., Declerck, M., Christ, A., Blein, T., Ma, L., Lelandais-Brière, C., ... Hartmann, C. (2014). A miR169 isoform regulates specific NF-YA targets and root architecture in *Arabidopsis*. *New Phytologist*, 202(4), 1197–1211. http://doi.org/10.1111/nph.12735.

Sternes, P. R., and Moyle, R. L. (2014). Deep sequencing reveals divergent expression patterns within the small RNA transcriptomes of cultured and vegetative tissues of sugarcane. *Plant Molecular Biology Reporter*. http://doi.org/10.1007/s11105-014-0787-0.

Studholme, D. J. (2012). Deep sequencing of small RNAs in plants: Applied bioinformatics. *Briefings in Functional Genomics*, 11(1), 71–85. http://doi.org/10.1093/bfgp/elr039.

Sunkar, R., and Jagadeeswaran, G. (2008). In silico identification of conserved microRNAs in large number of diverse plant species, *13*, 1–13. http://doi .org/10.1186/1471-2229-8-37.

Sunkar, R., Li, Y.-F., and Jagadeeswaran, G. (2012). Functions of microRNAs in plant stress responses. *Trends in Plant Science, 17*(4), 196–203, Table 3.1. http:// doi.org/10.1016/j.tplants.2012.01.010.

Sunkar, R., and Zhu, J. (2004). Novel and stress-regulated microRNAs and other small RNAs from *Arabidopsis. The Plant Cell, 16*(8), 2001–2019. http://doi .org/10.1105/tpc.104.022830.The.

Thiebaut, F., Grativol, C., Carnavale-Bottino, M., Rojas, C. A., Tanurdzic, M., Farinelli, L., ... Ferreira, P. C. G. (2012). Computational identification and analysis of novel sugarcane microRNAs. *BMC Genomics, 13*, 290. http://doi .org/10.1186/1471-2164-13-290.

Thiebaut, F., Grativol, C., Tanurdzic, M., Carnavale-Bottino, M., Vieira, T., Motta, M. R., ... Ferreira, P. C. G. (2014). Differential sRNA regulation in leaves and roots of sugarcane under water depletion. *PLoS ONE, 9*(4). http://doi .org/10.1371/journal.pone.0093822.

Thiebaut, F., Rojas, C. a, Almeida, K. L., Grativol, C., Domiciano, G. C., Lamb, C. R. C., ... Ferreira, P. C. G. (2012a). Regulation of miR319 during cold stress in sugarcane. *Plant, Cell and Environment, 35*(3), 502–512. http://doi .org/10.1111/j.1365-3040.2011.02430.x.

Turner, M., Yu, O., and Subramanian, S. (2012). Genome organization and char- acteristics of soybean microRNAs. *BMC Genomics, 13*(1), 169. http://doi .org/10.1186/1471-2164-13-169.

Varkonyi-Gasic, E., Wu, R., Wood, M., Walton, E. F., and Hellens, R. P. (2007). Protocol: A highly sensitive RT-PCR method for detection and quantification of microRNAs. *Plant Methods, 3*, 12. http://doi.org/10.1186/1746-4811-3-12.

Vazquez, F., Gasciolli, V., Crété, P., and Vaucheret, H. (2004). The nuclear dsRNA binding protein HYL1 is required for microRNA accumulation and plant development, but not posttranscriptional transgene silencing. *Current Biology, 14*(4), 346–351. http://doi.org/10.1016/j.cub.2004.01.035.

Wang, J., Wang, L., Mao, Y., Cai, W., Xue, H., and Chen, X. (2005). Control of root cap formation by MicroRNA-targeted auxin response factors in *Arabidopsis. The Plant Cell, 17*(8), 2204–2216. http://doi.org/10.1105/tpc.105.033076.

Wang, L., Song, X., Gu, L., Li, X., Cao, S., Chu, C., ... Cao, X. (2013). NOT2 proteins promote Pol II-dependent transcription and interact with multiple miRNA biogenesis factors in *Arabidopsis. The Plant Cell, 25*, 715–727.

Wassenegger, M., and Krczal, G. (2006). Nomenclature and functions of RNA- directed RNA polymerases. *Trends in Plant Science, 11*(3), 142–151. http://doi .org/10.1016/j.tplants.2006.01.003.

Wei, L. Q., Yan, L. F., and Wang, T. (2011). Deep sequencing on genome-wide scale reveals the unique composition and expression patterns of microRNAs in developing pollen of *Oryza sativa. Genome Biology, 12*(6), R53. http://doi .org/10.1186/gb-2011-12-6-r53.

Willmann, M. R., Endres, M. W., Cook, R. T., and Gregory, B. D. (2011). The func- tions of RNA-dependent RNA polymerases in *Arabidopsis. The Arabidopsis Book/American Society of Plant Biologists, 9*, e0146. http://doi.org/10.1199 /tab.0146.

Wu, G. (2013). Plant microRNAs and development. *Journal of Genetics and Genomics = Yi Chuan Xue Bao, 40*(5), 217–230. http://doi.org/10.1016/j.jgg.2013.04.002.

Wu, L., Zhou, H., Zhang, Q., Zhang, J., Ni, F., Liu, C., and Qi, Y. (2010). DNA methylation mediated by a microRNA pathway. *Molecular Cell, 38*(3), 465–475. http://doi.org/10.1016/j.molcel.2010.03.008.

Xie, Z., Kasschau, K. D., and Carrington, J. C. (2003). Negative feedback regulation of Dicer-like1 in *Arabidopsis* by microRNA-guided mRNA degradation. *Current Biology, 13*, 784–789. http://doi.org/10.1016/S.

Xie, Z., and Qi, X. (2008). Diverse small RNA-directed silencing pathways in plants. *Biochimica et Biophysica Acta, 1779*(11), 720–724. http://doi.org/10.1016/j.bbagrm.2008.02.009.

Xu, L., Lin, Z., Tao, Q., Liang, M., Zhao, G., Yin, X., and Fu, R. (2014). Multiple nuclear factor Y transcription factors respond to abiotic stress in *Brassica napus* L. *PLoS ONE, 9*(10), e111354. http://doi.org/10.1371/journal.pone.0111354.

Yang, C., Li, D., Mao, D., Liu, X., Ji, C., Li, X., ... Zhu, L. (2013). Overexpression of microRNA319 impacts leaf morphogenesis and leads to enhanced cold tolerance in rice (*Oryza sativa* L.). *Plant, Cell and Environment, 36*(12), 2207–2218.

Yang, X., and Li, L. (2011). miRDeep-P: A computational tool for analyzing the microRNA transcriptome in plants. *Bioinformatics (Oxford, England), 27*(18), 2614–2615. http://doi.org/10.1093/bioinformatics/btr430.

Yang, X., Wang, L., Yuan, D., Lindsey, K., and Zhang, X. (2013). Small RNA and degradome sequencing reveal complex miRNA regulation during cotton somatic embryogenesis. *Journal of Experimental Botany, 64*(6), 1521–1536. http://doi.org/10.1093/jxb/ert013.

Yu, B., Yang, Z., Li, J., Minakhina, S., Yang, M., Padgett, R. W., ... Chen, X. (2005). Methylation as a crucial step in plant microRNA biogenesis. *Science (New York), 307*(5711), 932–935. http://doi.org/10.1126/science.1107130.

Zanca, A. S., Vicentini, R., Ortiz-Morea, F. a, Del Bem, L. E. V, da Silva, M. J., Vincentz, M., and Nogueira, F. T. S. (2010). Identification and expression analysis of microRNAs and targets in the biofuel crop sugarcane. *BMC Plant Biology, 10*(1), 260. http://doi.org/10.1186/1471-2229-10-260.

Zhai, J., Arikit, S., Simon, S. A., Kingham, B. F., and Meyers, B. C. (2013). Rapid construction of parallel analysis of RNA end (PARE) libraries for Illumina sequencing. *Methods, 67*(1), 84–90.

Zhai, J., Jeong, D.-H., De Paoli, E., Park, S., Rosen, B. D., Li, Y., ... Meyers, B. C. (2011). MicroRNAs as master regulators of the plant NB-LRR defense gene family via the production of phased, trans-acting siRNAs. *Genes and Development, 25*(23), 2540–2553. http://doi.org/10.1101/gad.177527.111.

Zhang, B., Pan, X., Wang, Q., Cobb, G. P., and Anderson, T. A. (2006). Computational identification of microRNAs and their targets. *Computational Biology, 30*, 395–407. http://doi.org/10.1007/s13258-013-0070-z.

Zhang, B., Pan, X., Wang, Q., Cobb, G., and Anderson, T. A. (2005). Identification and characterization of new plant microRNAs using EST analysis. *Cell Research, 15*(5), 336–360. http://doi.org/10.1038/sj.cr.7290302.

Zhao, B., Ge, L., Liang, R., Li, W., Ruan, K., Lin, H., and Jin, Y. (2009). Members of miR-169 family are induced by high salinity and transiently inhibit the NF-YA transcription factor. *BMC Molecular Biology, 10*, 29. http://doi.org/10.1186/1471-2199-10-29.

Zhou, M., and Luo, H. (2013). MicroRNA-mediated gene regulation: Potential applications for plant genetic engineering. *Plant Molecular Biology, 83*(1–2), 59–75. http://doi.org/10.1007/s11103-013-0089-1.

chapter four

Biotechniques
Quest for stress-tolerant sugarcane

Bushra Tabassum, Idrees Ahmad Nasir
and Tayyab Husnain

Contents

Background

Plants, including sugarcane (*Saccharum officinarum*), cannot escape from their environment due to their sessile nature. They are exposed to a variety of environmental conditions, including extreme cold, heat and drought. Their livelihood, hence, depends on their innate ability to cope with changing extreme environments or whatever challenges they may face. With global climate change, environmental conditions are constantly in flux, which means that response to abiotic stress has become crucial to plant survival. Global warming is predicted to most severely affect developing countries, where agricultural systems are most vulnerable to climatic conditions and where small increases in temperature are very detrimental to productivity. Any environmental factor which limits the crop productivity or destroys biomass is termed as *stress* or *disturbance* (Grime, 1979). Abiotic stress is the primary cause of sugarcane crop loss worldwide. Drought, temperature and nutrient stress are common stress conditions that adversely affect sugarcane growth and ultimately crop production (Xiong et al., 2002). Minor abiotic stresses of sugarcane include fire, waterlogging, aluminium tolerance, salt tolerance, altitude, wind, soil pH and other metals.

Abiotic stress also determines the distribution of a particular plant species, as per soil and climatic conditions. It is noteworthy that some environmental factors like temperature can become stressful in just a few minutes, while others such as soil water content may take days to weeks and factors such as mineral deficiencies may take months to become stressful. Therefore, it is immensely important to understand the physiological processes that underlie stress injury and the adaptation and acclimatization mechanism of plants to environmental stress. Abiotic stresses that significantly affect sugarcane yield include nutrient stress, temperature stress (low temperatures and hot temperatures) and water stress. However, a few other factors including fire, waterlogging, altitude, wind, soil pH, and salt tolerance, also contribute in abiotic stress for sugarcane. Culm elongation is the most sensitive morphological change in sugarcane affected by water stress. Plant adaptation to salinity stress includes both osmotic and ionic stress tolerance. With increasing sodicity, soil physical properties deteriorate causing poor drainage. In sugarcane, low temperature stress has been studied in relation to bud sprouting, tillering, photosynthesis, culm growth, dry matter partitioning, juice quality, and gene expression and introgression to improve tolerance to chilling and freezing.

During abiotic stress, plants respond not only with changes in certain morphological, physiological characters, but changes in their pattern of gene expression and protein products also happen. Thus the ability of a certain plant to adapt has direct impact on its distribution and survival and on crop yields as well. In contrast, stress-tolerant plant species are able to grow in these conditions.

Stress induces a cascade of signaling events that activate kinase cascade, hormone biosynthesis and ion channels along with production of reactive oxygen species (Cheong et al., 2002). The signals in turn induce expression of stress responsive genes which define the overall plant defence. Abiotic stress in plants is, generally, a qualitative trait, which means that these are multigenic at genome level. Such traits are difficult to overcome and engineer.

Biotechniques to develop stress-tolerant sugarcane

Sustainable agriculture needs crop improvement for addressing critical problems including stress by employing available biotechniques. In recent decades, various biotechnological tools have proven to be most effective to overcome abiotic stress in plants, these include (1) crop breeding assisted with molecular markers, (2) genetic engineering (or in other words, transgenesis), (3) tissue culture-based *in vitro* selection, and (4) plant growth-promoting microbes (PGPR) (Ashraf et al., 2008; Sakhanokho et al., 2009) (Figure 4.1).

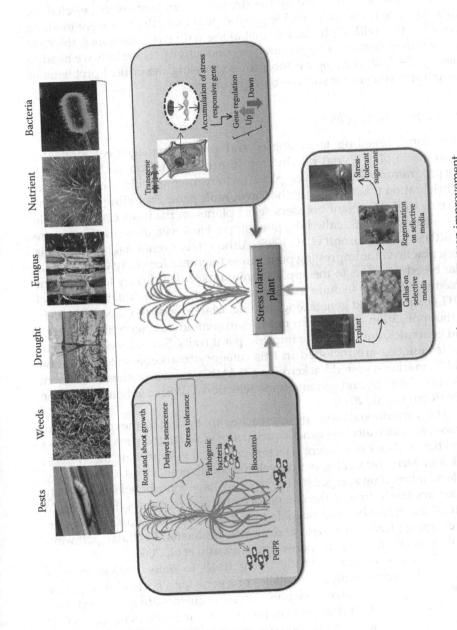

Figure 4.1 Biotechnological tools for biotic and abiotic stress for sugarcane crop improvement.

The basic aim underlying all available biotechniques is to make crop plants resistant/tolerant towards the abiotic stress and help plants overcome the pressure induced by the stress. Certain traits were selected for this purpose from the available genetic pool and attempts were made to combine this with high yield so that along with stress tolerance, the crop plant will maintain its vigour and/or productivity. Therefore, we need to deploy the biotechnological tools for addressing the critical problems of crop improvement for sustainable agriculture.

Marker-assisted selection

Traditional breeding technologies and proper management strategies continue to play a vital role in crop improvement. Conventional breeding programmes have been employed where inter-specific or inter-generic hybridization was opted to gather favourable genes including stress tolerance as well. The plant breeders select plants on the basis of their visible or measurable traits, called the phenotype. However, these methods were of little success (Purohit et al., 1998). Although conventional breeding has been practiced for improving plant stress tolerance for many years, molecular breeding refers to the application of molecular biology tools in plant breeding. Its areas include marker-assisted selection; qualitative trait loci (QTL) mapping and genetic engineering. Marker-assisted breeding refers to the application of molecular markers in combination with linkage maps and genomics, and is used to improve plant traits. Several modern breeding techniques are included in this category: marker-assisted selection (MAS), marker-assisted backcrossing (MABC), marker-assisted recurrent selection (MARS), and genome-wide selection (GWS) or genomic selection (GS) (Ribaut et al., 2010).

The conventional breeding approaches are limited by the complexity of stress-tolerance traits, low genetic variance of yield components under stress condition and lack of efficient selection criteria. It is important, therefore, to look for alternative strategies to develop stress-tolerant crops. Conclusively, molecular breeding and genetic engineering have contributed substantially to our understanding of the complexity of stress responses (Sreenivasulu et al., 2006). It has become possible to examine the mechanisms that perceive signals, transduce them and also to understand cellular pathways involved in abiotic stress response (Sreenivasulu et al., 2006).

DNA markers are currently being used to detect polymorphism between different genotypes or alleles of a gene for a particular sequence of DNA in a population or gene pool. These fragments are associated with a certain locus within the genome. The molecular markers include RFLP (restriction fragment length polymorphism) (Konieczny and Ausubel, 1993), RAPD (randomly amplified polymorphic DNA), AFLP (amplified fragment length polymorphism), SSR (simple sequence repeats) and SNP (single-nucleotide polymorphism) (Collard et al., 2005; Semagn et al., 2006).

In crop plants, traits are often of mono- or multigenic in nature. Even for some quality traits, one or a few major QTLs or genes can account for a very high proportion of the phenotypic variation of the trait (Bilyeu et al., 2006; Pham et al., 2012). Mostly, resistance to diseases/pests, male sterility, self-incompatibility and others related to shape, colour and architecture of whole plants and/or plant parts are multigenic in inheritance. Identification of such genes help breeders to assemble quality traits in a single variety. Additionally, these genes can be transformed in any of the crop plants to get transgenic progeny with improved character. Resistance to *Fusarium* head blight (FHB) in both wheat and barley is quantitatively inherited, and many QTLs have been identified from different resources of germplasm (Buerstmayr et al., 2009). Pumphrey et al. (2007) compared 19 pairs of near-isogenic lines (NILs) for *Fhb1* derived from an ongoing breeding program and found that the average reduction in disease severity between NIL pairs was 23% for disease severity and 27% for kernel infection. Later investigation from the group also demonstrated successful implementation of MAS for this QTL (Anderson et al. 2007).

Transgene technology

Transgenesis refers to a process where an exogenous gene, termed as *transgene*, is introduced into a host organism so that the organism will exhibit a new property. Biotechnology has offered multiple strategies to develop transgenic crop plants with improved tolerance to abiotic stress. Rapid advances in recombinant DNA technology in combination with efficient gene transfer protocols have resulted in efficient transformation and generation of transgenic lines in a number of crop plants (Gosal et al., 2009; Wani et al., 2008, 2011). Transgenic approach is currently being pursued actively throughout the world to improve traits including tolerance to various abiotic stresses in a number of crop plants (Ashraf et al., 2008). With the help of molecular tools, it is possible to directly select the desired gene from any organism.

Understanding the changes in cellular, biochemical and molecular mechanisms occurring in stressed plants has provided new tools and strategies to improve the environmental stress tolerance of crops. Since stress tolerance is a multigenic trait in plants, transformation of a single gene appears to have a limited effect. Most of the stress adaptation processes are under transcriptional control, therefore many transcription factors were chosen for the development of stress-tolerant plants. Additionally, stress responsive genes were taken from the wild relative of the crop in some cases. Transgenic technology offers the possibility to improve the existing genetic potential of the plant and develop new crop varieties tolerant towards various biotic and abiotic stresses.

A large number of genes have been transformed in different crop plants to make them resistant for one or the other stresses. The genes selected for transformation could be involved in encoding enzymes which are required for the biosynthesis of various osmoprotectants. Other classes of genes that have been selected for transformation include those that encoded enzymes for modifying membrane lipids; transcription factors; stress responsive genes from wild relative, late embryogenesis abundant (LEA) protein; and detoxification enzymes. These genes were expressed singly or in various combinations under the control of specific promoters that define the particular or localized expression of the transgene. The genes encoding protein factors regulate gene expression and signal transduction and thus prove to be helpful in inducing stress responsive genes during stressed conditions.

To date, several efforts have been made to enhance tolerance towards multiple stresses such as cold, drought and salt in crops including sugarcane. An increased tolerance to freezing and drought in *Arabidopsis* was achieved by overexpressing a transcriptional factor, *CBF4*, whose expression was induced during drought stress and by abscisic acid (ABA) treatment (Dai et al., 2007). Various reports are present where the transgene technique was employed to create resistance in transgenic sugarcane plants against abiotic stresses. These include Trujillo et al. (2009), Trujillo et al. (2008), Wahid and Close (2007), Patade et al. (2006), Patade et al. (2011) and Prabu et al. (2011).

In vitro *selection*

As an alternative biotechnique for development of stress-tolerant crop plants, tissue culture-based *in vitro* selection has gained much importance. *In vitro* culture techniques are an excellent tool to study the behaviour of undifferentiated cells and the whole plants in ambient stress under controlled conditions. The technique is a quite feasible and cost-effective tool for developing stress-tolerant plants (Ochatt et al., 1999; Queiros et al., 2007). In this biotechnique, *in vitro* culture of plant cells, tissues or organs is grown on a medium supplemented with specific selective agent followed by selection of regenerated plants with desirable characters. Somaclonal variation does exist in tissue culture which can be exploited for *in vitro* selection of cells and tissues against several stresses (Bajaj, 1987; Tal, 1996). However, the success also depends on that particular trait whether or not it is amenable to *in vitro* selection.

Plants tolerant to both the biotic and the abiotic stresses can be acquired by applying the selecting agents such as NaCl for salt tolerance, or PEG or mannitol for drought tolerance (Errabii et al. 2008). *In vitro* selection for cells exhibiting increased tolerance to drought stress has been reported (El-Haris and Barakat, 1998; Errabii et al., 2006; Hassan et al., 2004; Sabbah and Tal, 1990; Santos-Diaz and Ochoa-Alejo, 1994). The effect of NaCl and PEG on growth, osmolytes accumulation and antioxidant defence in cultured cells of sugarcane is reported by Patade et al. (2011, 2012); however, in their investigations no attempt is made to select PEG-tolerant callus lines, characterization of PEG-tolerant callus lines and regeneration of plantlets from PEG-tolerant callus lines. Musa (2011) used PEG as selection agent to screen sugarcane varieties for drought tolerance. Begum et al. (2011) has screened tissue culture-raised sugarcane somaclones for drought tolerance using PEG as the selection agent.

Due to the complex and variable nature of the abiotic stresses, it is difficult to analyse the response of plants to various abiotic stresses in the field or in greenhouse conditions. *In vitro* selection can allow a deeper understanding of the physiology and biochemistry in plants cultured under adverse environmental conditions (Benderradji et al., 2012). *In vitro* selection has advantages over natural conditions, including (1) unfavourable weather and climate conditions are avoided, (2) a large number of plantlets can be screened for a particular resistance and (3) manipulation of mutants and somaclones with high genomic variability became very easy.

Plant growth-promoting rhizobacteria (PGPR) help plants tolerate abiotic stress

Plant growth-promoting rhizobacteria (PGPR) are root-colonizing bacteria that form symbiotic relationships with many plants. These microbes enhance plant growth through various mechanisms that include phosphate solubilisation; nitrogen fixation; and production of siderophore, ACC (1-aminocyclopropane-1-carboxylate deaminase), phytohormone and volatile organic compounds (Grover et al., 2011). PGPR exhibit antifungal activity and induce systemic resistance, promote beneficial plant–microbe symbioses and interfere with pathogen toxin production. Briefly, PGPR produce growth-promoting substances in the rhizosphere that have a direct influence on the morphology of the plant (Dimkpa et al., 2009; Kim et al., 2013; Timmusk and Nevo, 2011; Timmusk et al., 2014; Yang et al., 2009).

PGPR reside below the ground and influence the selection on plant traits by modifying the effects of abiotic stress (Lau and Lennon, 2011). Their role in biocontrol activity, plant growth and nutrient management is well established. Additionally, these microbes are being used in the management of biotic and abiotic stresses where they employ a variety of mechanisms to alleviate stress on plants (Berg et al., 2013; Rolli et al., 2014). In the direct mechanism, PGPR can enhance uptake of micronutrients by plants and help maintain phytohormone homeostasis, while indirectly,

PGPR stimulate the plant immune system to fight against potential pathogens (Balloi et al., 2010). During stress conditions, ethylene regulates the plant homeostasis resulting in reduced root and shoot growth (Glick et al., 2007). However, bacteria carrying ACC deaminase enzyme cleave the ethylene precursor in stressed plants that ultimately lowers the ethylene level in them (Glick, 2004) rendering plants tolerant towards the stress.

In plants, PGPR selectively adapt to a certain ecological niche (Gray and Smith, 2005). PGP bacteria surpass the plant's barrier (endodermis) and travel through the root cortex to the vascular system and reside in the plant stem, leaves, tubers and other organs as endophytes (Compant et al., 2005; Gray and Smith, 2005). Endophytic bacteria induce resistance to abiotic stress in plant species by reducing water consumption and enhancing the growth rate and biomass of the stressed plant (Cherif et al., 2015; Theocharis et al., 2012). Such findings indicate that the symbiotic relationship between PGPR and plants may be useful in alleviating the impact of drastic climatic change and expanding productive agricultural land (Redman et al., 2011).

Various studies have documented that drought-stressed plants when inoculated with PGPR exhibit drought tolerance (Figueiredo et al. 2008; Glick et al., 2007; Kohler et al., 2008; Mayak et al., 2004; Timmusk and Wagner, 1999). Similarly, PGPR inoculation renders plants tolerant towards excess salt (Mayak et al., 2004; Zhang et al., 2008), and helps plants with fertility and nutrient uptake (Adesemoye et al., 2008; Gyaneshwar et al., 2002; Mantelin and Touraine, 2004; Malakoff, 1998) and plant architecture (Kloepper et al., 2007).

PGPR plays an important role in conferring resistance and adaptation of plants to drought stresses and have the potential role in solving future food security issues. The interaction between plants and PGPR under drought conditions affects not only the plant but also changes the soil properties. The mechanisms elicited by PGPR such as triggering osmotic responses and induction of novel genes play a vital role in ensuring plant survival under drought stress. The development of drought-tolerant crop varieties through genetic engineering and plant breeding is essential, but it is a long drawn process, whereas PGPR inoculation to alleviate drought stresses in plants opens a new chapter in the application of microorganisms in dry land agriculture. Taking the current leads available, concerted future research is needed in terms of identification of the right kind of microbes and addressing the issue of delivery systems and field evaluation of potential organisms.

Conclusion

Keeping in view the economic importance of sugarcane, development of stress-tolerant sugarcane varieties by applying any of the available biotechniques is mandatory. Transgenics have provided a variety of improvements in sugarcane breeding programmes through development of abiotic stress-tolerant sugarcane varieties. In this venture, molecular markers complement transgenics and classical breeding for developing varieties that can tolerate abiotic stresses. Molecular breeding and genetic engineering have contributed substantially to our understanding of the complexity of stress responses (Sreenivasulu et al., 2006). It has become possible to examine the mechanisms that perceive signals, transduce them and also to understand cellular pathways involved in abiotic stress response (Sreenivasulu et al., 2006).

References

Adesemoye, A.O., Torbert, H.A., Kloepper, J.W. 2008. Enhanced plant nutrient use efficiency with PGPR and AMF in an integrated nutrient management system. *Can. J. Microbiol.* 54, 876–886.

Anderson, J.A., Chao, S., Liu, S. 2007. Molecular breeding using a major QTL for *Fusarium* head blight resistance in wheat. *Crop Sci.* 47, S-112–119.

Ashraf, M., Athar, H.R., Harris, P.J.C., Kwon, T.R. 2008. Some prospective strategies for improving crop salt tolerance. *Adv. Agron.* 97, 45–110.

Balloi, A., Rolli, E., Marasco, R., Mapelli, F., Tamagnini, I., Cappitelli, F. et al. 2010. The role of microorganisms in bioremediation and phytoremediation of polluted and stressed soils. *Agrochimica* 54, 353–369.

Benderradji, L., Brini, F., Kellou, K., Ykhelf, N., Djekoun, A., Masmoudi, K., Bouzerour, H. 2012. Callus induction, proliferation, and plantlets regeneration of two bread wheat (*Triticum aestivum* L.) genotypes under saline and heat stress conditions. *ISRN Agronomy*, Article ID 367851.

Berg, G., Zachow, C., Müller, H., Philipps, J., Tilcher, R. 2013. Next-generation bioproducts sowing the seeds of success for sustainable agriculture. *Agron.* 3, 648–656.

Buerstmayr, H., Ban, T., Anderson, J.A. 2009. QTL mapping and marker-assisted selection for *Fusarium* head blight resistance in wheat: A review. *Plant Breeding* 128, 1–26.

Cheong Y.H., Chang H.S., Gupta R., Wang X., Zhu T., Luan S. 2002. Transcriptional profiling reveals novel interactions between wounding, pathogen, abiotic stress, and hormonal responses in *Arabidopsis*. *Plant Physiol.* 129, 661–677.

Cherif, H., Marasco, R., Rolli, E., Ferjani, R., Fusi, M., Souss, A. et al. 2015. Oasis desert farming selects environment-specific date palm root endophytic communities and cultivable bacteria that promote resistance to drought. *Environ. Microbiol. Rep.* 7, 668–678.

Collard, B.C.Y., Jahufer, M.Z.Z., Brouwer, J.B., Pang, E.C.K. 2005. An introduction to markers, quantitative trait loci (QTL) mapping and marker-assisted selection for crop improvement: The basic concepts. *Euphytica* 142, 169–196.

Compant, S., Reiter, B., Sessitsch, A., Nowak, J., Clément, C., Ait Barka, E. 2005. Endophytic colonization of *Vitis vinifera* L. by plant growth-promoting bacterium *Burkholderia* sp. strain PsJN. *Appl. Environ. Microbiol.* 71, 1685–1693.

Dai, X., Xu, Y., Ma, Q., Xu, W., Wang, T., Xue, Y., Chong, K. 2007. Overexpression of an R1R2R3 MYB gene, OsMYB3R-2, increases tolerance to freezing, drought, and salt stress in transgenic *Arabidopsis*. *Plant Physiol.* 143, 1739–1751.

Dimkpa, C., Weinand, T., Asch, F. 2009. Plant-rhizobacteria interactions alleviate abiotic stress conditions. *Plant Cell Environ.* 32, 1682–1694.

Figueiredo, V.B., Burity, H.A., Martinez, C.R., Chanway, C.P. 2008. Alleviation of drought stress in the common bean (*Phaseolus vulgaris* L.) by co-inoculation with *Paenibacillus polymyxa* and *Rhizobium tropici. Appl. Soil Ecol.* 40, 182–188.

Glick, B.R. 2004. Bacterial ACC deaminase and the alleviation of plant stress. *Adv. Appl. Microbiol.* 56, 291–312.

Glick, B.R., Cheng, Z., Czarny, J., Duan, J. 2007. Promotion of plant growth by ACC deaminase-producing soil bacteria. In: Bakker, P.A.H., Raaijmakers, J., Lemanceau, P., Bloemberg, G. (Eds.), *New perspectives and approaches in plant growth-promoting rhizobacteria research*, 329–339. Netherlands: Springer.

Glick, B.R., Todorovic, B., Czarny, J., Cheng, Z., Duan, J., McConkey, B. 2007. Promotion of plant growth by bacterial ACC deaminase. *Crit. Rev. Plant Sci.* 26, 227–242.

Gosal, S.S., Wani, S.H., Kang, M.S. 2009. Biotechnology and drought tolerance. *J. Crop Improv.* 23, 19–54.

Gray, E.J., Smith, D.L. 2005. Intracellular and extracellular PGPR: Commonalities and distinctions in the plant-bacterium signaling processes. *Soil Biol. Biochem.* 37, 395–412.

Grover, M., Ali SkZ. Sandhya, V., Venkateswarlu, B. 2011. Role of microorganisms in adaptation of agricultural crops to abiotic stresses. *World J. Microbiol. Biotechnol.* 27, 1231–1240.

Gyaneshwar, P., Naresh Kumar, G., Parekh, L.J., Poole, P.S. 2002. Role of soil microorganisms in improving P nutrition of plants. *Plant Soil* 245, 83–93.

Kim, Y.C., Glick, B., Bashan, Y., Ryu, C.M. 2013. Enhancement of plant drought tolerance by microbes. In: Aroca, R. (Ed.), *Plant responses to drought stress.* Berlin: Springer Verlag.

Kloepper, J.W., Gutierrez-Estrada, A., McInroy, J.A. 2007. Photoperiod regulates elicitation of growth promotion but not induced resistance by plant growth-promoting rhizobacteria. *Can. J. Microbiol.* 53, 159–167.

Kohler, J., Hernandez, J.A., Caravaca, F., Roldan, A. 2008. Plant-growth-promoting rhizobacteria and arbuscular mycorrhizal fungi modify alleviation biochemical mechanisms in water-stressed plants. *Funct. Plant Biol.* 35, 141–151.

Konieczny, A., Ausubel, F. 1993. A procedure for mapping *Arabidopsis* mutations using co-dominant ecotype-specific PCR based markers. *The Plant Journal* 4, 403–410.

Lau, J.A., Lennon, J.T. 2011. Evolutionary ecology of plant–microbe interactions: Soil microbial structure alters selection on plant traits. *New Phytol.* 192, 215–224.

Malakoff, D. 1998. Coastal ecology: Death by suffocation in the Gulf of Mexico. *Science* 281, 190–192.

Mantelin, S., Touraine, B. 2004. Plant growth-promoting bacteria and nitrate availability impacts on root development and nitrate uptake. *J. Exp. Bot.* 55, 27–34.

Mayak, S. Tirosh, T., Glick, B.R. 2004. Plant growth-promoting bacteria that confer resistance to water stress in tomatoes and peppers. *Plant Sci.* 166, 525–530.

Patade, V.Y., Bhargava, S., Suprasanna, P. 2011. Salt and drought tolerance of sugarcane under iso-osmotic salt and water stress: Growth, osmolytes accumulation, and antioxidant defense. *J. Plant Interact.* 6(4), 275–282.

Patade, V.Y., Suprasanna, P., Bapat, V.A. 2006. Selection for abiotic (salinity and drought) stress tolerance and molecular characterization of tolerant lines in sugarcane. *BARC Newsletter* 273, 244.

Prabu, G., Kawar, P.G., Pagariya, M.C., Prasad, D.T. 2011. Identification of water deficit stress upregulated genes in sugarcane. *Plant Mol. Bio. Rep.* 29(2), 291–304.

Pumphrey, M.O., Bernardo, R., Anderson, J.A. 2007. Validating the *Fhb1* QTL for Fusarium head blight resistance in near-isogenic wheat lines developed from breeding populations. *Crop Sci.* 47, 200–206.

Redman, R.S., Kim, Y.O., Woodward, C.J.D.A., Greer, C., Espino, L. et al. 2011. Increased fitness of rice plants to abiotic stress via habitat adapted symbiosis: A strategy for mitigating impacts of climate change. *PLoS One* 6, e14823.

Rolli, E., Marasco, R., Vigani, G., Ettoumi, B., Mapelli, F., Deangelis, M.L. et al. 2014. Improved plant resistance to drought is promoted by the root-associated microbiome as a water stress-dependent trait. *Environ. Microbiol.* 17, 316–331.

Sakhanokho, H.F., Kelley, R.Y. 2009. Influence of salicylic acid on *in vitro* propagation and salt tolerance in *Hibiscus acetosella* and *Hibiscus moscheutos* (cv 'Luna Red'). *Afr. J. Biotechnol.* 8, 1474–1481.

Semagn, K., Bjornstad, A., Ndjiondjop, M.N. 2006a. An overview of molecular marker methods for plants. *Afr. J. Biotechnol.* 5, 2540–2568.

Theocharis, A., Bordiec, S., Fernandez, O., Paquis, S., Dhondt-Cordelier, S., Baillieul, F. et al. 2012. *Burkholderia phytofirmans* PsJN primes *Vitis vinifera* L. and confers a better tolerance to low nonfreezing temperatures. *Mol. Plant Interact.* 25, 241–249.

Timmusk, S., Islam, A., Abd El, D., Lucian, C., Tanilas, T., Kannaste, A. et al. 2014. Drought-tolerance of wheat improved by rhizosphere bacteria from harsh environments: Enhanced biomass production and reduced emissions of stress volatiles. *PLoS One* 9, 1–13.

Timmusk, S., Nevo, E., 2011. Plant root associated biofilms. In: Maheshwari, D.K. (Ed.), *Bacteria in agrobiology*, 285–300. Berlin: Springer Verlag.

Timmusk, S., Wagner, E.G.H. 1999. The plant-growth-promoting rhizobacterium *Paenibacillus polymyxa* induces changes in *Arabidopsis thaliana* gene expression: A possible connection between biotic and abiotic stress responses. *Mol. Plant Microb. Interact.* 12, 951–959.

Trujillo, L.E., Sotolongo, M., Menendez, C., Ochogavia, M.E., Coll, Y., Hernandez, I. et al. 2008. SodERF3, a novel sugarcane ethylene responsive factor (ERF), enhances salt and drought tolerance when overexpressed in tobacco plants. *Plant Cell Physiol.* 49(4), 512–525.

Trujillo, L.E., Menendez, C., Ochogavia, M.E., Hernandez, I., Borras, O., Rodriguez, R. et al. 2009. Engineering drought and salt tolerance in plants using SodERF3, a novel sugarcane ethylene responsive factor. *Biotecnología Aplicada* 26(2), 168–171.

Wahid, A., Close, T.J. 2007. Expression of dehydrins under heat stress and their relationship with water relations of sugarcane leaves. *Biologia Plantarum* 51(1), 104–109.

Wani, S.H., Gosal, S.S. In press. Introduction of *OsglyII* gene into *Indica* rice through particle bombardment for increased salinity tolerance. *Biol. Plant.*

Wani, S.H., Sandhu, J.S., Gosal, S.S. 2008. Genetic engineering of crop plants for abiotic stress tolerance. In: C.P. Malik, B. Kaur, C. Wadhwani (Eds.), *Advanced topics in plant biotechnology and plant biology*, 149–183. New Delhi: MD Publications.

Yang, J., Kloepper, J.W., Ryu, C.M. 2009. Rhizosphere bacteria help plants tolerate abiotic stress. *Trends Plant Sci.* 14, 1–4.

Zhang, H., Kim, M.S., Sun, Y., Dowd, S.E., Shi, H., Pare, P.W. 2008. Soil bacteria confer plant salt tolerance by tissue-specific regulation of the sodium transporter HKT1. *Mol. Plant Microbe Interact.* 21, 737–744.

chapter five

Biotechnological interventions to overcome the effect of climate change in sugarcane production

Yogendra Singh

Contents

Introduction

Sugarcane originated in New Guinea, where it has been known since 6000 B.C. In 1000 B.C., its cultivation slowly spread along with human migration routes to Southern Asia and India. Some literature reveals that it hybridized with wild sugarcane from India and China to produce the thin canes. Sugarcane has an incredibly long history of cultivation in the Indian subcontinent. The earliest reference to it is in the Atharva Veda (1500–800 B.C.), where sugarcane was called *Ikshu* and has mention as an offering during sacrificial rites. The Atharva Veda also mentioned it as the source and symbol of sweet pleasant appearance. The word 'sugar' is derived from the ancient Sanskrit word *sharkara*. By 6th century B.C., sharkara was frequently referred to in Sanskrit texts which even distinguished superior and inferior varieties of sugarcane. A Persian description from 6th century B.C. gives the first account of solid sugar and describes it as coming from the Indus Valley. This early sugar would have resembled raw sugar, traditional dark brown sugar/Indian jaggery/gur. At present, 115 countries of the world cultivate sugarcane for sugar production and produce about three-fourths of the total sugar production of the world. The remaining sugar comes from sugar beet. Sugar beet cultivation and processing is highly subsidized in the European Union, which contributes nearly 21.5% of the world's sugar. With a World Trade Organization agreement in place, sugar beet cultivation may not remain a profitable proposition in the European Union due to drastic reduction in subsidies on its production and processing. Sugarcane is the world's largest crop by production quantity. In 2012, the Food and Agriculture Organization of the United Nations (FAO) estimates it was cultivated on about 26 million hectares, in more than 90 countries, with a worldwide harvest of 1.83 billion tons. Brazil was the largest producer of sugarcane in the world. The next five major producers, in decreasing amounts of production, were India, China, Thailand, Pakistan and Mexico. Brazil led the world in sugarcane production in 2013 with a 739,267 TMT harvest. India was the second largest producer with 341,200 TMT, and China the third largest producer with 125,536 TMT harvest (Table 5.1 and Figure 5.1). The average worldwide yield of sugarcane crops in 2013 was 70.77 tons per hectare. The most productive farms in the world were in Peru with a nationwide average sugarcane crop yield of 133.71 tons per hectare.

Table 5.1 Top 10 sugarcane producers 2013

Ranking	Country	Production (thousand metric tons, TMT)
1	Brazil	739,267
2	India	341,200
3	China	125,536[a]
4	Thailand	100,096
5	Pakistan	63,750
6	Mexico	61,182
7	Colombia	34,876
8	Indonesia	33,700[a]
9	Philippines	31,874
10	United States	27,906
	World	1,877,105

Source: Food and Agricultural Organization of United Nations, Economic and Social Department, Statistical Division.

[a] Unofficial/semi-official/mirror data.

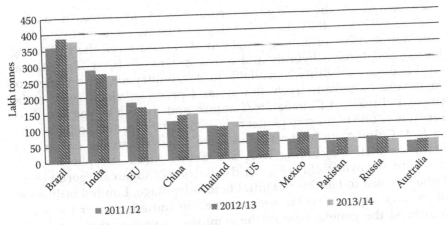

Figure 5.1 Major sugar producing countries of the world. India was the second largest producer of sugar and Brazil was the leading producer in 2013–2014. India's share in the world production of sugar was 15.39% in 2013–2014. (From the United States Department of Agriculture.)

Other than sugar, products derived from sugarcane include falernum, molasses, rum, *cachaça* (a traditional spirit from Brazil), bagasse and ethanol. In some regions, people use sugarcane reeds to make pens, mats, screens and thatch. The young, unexpanded inflorescence of *tebu telor* is eaten raw, steamed or toasted, and prepared in various ways in certain island communities of Indonesia (Dahlia et al., 2010). A natural nutrient, sugar, also known as sucrose, is a vital ingredient in our daily diet. To meet the body's energy requirements, carbohydrates should account for between 50% and 55% of a balanced diet. Sugar, whatever its form, is a source of the carbohydrates essential to our health and well-being. Sugar displays a whole range of characteristics and tastes that affect the way it behaves when used. It is a sweetening, colouring and bulking agent, and a preservative. It can alter boiling and freezing points, affect the flavour and smell of foods, and add bulk to foods.

Sugarcane germplasm

Natural variability in the form of germplasm is one of the most important and basic raw materials at the disposal of the sugarcane breeders for meeting the future needs of crop improvement. India has the world collection of sugarcane germplasm and it is being maintained by the Sugarcane Breeding Institute, Coimbatore. The cultivated and wild species of *Saccharum*, namely, *S. officinarum*, *S. barberi*, *S. sinensis*, *S. spontaneum* and *S. robustum* along with related genera which can hybridize with sugarcane, namely, *Erianthus*, *Miscanthus*, *Narenga* and

Sclerostachya, are the basic genetic resources of sugarcane. In India, a total collection of 4803 clones is being maintained. Out of this collection, about 2070 are basic germplasm at species level; the rest being Indian and foreign hybrids and allied genera. Thus, at the national level enough variability is available in the country for improvement of sugarcane. Sugarcane freely flowers and produces viable seed in the tropical climate of Coimbatore/Kannur, and the Sugarcane Breeding Institute is one of the two pioneers in sugarcane breeding in the world. In fact, Coimbatore-bred varieties have ruled sugarcane scenario of many countries, including USA and South Africa. Despite the rich diversity and availability of a world collection of sugarcane germplasm, India has still to tap this potential to its advantage. Limited utilisation of basic germplasm (at the species level) in India has, over the years, narrowed the genetic base of the commercial cane varieties, and the plateauing effect has become apparent in terms of sugar recovery and cane productivity. Out of 2070 clones available at species level, hardly 50 have been utilised in the breeding programme. Further, out of 764 *S. officinarum* (noble cane, the main contributor of genes for sugar) clones available at the Sugarcane Breeding Institute, hardly 100 accessions flower in nature, and out of which only 23 have been utilised in breeding programmes so far. It is imperative to tap the huge available genetic potential to break both sugar and yield barriers.

Sugarcane production system

In India, more than 6 million farmers are engaged in sugarcane cultivation, and the majority of them are small and marginal with very small land holdings. About 50% of the total area under sugarcane is comprised of holdings between 0.5 to 5 ha. For 20.7% of the area, holding size ranges between 5 and 10 ha (Anonymous, 2011). This has provided a unique advantage for better land use through intercropping and increases in the input use efficiency. High value and remunerative crops like vegetables, potato, oilseeds and pulses offer great scope for growing as intercrops, and in further providing additional income and reducing risks in the long duration crop of sugarcane as well as in improving land-use efficiency. Monoculture of cane has resulted in substantial reduction in productivity. The proper sequence of cropping, such as sugarcane and leguminous crops, as a component of system approach of nutrient management is suitable for sustainable productivity. Sugarcane-based cropping systems generally are at least of 3 to 4 years of duration. The plant crop of sugarcane is invariably followed by its ratoon crop. Mostly one to two ratoons are taken in succession. However, the crop preceding sugarcane and succeeding its ratoon crop varies in different agro-climatic conditions and under socioeconomic situations. Sugarcane-based systems are well integrated

Table 5.2 Production, area under cultivation, and yield of sugarcane and sugar

Crop/marking year	Area (lakh hectares)	Production (lakh tonnes)		Sugarcane yield (tonnes/ hectares)
		Sugarcane	Sugar	
2005–06	42.0	2811.7	193.2	66.92
2006–07	51.5	3555.2	282.0	69.02
2007–08	50.6	3481.9	263.0	68.88
2008–09	44.2	2850.3	146.8	64.55
2009–10	41.7	2923.0	188.0	70.02
2010–11	48.8	3423.8	243.5	70.09
2011–12	50.4	3610.4	263.4	71.67
2012–13	49.99	3412.0	258.5	68.25
2013–14	50.12	3521.4	245.5	69.84
2014–15	N.A.	3549.5[a]	250.46	N.A.

Source: India Department of Food and Public Distribution (for sugar production) and Agricultural Statistics (for production and area of sugarcane).

Note: India's sugar production has increased in the last 10 years at a CAGR (compound annual growth rate) of 6.04%. During the same period, India's sugarcane production has increased at a CAGR of 3.97% and area under cultivation at a CAGR of 3.19%.

[a] As per second advance estimate (2014–2015) of DAC released on 18 February 2015.

with rice–wheat and rice-based cropping systems. The major sugarcane-based cropping systems are rice–wheat/mustard–sugarcane plant–first ratoon wheat (rice–wheat–sugarcane) and maize–wheat–sugarcane plant–first ratoon wheat (sugarcane–wheat) in Uttar Pradesh, cotton–sugarcane plant–first ratoon–sorghum in Maharashtra, groundnut–sugarcane in Gujarat and rice–sugarcane plant–first ratoon (sugarcane–rice) in Western and South India. Sugarcane-based cropping systems are better in terms of infrastructure and socioeconomic development indicators such as the value of agricultural output per hectare compared to other cropping systems. Thus, it offers great opportunity to study and quantify the beneficial effects of the sugarcane-based systems on soil fertility and productivity. According to the Indian Department of Food and Public Distribution (Table 5.2), the total sugarcane production was 3549.5 lakh tones in 2014–2015.

Stresses in sugarcane

Stress can be defined as any condition that hampers the expression of the full genetic potential of a living being. As sessile organisms, plants face many stressful conditions that required the evolutionary establishment of diverse developmental and physiological strategies to cope or avoid the stress condition.

Abiotic stresses

Water scarcity is a major abiotic stress for sugarcane. Elucidating tolerance mechanisms would enable the development of cultivars more tolerant to drought, allowing cultivation in marginal areas, while ensuring the sustainability and viability of the industry in such drought-prone areas. Plant irrigation is a good option for agriculture, but it also increases salinity on soil. Besides, it corresponds to 65% of global water demand, and considering the expansion of cultivation to areas without fresh water, tolerance to drought will become increasingly important. Drought tolerance would also contribute to reducing irrigation and water use (Rocha et al., 2007). Molecular marker technology is a boon for abiotic stress management (Singh, 2013). Even though sugarcane can survive long dry periods, it demands a fair amount of water for optimal yield, leading to the use of irrigation in many areas. Whereas irrigation of sugarcane plantations in Brazil is minimal, 60% of Australia fields and 40% of South African cultivation are irrigated (Inman-Bamber and Smith, 2005). However, the lack of genetic and molecular information about drought tolerance mechanisms and inheritance in sugarcane has limited the development of improved cultivars. There is a need to distinguish genes definitely associated with the response to water deficit, which hold an adaptive function to water deprivation and in stress environments. Genes associated with regulation of expression under water deficit or during the establishment of drought tolerance are potential candidates to evaluate differential expression between contrasting sugarcane genotypes.

Biotic stresses

Genetic resistance to pests and diseases is an indispensable and essential condition in crop improvement. Pests and pathogens often overcome new territory and are well known to dynamically evolve towards breaking resistances, always posing new challenges. Molecular markers may help in diagnosis and management of fungal pathogens (Singh et al., 2013b). Similarly, genomic fingerprinting of these pathogens may assist in their management (Singh and Kumar, 2010). Indeed, biotic stresses are of special concern in sugarcane breeding programs, because they may cause great economical impact in plantations with susceptible cultivars. Examples of biotic stresses to which sugarcane breeders, geneticists, pathologists and entomologists have been paying attention, depending on the location of the breeding programme, are fungal diseases such as rusts, especially the brown rust (*Puccinia melanocephala*) and the orange rust (*Puccinia kuehnii*) that recently invaded the American continent, as well as 'smut' (*Ustilago scitaminea*). The main bacterial diseases are ratoon stunting disease (*Leifsonia xyli*) and leaf scald (*Xanthomonas albilineans*), and important viral diseases are sugarcane mosaic virus (SCMV) and sugarcane yellow leaf virus (SCYLV). Additional diseases with constrained proliferation or of potentially less significant economical impact comprise the fungal diseases red rot (*Glomerella tucumanensis*), eyespot (*Helminthosporium sacchari*), pokka boeng (*Fusarium moniloforme*) and pineapple disease (*Ceratocystis paradoxa*). Identification of genetic resistance for these diseases is important to allow incorporation of resistance traits as goals in breeding programmes to reduce production threats (current and potential), as well as reduce fungicide spraying. In the same context, insects are also potential threats to sugarcane production, either directly or as disease vectors. The main sugarcane pests include root froghopper (*Mahanarva fimbriolata*, Hemiptera: Cercopidae), the sugarcane weevil (*Sphenophorus levis*, Coleoptera: Curculionidade), longhorn beetle (*Migdolus fryanus*, Coleoptera: Cerambycidae), sugarcane borer (*Diatreae saccharallis*, Lepdoptera: Pyralidae) and the stem borer (*Telchin licus*, Lepdoptera: Castniidae). Whereas aphids are of little concern as pests per se, two species (*Melanaphis sacchari* and *Sipha flava*, Hemiptera: Aphididae) are SCYLV vectors. Sugarcane resistance against these insects is beneficial where the virus is a potential danger. Biological control of sugarcane pests by using natural enemies is a viable crop management technique in some cases, such as the fungus *Metarhizium anisopliae* that controls the root froghopper. However, the incorporation of genetic resistances against pests of economical or potential impact is indisputably the best option, when available.

Biotechnological interventions in sugarcane

Plant tissue cultures

Although several graminaceous crop plants and forage grasses have been successfully regenerated from tissue culture, they lack the multiplication procedure that has long been a serious problem in sugarcane cultivation. A newly identified variety with desirable character, that is, pest/disease resistant, high sugar content, and stress resistant, can be propagated through tissue culture and made available to the farmer for cultivation. Micropropagation can also be used for large-scale production of disease-/pathogen-free seed stock. *In vitro* multiplication of sugarcane has received considerable research attention because of its economic importance as a cash crop. Micropropagation is currently the only realistic means of achieving rapid, large-scale production of disease-free quality planting material as seed canes of newly developed varieties in order to speed up the breeding and commercialization process in sugarcane (Feldmenn et al., 1994; Lal and Krishna, 1994; Lorenzo et. al., 2001). Micropropagation is a part of plant tissue culture. Basically it is the branch of biotechnology used to clone plants at a very high speed without the restriction of the season (Bhojwani and Razdan, 1983). Plant tissue culture research is multidimensional, and it has direct commercial applications as well as value in basic research into cell biology, genetics and biochemistry. Instead of being a tool for rapid propagation, tissue culture became an alternating tool of breeding by selecting and testing for useful variants (Babra et al., 1978; Heinz et al., 1971). Crop production through micropropagation also eliminates the possibility of any interruption in the growing season because it can be carried out inside the carefully regulated environment of a greenhouse (Figure 5.2). Because the chemical and physical environment inside a greenhouse can be closely monitored, any lull in production that might typically occur as a result of seasonal change can be avoided (Lineberger, 2002). Multiple shoot formation is preferred in micropropagation. Bhuria et al. (2014) reported multiple shoot formation during micropropagation.

Sugarcane is an economically important, polysomatic, highly heterozygous, clonally propagated crop that accounts for more than 60% of the world's sugar production (Guimarces and Sobral, 1998). Lack of rapid multiplication has been a serious problem in sugarcane breeding (Ali and Afghan, 2001). The technique of plant tissue culture may play a key role in the Second Green Revolution in which biotechnology and gene modification are being used to improve crop yield and quality. Micropropagation can also be used for production of disease-/pathogen-free stock material (widely used in horticulture industry), as through meristem culture. Apart from these conventional benefits, research on various aspects has widened the scope of plant tissue culture. Secondary metabolite production through tissue culture technology has been proved more efficient as compared to conventional extraction from field grown/naturally cultivated plants e.g. commercial production of the napthoquinone pigment using plant tissue culture technique is more economical. Even though tissue culture has tremendous applications in various fields, it has some limitations also. Tissue culture requires advanced manpower skill, specialized equipment and capital-intensive facilities. Tissue culture techniques are usually specific and hence success rates vary widely. Sengar et al. (2011) and Srivastava et al. (2014) have also supported micropropagation for improvement of sugarcane. Genetic fidelity study removes chances of somaclonal variation in micropropagation of sugarcane (Singh et al., 2014).

Molecular marker technology

A molecular marker, also called a DNA marker, is a DNA sequence that is readily detected and whose inheritance can be easily monitored. The use

Figure 5.2 Regeneration of sugarcane through plant tissue culture. (a) Initiation of regeneration from shoot tip. (b, c) Multiplication of shoots in initial stage. (d, e) Multiplication of shoots in final stage. (f, g) Rooting of regenerated shoots. (h) Rooted plants transferred in soil. (i) Plants in field condition.

of molecular markers is based on a naturally occurring DNA polymorphism, which forms the basis for designing strategies to exploit applied purpose. A molecular marker has some desirable properties:

- Must be polymorphic
- Co-dominant inheritance
- Should be evenly and frequently distributed
- Should be reproducible
- Should be easy, fast and cheap to detect

Molecular markers have the potential to speed breeding, and their main contribution in crops breeding relies on marker-assisted selection (MAS). MAS is a possible tool for the Second Green Revolution (Singh, 2010a). PCR-based molecular markers are considered as a boon for agriculture development (Singh, 2010b). There are many breeding challenges posed by sugarcane genetics, which consequently affect breeding programs, and so far these have very little benefit for molecular tools generated for sugarcane. The first molecular markers may be explored for genetic divergence study and analysis (Singh, 2011; Singh and Singh, 2012; Singh et al., 2013a). The sugarcane genetic map has more than 1100 molecular markers (considering diverse marker types) with a total map length of 2600 cM and a marker density of 7.3 cM (Garcia et al., 2006), which includes DNA regions (Tabasum et al., 2010) as well as expressed RNA sequences (Wei et al., 2010). Investigation of diversity within sugarcane cultivars using RFLP (restriction fragment length polymorphism) has shown that modern sugarcane cultivars are highly heterozygous with many distinct alleles at a locus (Jannoo et al., 1999). This has also been demonstrated with simple sequence repeats (SSRs) (Selvi et al., 2003). The majority of the diversity is due to the more polymorphic and smaller contribution of the *S. spontaneum* portion of the genome (Jannoo et al., 1999). Recently, AFLP (amplified fragment length polymorphism) markers have shown that they can be used to determine genetic similarity among sugarcane cultivars (Lima et al., 2002).

Introgression of genes from these types of hybrids would be greatly facilitated by identification of molecular markers linked to genes of interest. Molecular genetic mapping of Saccharum species has been limited due to its polyploidy nature and the resulting mix of single-dose and multidose alleles. But using single-dose markers (Wu et al., 1992) linkage maps have been constructed with RAPD (random amplified polymorphic DNA) (Mudge et al., 1996) and RFLP (Ming et al., 2002b) markers for *S. spontaneum*, *S. officinarum* and *S. robustum*. Single-dose linkage maps have also been constructed in cultivars using AFLP (Hoarau et al., 2001), and AFLP and SSR markers (Rossi et al., 2003) Aitken et al. (2005) used over 1000 SSR and AFLP markers to cluster 123 linkage groups into eight homology groups which correspond to the basic chromosome number of *S. spontaneum.* Comparative analysis of quantitative trait loci (QTL) affecting plant height and flowering between sorghum and sugarcane identified QTL clusters for both these traits in sugarcane, which corresponded closely to QTL previously mapped in sorghum (Ming et al., 2002a). The colinearity between sugarcane and sorghum chromosomes (Ming et al., 1998) means that genes identified in this diploid relative could be of use in locating the same genes in sugarcane.

Genomics

The complex genome of the cultivated sugarcane is currently being sequenced (Sugarcane Genome Sequencing Initiative, http://sugarcanegenome.org), and more effort is supposed to accelerate the discovery of genes responsible for most desirable traits. It will allow the identification of regulatory regions, comparative studies of grass chromosomes, segment evolution, and detection of intra- and inter-cultivar allelic loci. Transcriptome efforts in sugarcane had a landmark in the late 1990s, when the large-scale cDNA libraries sequencing project SUCEST was set (Vettore et al., 2003), and from which almost 300,000 ESTs (expressed sequence tags) were obtained and assembled into ~43,000 unique transcribed sequences, the closest picture of sugarcane transcriptional units. Most functional genomics projects performed in the 1990s focused on sucrose content, disease resistance and stress tolerance, and involved several techniques, such as EST characterization, microarray and SAGE (serial analysis of gene expression) (Iskandar et al., 2011; Papini-Terzi et al., 2009). The post-genomic era comprises the use of this information into breeding programmes, with the identified markers that reveal expression profile of genes in different environmental conditions (Khan et al., 2011). Modern genomics offers the knowledge needed to assign a physiological function to a gene. However, the distance from genotype to phenotype still requires a more integrated approach from molecular data to sugarcane physiology and production, thus setting the basis for modeling the regulatory pathways that link genes, metabolites and physiological processes.

Genetic engineering

Sugarcane is an important food and energy crop, and there are so many reasons that make this crop an appropriate candidate for improvement via genetic engineering. Plant biotechnology has made significant strides in the past two decades or so, encompassing within its folds spectacular developments in plant genetic engineering.

Genetic transformation using agrobacterium

Agrobacterium tumefaciens is a soil dwelling, gram negative bacterium that has a natural ability to mobilize and integrate a part of its large tumour inducing (Ti) plasmid called transfer DNA (T-DNA) into the nucleus of infected plant cells (Chilton, 1977). As soon as this fact was discovered, scientists started using modified (disarmed) *Agrobacterium* strains developed by Fraley (Fraley, 1985) lacking tumour-forming genes to transfer useful genes into plants. This system had a number of advantages like higher transformation efficiency, integration of defined DNA pieces frequently as a single copy, Mendelian transmission to the next generations and lower cost of equipment than used in biolistic method. Genetic transformation using the *A. tumefaciens* system has significant advantages over biolistic technology. The *Agrobacterium* system has a stable expression and higher transformation efficiency. Moreover, fewer transgene integrations result in lower frequency of transgene silencing (Dai et al., 2001).

   A. tumefaciens-mediated genetic transformation is a routine method of gene transfer in dicotyledonous plants. Monocots as a group were earlier considered outside the host range of *Agrobacterium*. This is because monocots, particularly the grasses, secrete little or no phenolic compounds, lack receptor sites on their cells for *A. tumefaciens* attachment, and lack tumour formation or reduced activity of T-DNA promoters in monocots. But now monocots are also being transformed using *Agrobacterium*. This methodology presents several advantages over other approaches, including the ability to transfer large segments of DNA with minimal rearrangement of DNA, fewer copy gene insertion, higher efficiency, minimal cost and reduction in the occurrence of transgene silencing (Dai et al., 2001). An *Agrobacterium*-mediated genetic transformation system is illustrated in Figure 5.3.

   Viable protocols for transformation using *A. tumefaciens*-mediated transformation were assessed by Enriquez et al. (2000). A sugarcane plant resistant to the herbicide BASTA was generated. A high regeneration rate and transformation frequencies 10% to 35% were recorded. A series of improved vectors useful for monocotyledonous genetic transformation by the *A. tumefaciens* system having fruitful features like selectable marker genes, the *hpt* gene for hygromycin resistance and the *bar* gene for phosphinothricin resistance, a polylinker sequence for transgene insertion and a number of origins of replication. Sugarcane cultivars Ja60-5 and B4362 were transformed (Enriquez et al., 2000), with *A. tumefaciens* harbouring the binary plasmid pGT GUSBAR (glufosinate resistance). Manickavasagam et al. (2004) developed herbicide-resistant sugarcane plants using axillary buds of sugarcane cultivars Co 92061 and Co 671 with *Agrobacterium* strains LBA 4404 and EHA 105. Zhangsun et al. (2007) reported the most useful nptII gene (selectable marker) for selection of

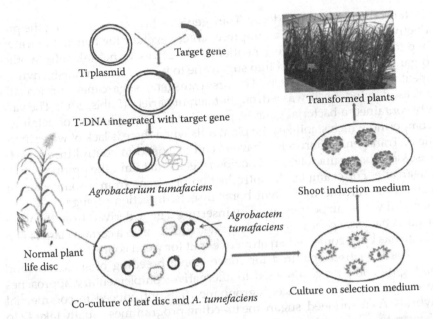

Figure 5.3 Flow diagram of *Agrobacterium* mediated transformation in sugarcane.

sugarcane callus transformation by the *A. tumefaciens* system. Genetic manipulation is being conducted to increase sucrose content of sugarcane. This work requires an understanding of the many interacting processes involved in accumulation of sucrose in sugar-storing stems. Scientists have identified the key enzymes that set in motion these processes, which can be hastened or slowed by genetic engineering towards more efficient buildup of sucrose in stems (Patrick et al., 2013).

In sugarcane, genetic modification is being carried out one step at a time to boost the sucrose yield. For example, as a first step, South African scientists genetically knocked down a particular enzyme. This raised the amount of sucrose in young stems of the engineered sugarcane plants. In Australia, researchers have inserted microbial genes into sugarcane, creating transgenic plants that can make cellulose-degrading enzymes precisely engineered to operate in the leaves of mature plants (Harrison et al., 2011). Genetically engineered sugarcane plants were shown to produce high-value chemicals like therapeutic proteins and natural precursors of biopolymers. A remarkable achievement in this area is the production of an alternative sweetener called isomaltulose in transgenic sugarcane. This was achieved by inserting a bacterial gene for making an enzyme that transforms sucrose into isomaltulose. If used as a sweetener, isomaltulose may bring certain health benefits because it is digested more slowly than sucrose, which is good for diabetics, and it does not support the growth of bacteria that cause tooth decay. Transgenic technology may bring the productivity of sugarcane to an unprecedented level for the benefit of farmers and to complement the aforementioned objectives. Genes taken from other organisms can be inserted into sugarcane to protect it from harsh environmental conditions and pests. The first transgenic sugarcane commercially released in Indonesia was a drought-tolerant variety (Lubis, 2013). The variety contained a bacterial gene responsible for the production of betaine, a compound which stabilizes the plant cells when there is lack of water in the field. Transgenic approaches have been developed to control insect pests, disease-causing microbes and noxious weeds that limit the productivity of sugarcane. For example, the introduction of a gene from a soil bacterium protects sugarcane from stem borer insects. Infection of sugarcane by a harmful virus can be prevented by inserting a gene derived from the virus itself. A bacterial gene responsible for detoxification of a certain class of herbicide has been conferred an attractive trait for weed control.

The long time required for conventional breeding of sugarcane and its highly complex genome led to alternative complementary approaches to obtain novel or enhanced agronomic traits introduced in commercial hybrids. As mentioned, sugarcane breeding programmes usually take 12 to 15 years to carry out, test and launch a new variety. The transgenic approach using candidate genes for targeted traits is an alternative to significantly shorten breeding time. Sugarcane is a recalcitrant species regarding genetic transformation, and several parameters usually need optimization at the variety level to reach higher transformation efficiencies. Genetic transformation of sugarcane first relied on particle bombardment (biolistic) of cell suspension, embryogenic callus or meristem. Genes associated with sucrose content were identified and validated *in vivo* via genetic transformation, resulting in higher sucrose concentration in transgenic plants (Papini-Terzi et al., 2009). Another important application of genetic transformation is the development of resistance to pests and pathogens, including constructs against bacteria and viruses, as seen after biolistic transformation with the capsid gene of the yellow leaf virus. Beyond the use of herbicide-resistant genes (e.g., *bar* and *pat*) as selective markers, they also confer an attractive trait to reduce production costs (Manickavasagam et al., 2004). Increase in drought tolerance was correlated with proline accumulation in transgenic sugarcane (Molinari et al., 2007).

Conclusion and perspective

Sugarcane (*Saccharum officinarum* L.) is raised as an important industrial cash crop worldwide. It is cultivated in tropical and subtropical regions of the world in a range of climates from hot and dry near sea level to cool and moist at higher elevations for the production of sugar and bioethanol. Traditionally, the main focus on sugarcane breeding had been on sugar yield. However, recently, a new sugarcane genotype concept is emerging, focusing on biomass production to enable better exploration of ethanol or energy production. Within this new concept, breeding programs must be reoriented to strengthen its efforts on the development of new cultivars that fit this new variety profile. For this, it is essential to quickly answer questions related to biometrics (stalk number, diameter, height) and processing (sucrose content, reducing sugars, fibre content). Surely, new germplasm resources must be explored by sugarcane breeding programs. The implementation of a parallel introgression program, aimed at broadening the genetic base of sugarcane cultivars for sugar content and/or biomass production, will bring great contributions for increases of yield, ensuring a more sustainable cultivation of sugarcane. Gains on important traits, such as vigour (robustness), will contribute to biomass production and may be found within *S. spontaneum* accessions and related genera, such as *Miscanthus* and *Erianthus*. New resources and tools are constantly been made available for sugarcane, such as better understanding of its genome, genetics, physiology, molecular biology, new markers associated with traits of agronomical relevance and new analysis tools. Breeding programs should take advantage of these tools and incorporate in their selection pipelines to generate superior new cultivars that respond to current and future needs of the industry and the hope of the general society. In short we can say that proper use of agriculture biotechnology may provide food security for every one (Singh, 2009).

References

Aitken, K.S., Jackson, P.A., and McIntyre, C.L. 2005. A combination of AFLP and SSR markers provides extensive map coverage and identification of homo(eo)logous linkage groups in a sugarcane cultivar. *Theoretical and Applied Genetics* 110: 789–801.

Ali, K., and Afghan S. 2001. Rapid multiplication of sugarcane through micropropagation technique. *Pakistan Sugar Journal* 16(6): 11–14.

Anonymous. 2011. Vision 2030. Indian Council of Agriculture Research (ICAR), New Delhi.

Barba, R.C., Zamora, A.B., Malion A.K., and Linga C.K. 1978. Sugarcane tissue culture research. *Proceedings of the International Society of Sugarcane Technologists* 16: 1843–1863.

Bhojwani, S.S., and Razdan, M.K. 1992. *Plant tissue culture: Theory and practice.* Amsterdam: Elsevier Science Publishers B.V.

Bhuria, P., Singh, Y., Tiwari, S., and Chaturvedi, M. 2014. Multiple shoot regeneration from nodal explant of Ashwagandha (Withania somnifera). *Indian Research Journal of Genetics and Biotechnology* 6(3): 521–525.

Chilton, M.D. 1977. Successful integration of T-DNA in plants. *Cell* 11: 263–271.

Dahlia, L., Kurniawan, I., Anggakusuma, D., and Roshetko, J.M. 2010. *Consumer preference for indigenous vegetables: Asia.* Bogor, Indonesia: World Agroforestry Centre (ICRAF).

Dai, S.H., Zheng, P., Marmey, P., Zhang, S.P., Tian, W.Z., Chen, S., Beachy, R.N., and Fauquet, C. 2001. Comparative analysis of transgenic rice plants obtained by Agrobacterium-mediated transformation and particle bombardment. *Molecular Breeding* 7: 25–33.

Enriquez, G.A., Trujillo, L.E., Menendez, C., Vazquez, R.I., Tiel, K. et al. 2000. Sugarcane (*Saccharum* hybrid) genetic transformation mediated by Agro-bacterium tumefaciens: Production of transgenic plants expressing proteins with agronomic and industrial value. Proceedings of the International Symposium on 'Plant Genetic Engineering: Towards the Third Millennium', Havana, Cuba.

Feldmenn, P., Sapotille, J., Gredoire, P., and Rott, P. 1994. Micro propagation of sugarcane. In: C. Teisson, ed., *In vitro culture of tropical plants*, 15–17. France: CIRAD.

Fraley, R.T. 1985. Development of disarmed Ti plasmid vector system for plant transformation. *Biotechnology* 3: 629–635.

Garcia, A.A., Kido, E.A., Meza, A.N., Souza, H.M., Pinto, L.R., Pastina, M.M. et al. 2006. Development of an integrated genetic map of a sugarcane (*Saccharum* spp.) commercial cross, based on a maximum-likelihood approach for estimation of linkage and linkage phases. *Theoretical and Applied Genetics* 112(2): 298–314.

Guimarces, C.T., and Sobral, W.S. 1998. The *Saccharum* complex: Relation to other andropogoneae. *Plant Breeding Reviews* 16: 269–288.

Harrison, M.D. 2011. Accumulation of recombinant cellobiohydrolase and endo-glucanase in the leaves of mature transgenic sugar cane. *Plant Biotechnology Journal* 9: 884–896.

Heinz, D.J., and Mee, G.W.P. 1971. Plant differentiation from callus tissue of *Saccharum* sp. *Crop Science* 9: 346–348.

Hoarau, J.-Y., Offmann, B., D'Hont, A., Risterucci, A.M., Roques, D., Glaszmann, J.-C., and Grivett, L. 2001. Genetic dissection of a modern sugarcane cultivar (*Saccharum* spp.) I. Genome mapping with AFLP markers. *Theoretical and Applied Genetics* 103: 84–97.

Inman-Bamber, N.G., and Smith, D.M. 2005. Water relations in sugarcane and response to water deficits. *Field Crops Research* 92(2–3): 185–202.

Iskandar, H.M., Casu, R.E., Fletcher, A.T., Schmidt, S., Xu, J., Maclean, D.J., Manners, J.M., and Bonnett, G.D. 2011. Identification of drought response genes and a study of their expression during sucrose accumulation and water deficit in sugarcane culms. *BMC Plant Biology* 11: 12.

Jannoo, N., Grivet, L., Seguin, M., Paulet, F., Domaingue, R., Rao, P.S., Dookun, A., D'Hont, A., Glaszmann, J.-C. 1999. Molecular investigation of the genetic base of sugarcane cultivars. *Theoretical and Applied Genetics* 99: 171–184.

Khan, M.S., Yadav, S., Srivastava, S., Swapna, M., Chandra, A., and Singh, R.K. 2011. Development and utilization of conserved intron scanning marker in sugarcane. *Australian Journal of Botany* 59(1): 38–45.

Lal, N., and Krishna, R. 1994. Sugarcane and its problems: Tissue culture for pure and disease free seed production in sugarcane. *Ind. Sugar* 44: 847–848.

Lima, M.L.A., Garcia, A.A.F., Oliveira, K.M., Matsuoka, S., Arizono, H., De Souza, C.L., and De Souza, A.P. 2002. Analysis of genetic similarity detected by AFLP and coefficient of parentage among genotypes of sugar cane (*Saccharum* spp.). *Theoretical and Applied Genetics* 104: 30–38.

Lineberger, R.D. 2002. Horticulture in the post-land grant era. *HortScience* 7:1147–1149.

Lorenzo, J.C., Ojeda, E., Espinosa, A., and Borroto, C. 2001. Field performance of temporary immersion bioreactor derived sugarcane plants. *In Vitro Cellular and Developmental Biology – Plant* 37: 803–806.

Lubis, A.M. 2013. Development underway for first transgenic sugarcane plantation. *The Jakarta Post*, May 20. http://www.thejakartapost.com/news/2013/05/20 /development-underway-first-transgenic-sugarcane-plantation.html.

Manickavasagam, M., Ganapathi, A., Anbazhagan, V.R., Sudhakar, B., Selvaraj, N., Vasudevan, A., and Kasthurirengan, S. 2004. *Agrobacterium* mediated genetic transformation and development of herbicide-resistant sugarcane (*Saccharum* species hybrids) using axillary buds. *Plant Cell Reports* 23(3): 134–143.

Ming, R., Del Monte, T.A., Hernandez, E., Moore, P.H., Irvine, J.E., and Paterson, A.H. 2002a. Comparative analysis of QTLs affecting plant height and flowering among closely-related diploid and polyploidy genomes. *Genome* 45: 794–803.

Ming, R., Liu, S.C., Bowers, J.E., Moore, P.H., Irvine, J.E., and Paterson, A.H. 2002b. Construction of a *Saccharum* consensus genetic map from two interspecific crosses. *Crop Science* 42: 570–583.

Ming, R., Liu, S.C., Lin, Y.R., Da Silva, J., Wilson, W., Braga, D. et al. 1998. Detailed alignment of Saccharum and Sorghum chromosomes: Comparative organization of closely related diploid and polyploid genomes. *Genetics* 150: 1663–1682.

Molinari, H.B.C., Marur, C.J., Daros, E., Campos, M.K.F., Carvalho, J.F.R.P., Bespalhok-Filho, J.C., Pereira, L.F.P., and Vieira, L.G.E. 2007. Evaluation of the stress-inducible production of 6-proline in transgenic sugarcane (*Saccharum* spp.): Osmotic adjustment, chlorophyll fluorescence and oxidative stress. *Physiologia Plantarum* 130(2): 218–229.

Mudge, J., Andersen, W.R., Kehrer, R.L., and Fairbanks, D.J. 1996. *A RAPD genetic map of Saccharum officinarum. Crop Science* 36: 1362–1366.

Papini-Terzi, F.S., Rocha, F.R., Vêncio, R.Z.N., Felix, J.M., Branco, D.S., Waclawovsky, A.J. et al. 2009. Sugarcane genes associated with sucrose content. *BMC Genomics* 10: 120.

Patrick, J.W. 2013. Metabolic engineering of sugars and simple sugar derivatives in plants. *Plant Biotechnology Journal* 11: 142–156.

Rocha, F.R., Papini-Terzi, F.S., Nishiyama Jr., M.Y., Vêncio, R.Z., Vicentini, R., Duarte, R.D. et al. 2007. Signal transduction related responses to phytohormones and environmental challenges in sugarcane. *BMC Genomics* 8: 71.

Rossi, M., Araujo, P.G., Paulet, F., Garsmeur, O., Dias, V.M., Chen, H., Van Sluys, M.-A., and D'Hont, A. 2003. Genomic distribution and characterization of EST-derived resistance gene analogs (RGAs) in sugarcane. *Molecular Genetics and Genomics* 269: 406–409.

Selvi, A., Nair, N.V., Balasundaram, N., and Mohapatra, T. 2003. Evaluation of maize microsatellite markers for genetic diversity analysis for fingerprinting in sugarcane. *Genome* 46: 394–403.

Sengar, R.S., Sengar K., and Garg, S.K. 2011. Biotechnological approaches for high sugarcane yield. *Plant Sciences Feed* 1(7): 101–111.

Singh, Y. 2009. Food security for every one through agriculture biotechnology. *AGROBIOS Newsletters* 8(6): 7–9.

Singh, Y. 2010a. Marker assisted selection: A possible tool for second green revolution. *Indian Farmer's Digest* 43(7): 15–17.

Singh, Y. 2010b. PCR based molecular markers: Boon for agriculture development. *AGROBIOS Newsletters* 8(10): 9–12.

Singh, Y. 2011. Molecular approaches to assess genetic divergence in Rice. *GERF Bulletin of Biosciences* 2(1): 41–48.

Singh, Y. 2013. Molecular markers technology for abiotic stress management in rice. *International Journal of Advanced Biotechnology and Research* 4(4): 542–552.

Singh, Y., and Kumar J. 2010. Study of genomic fingerprints of *Magnaporthe grisea* from Finger millet (*Eleusine Coracona*) by RAPD-PCR. *African Journal of Biotechnology* 9(46): 7798–7804.

Singh, Y., Pani, D.R., Khokhar, D., and Singh, U.S. 2013a. Agro-morphological characterization and molecular diversity analysis of aromatic rice germplasm using RAPD markers. *ORYZA* 50(1): 26–34.

Singh, Y., Sapre, S., Prakash, V., Indhane, S.S., and Tiwari, S. 2014. Assessment of genetic fidelity of micropropagated sugarcane plants using RAPD markers in souvenir of. National Conference on Biotechnology for Sustainable Agriculture, Jabalpur, India. p. 42.

Singh, Y., Singh, J., and Pandey, A.K. 2013b. Molecular markers in diagnosis and management of fungal pathogens: A review. *International Journal of Advanced Biotechnology and Research* 4(2): 180–188.

Singh, Y., and Singh U.S. 2012. Simple sequence repeats markers and grain quality characteristics for genetic divergence and selective identification of aromatic rices. *International Journal of Advanced Biotechnology and Research* 3(3): 711–719

Srivastava, B., Singh, Y., Tiwari, S., and Sapre, S. 2014. Studies on micropropagation of sugarcane from leaf roll, eye bud and meristem tip. *Indian Research Journal of Genetics and Biotechnology* 6(1): 319–323.

Tabasum, S., Khan, F.A., Nawaz, S., Iqbal, M.Z., and Saeed, A. 2010. DNA profiling of sugarcane genotypes using randomly amplified polymorphic DNA. *Genetics and Molecular Research* 9(1): 471–483.

Vettore, A.L., da Silva, F.R., and Kemper, E.L. 2003. Analysis and functional annotation of an expressed sequence tag collection for tropical crop sugarcane. *Genome Research* 13(12): 2725–2735.

Wei, X., Jackson, P.A., Hermann, S., Kilian, A., Heller-Uszynska, K., and Deomano, E. 2010. Simultaneously accounting for population structure, genotype by environment interaction, and spatial variation in marker-trait associations in sugarcane. *Genome* 53(11): 973–981.

Wu, K.K., Burnquist, W., Sorrels, M.E., Tew, T.L., Moore, P.H., and Tanksley, S.D. 1992. The detection and estimation of linkage in polyploids using single dose restriction fragments. *Theoretical and Applied Genetics* 83: 294–300.

chapter six

Abiotic stress tolerance in sugarcane using genomics and proteomics techniques

Naresh Pratap Singh and Vaishali Shami

Contents

Sugarcane (*Saccharum* spp. family Gramineae) is an important industrial crop, ranking among the 10 most planted crops in the world. Besides being the major sugar contributor with more than 70% of the world's sugar, sugarcane is important as the raw material for sugar and allied industries. Worldwide, sugarcane occupies an area of 20.42 million ha with a total production of 1333 million metric tons (Food and Agriculture Organization of the United Nations [FAO], 2003). India is the largest single producer of sugar including traditional cane sugar sweeteners, khandsari and gur equivalent to 26 million tonnes in 2015–16 (Indian Sugar, 2008). Out of 121 sugarcane-producing countries, 15 countries (Brazil, India, China, Thailand, Pakistan, Mexico, Cuba, Columbia, Australia, United States, Philippines, South Africa, Argentina, Myanmar, Bangladesh) use about 86% of area and gives 87.1% of production. About 50 million farmers and an equal number of agricultural labourers worldwide depend on sugarcane for their livelihood. Half a million skilled workers are also engaged in the sugar industry. A total of 453 sugar factories crushed 194.4 million tonnes of cane (69% of total cane production) producing 20.14 million tonnes of sugar.

In India sugarcane is cultivated under a wide range of agro-climatic conditions, both in tropical and subtropical regions between 0–10 and 10–30 latitudes, respectively. In a tropical climate, the cultivation of sugarcane is more successful in terms of cane yield and sugar recovery throughout the year. Among the subtropical states, Uttar Pradesh is a major sugarcane growing state. The total area under sugarcane cultivation is about 22.47 lakh hectares with an average yield of about 59.6 tonnes per hectare and sugar production of about 84.75 lakh tonnes (Anonymous, 2008). However, the average yield per hectare and sugar recovery are lower than the national average. Thus, sugarcane production offers continuing challenges to the development of high yielding, high sugar content and disease resistant.

In the current situation where dramatic changes in environmental conditions are very common, finding alternate options for increasing crop productivity is the main area that needs emphasizing to feed overwhelming populations and ensure food security. Plants, as sessile organisms, have developed in the course of their evolution, efficient strategies of response to avoid, tolerate or adapt to different types of stress situations. The diverse stress factors that plants have to face often activate similar cell signaling pathways and cellular responses, such as the production of stress proteins, upregulation of the antioxidant machinery and accumulation of compatible solutes, so the cellular and molecular responses of plants to environmental stress have been intensively studied (Hasegawa et al., 2000). Among environmental conditions, various abiotic stresses are playing major roles in decreasing crop productivity. Abiotic stress includes high and low temperature, salinity, drought, flooding and heavy metals. These stresses reduce the yield of crops, depending on the type of crop and stress period. Drought and salinity are becoming particularly widespread in many regions, and may cause serious salinization of more than 50% of all arable lands by the year 2050 (Bray et al., 2000), which might severely affect plant growth and biomass production. About 25 million acres of land is lost each year due to salinity caused by unsustainable irrigation (Wang et al., 2003). Low temperature, drought and high salinity are common stress conditions that adversely affect plant growth and crop production (Xiong et al., 2002).

Biotechnology is changing the agricultural scenario in three major areas: (1) growth and development control (vegetative and reproduction/propagation); (2) protecting plants against the ever-increasing threats of abiotic and biotic stress; and (3) expanding the horizons by producing specialty foods, biochemicals and pharmaceuticals (Haggag and Mohamed, 2007). Sugarcane is a typical glycophyte and hence exhibits stunted growth or no growth under salinity, with its yield falling to 50% or less than its true potential. To sustain sugarcane production and to improve the productivity, tolerance to biotic and abiotic stresses, nutrient management, and improved sugar recovery are some of the concerns. Both the conventional and biotechnological methods have greatly contributed in solving some of these constraints. Plant responses to stress are mediated via profound changes in gene expression which result in changes in composition of plant transcriptome, proteome and metabolome (Pérez-Alfocea et al., 2011).

Biotechnology for sustainable production of sugarcane

Biotechnology has been contributing to sustainable production of sugarcane in the following ways:

- Increased resistance against abiotic stresses (salinity, drought, cold, flooding and problem soils)
- Bioremediation of polluted soils and biodetectors for monitoring pollution
- Increased productivity and quality
- Enhanced nitrogen fixation and increased nutrient uptake and use efficiency
- Improved fermentation technology
- Improved technologies for generating biomass-derived energy
- Generation of high nutrient levels in nutrient-deficient staple crops

Functional genomic approaches for sugarcane improvement

Till now, sugarcane genetics has received relatively little attention as compared to other crops, mainly due to its highly heterozygous polyploidy, complex genome, poor fertility and long breeding cycle (Gupta et al., 2010). However, advancement in modern technologies, including development of highly efficient DNA sequencing techniques, identification of single-nucleotide polymorphisms (SNPs) and genome mapping, DNA microarray technologies for gene expression analysis, RNA interference (RNAi) technology and the rapid improvement in data mining tools, can have a major influence on future sugarcane crop improvement programs. At present, both micro- and macroarrays are being used for the identification of genes expressed specifically in stems, disease resistance genes, and those involved in carbohydrate metabolism (Casu et al., 2005; Grivet et al., 2001; Ulian, 2000). Sequencing of sugarcane ESTs (expressed sequence tags) greatly contributed to the gene discovery process (Prabu et al., 2010) and the suppression subtractive hybridization (SSH) approach helps in identification of salt-induced genes in sugarcane leaves specifically to target rare transcripts, such as those participating in cell signaling and the regulation of gene expression (Patade et al., 2011b,c). The validated expression data by real-time polymerase chain reaction (PCR) can aid in assigning function for the sugarcane genes and characterization of regulatory sequences in sugarcane.

Genomics for abiotic stress tolerance in sugarcane

Genomics researches have opened new opportunities for improving crop plants and their productivity. The development of genomic technologies that can yield structural and functional information about key genes provides useful information through profiling experiments and/or through the candidate gene approach. The candidate gene approach is facilitated by the large number of sequences and freely available gene information found in plant genetic databases to identify potential candidate genes and pathways involved in stress tolerance. Functional genomics allows large-scale gene function analysis with high throughput technology and incorporates interaction of gene products at cellular and organism level. The information coming from sequencing programs is providing enormous input about genes to be analysed (Chain et al., 2009; Feuillet et al., 2010). Among all the abiotic stresses, salinity and drought are important environmental factors that limit crop productivity. Sugarcane, being a typical glycophyte, exhibits stunted growth or no growth under salinity due to alterations in water relations, metabolic perturbations, generation of reactive oxygen species (ROS), tissue damage (Patade et al., 2011a) and enzymes involved in sugar metabolism (Gomathi and Thandapani, 2005) resulting in declines in yield up to 50% or more as compared to its true potential (Akhtar et al., 2003; Wiedenfeld, 2008). There is a large collection of ESTs and cDNA (complementary DNA) sequences available (Marques et al., 2009). The basic interest behind these EST projects is to identify genes responsible for critical functions. Patade et al. (2011c) constructed a forward subtracted cDNA library from sugarcane plants stressed with NaCl (200 mM) for 0.5 to 18 h to find mRNA species that are differentially expressed in sugarcane in response to salinity stress. Therefore, construction of cDNA libraries enriched for differentially expressed tran.scripts is an important first step in attempting to study stress responsive genes. Sequencing the differentially expressed few cDNAs clones led to the identification of salinity-induced kinase (designated as sugarcane shaggy-like protein kinase-SuSk). The expression was induced by salt as well as polyethylene glycol (PEG) stress indicating that the induction of this gene occurred in response to the osmotic component rather than the ionic component. Gene expression profiling is an important tool to investigate responses to environmental changes at the transcriptional level. Transcriptomic study of short-term (up to 24 h) salt (NaCl, 200 mM) or iso-osmotic PEG 8000 (20% w/v) stress has revealed altered expression of representative stress responsive genes in sugarcane leaves (Patade et al., 2011b). Suprasanna et al. (2011) reported downregulation of a sugarcane homologue of NHX belonging to the family of Na/H and K/H antiporters in response to the salt stress. Though the NHX transcript levels increased transiently in sugarcane plants stressed with salt or PEG for 15 days and correlated to growth inhibition (Patade, 2009). The transcript levels of P5CS, an important gene in the proline biosynthesis pathway, did not significantly alter upon exposure to salt stress, but were severely reduced in response to PEG stress. On the other hand, expression of PDH, which plays a role in proline catabolism, was inhibited in response to salt stress and induced under PEG stress treatments (Patade et al., 2011b). On long-term exposure to salt or PEG stress the steady-state levels of both P5CS and PDH gene expression increased (Patade, 2009), which also correlated to proline accumulation under these stress conditions (Patade et al., 2011a). The steady-state transcript levels of CAT2 were also lower in non-primed stressed sugarcane plants in response to iso-osmotic salt or PEG stress for 15 days. To understand the molecular basis of salt stress response, Pagariya et al. (2010) carried out cDNA RAPD (random amplified polymorphic DNA)-based gene expression at early growth stage in tolerant sugarcane variety. Among 335 differentially expressed transcript-derived fragments (TDFs), 156 upregulated and 85 downregulated were sequenced. The 17% TDFs representing potential transcripts involved in stress tolerance and plant defence in sugarcane were reported. Recently, Prabu et al. (2010) based on sqRT-PCR analysis showed higher transcript expression of WRKY, 22-kDa drought-induced protein, MIPS and ornithine-oxo-acid amino transferase at initial stages of stress induction with a gradual decrease in advanced stages. Analysis of the expression of these stress responsive genes in sugarcane plants under water deficit stress revealed a different transcriptional profile compared with sucrose accumulation. Prabu et al. (2010) identified differentially expressed transcripts in response to water deficiency stress in sugarcane using PCR-based cDNA SSH. The 158 cDNA was cloned; out of them on the basis of Dot blot, 62% showed similarity with known functional genes, and 12% with hypothetical proteins of plant origin, while 26% represented new unknown sequences. Annotation of these differentially ESTs indicated their possible function in cellular organization, protein metabolism, signal transduction and transcription.

Role of miRNA in abiotic stress responses in sugarcane

The involvement of miRNAs in abiotic stress has been studied in plants in response to dehydration or NaCl by using expression analysis, suggesting stress specific regulation of expression of miRNA (Patade and

Suprasanna, 2010) in sugarcane. MicroRNAs (miRNAs) are small, single-stranded, non-coding, naturally occurring, highly conserved families of transcripts (18–25 nt in length). Several miRNAs are either upregulated or downregulated by abiotic stresses, which might be involved in adaptation to stress and its related responsive gene expressions (Sunkar and Zhu, 2004). Under the short-term iso-osmotic (–0.7 MPa) NaCl or PEG stress (up to 24 h), the transcript level of the mature miRNA increased to 112% of the control and it progressively increased with the stress exposure period (1.3-fold at 8 h treatment) but reported no change in long-term (15 days) stress. This indicated that expression of the miR159 gene was more responsive to osmotic stress than ionic stress. Suprasanna et al. (2011) also supported the aforementioned statement by predicting targeted MYB under the same stress (NaCl or PEG) conditions to study the changes in target gene expression in response to over- or underexpression of miR159. The results on the expression of specific miR159 and its targets could be useful in developing appropriate markers for selection of tolerant cultivars in sugarcane.

Tissue culture for abiotic stress tolerance in sugarcane

Tissue culture methods showed a great impact both on basic research and applied commercial interest. This *in vitro* culture include micropropagation of elite clones, production of disease-free planting material, generation of agronomically superior somaclones, screening methods for biotic and abiotic stress tolerance, and conservation of novel and useful germplasm. *In vitro* techniques for the mass propagation via direct and indirect regeneration pathways are well established for healthy sugarcane plantlets and efforts are in progress to improve sugarcane germplasm through genetic engineering (Snyman et al., 2011). In direct morphogenesis, plants are regenerated directly from tissues such as immature leaf roll discs and also from shoot tip culture, by which sugarcane is propagated commercially (Hendre et al., 1983). Indirect morphogenesis involves initial culturing of leaf roll sections or inflorescences on an auxin-containing medium to produce an undifferentiated mass of cells, or callus.

Somatic embryogenesis is one of the economically important methods which allows propagating a large number of uniform plants in less time, for obtaining resistant plants through somaclonal variation, mutagenesis and developing transgenic plants (Suprasanna and Bapat, 2006; Suprasanna et al., 2007). *In vitro* preservation of sugarcane germplasm has been explored using slow growth (Chandran, 2010) and cryopreservation techniques (Gonzales-Arnao and Engelmann, 2006). Tissue culture laboratories had successfully developed protocols for direct somatic embryogenesis (DSEM) and indirect somatic embryogenesis (ISEM) using young

leaf rolls and immature inflorescence segments of Indian sugarcane cultivars (Desai et al., 2004a). Both protocols use different media combinations
of coconut water (CW), kinetin, zeatin and thidiazuron (TDZ) to optimize callus growth and regeneration. CW and zeatin were more effective over other growth regulators for callus induction, whereas CW alone
was found effective for plant regeneration. The DSEM system is useful
for cost-effective and large-scale clonal propagation over ISEM, providing a new target explant source for genetic transformation (Suprasanna
et al., 2007). The plants derived through DSEM have been found to be
uniform in growth pattern with more vigour compared to plants derived
through the indirect somatic embryogenesis pathway. This method also
yielded a large number of plants (7–8 per explant) in a short span of
7 weeks. When these explants have been analysed at the molecular level
using RAPD markers, these plants exhibited no variations (Suprasanna et
al., 2006a). Subsequently, this technique of direct somatic embryogenesis
has been extended to several other Indian sugarcane cultivars, suggesting wide adaptability. Nowadays, researchers are refining and improving frequency of callus induction, proliferation and plant regeneration by
using different media, growth regulators and other additives (Snyman et
al., 2011). Alternative strategies, such as partial desiccation (Desai et al.,
2004a, 2004b), silver nitrate and copper sulphate, have also been attempted
for improving regeneration response (Patel et al., 2007). Cefotaxime has
been found to be beneficial for improving frequency of shoot multiplication and elongation (Kaur et al., 2008). The somatic embryos derived
through direct somatic embryogenesis of sugarcane (Desai et al., 2004a)
were encapsulated in sodium alginate and the beads showed maximum
percentage of germination (73%) on half strength mass spectrometry (MS)
media. Synchronous somatic embryo production combined with healthy
plant regeneration may be useful for synthetic seed technology in sugarcane. In sugarcane, there have been very few studies demonstrating the
synthetic seed research. Synthetic seed technology is emerging as a novel
tool in plant biology (Suprasanna et al., 2006b) and synthetic seeds are
potential delivery systems and provide an alternative to current high-cost
vegetative propagation and for storage of novel and important germplasm.

Proteomic approach for abiotic stress tolerance in sugarcane

Research on abiotic stress biology of plants has been enriched with a
broad range of transcriptomic and proteomic studies. These studies provide comprehensive information on alteration of gene expression and
proteome profile during and following stress conditions (Hakeem et al.,
2012; Mizoi et al., 2012). About 50% of the genes responsive to flood,
salinity and extreme temperatures were found to encode transcriptional

regulators (Kilian et al., 2007; Mizoi et al., 2012). Thus, transcription factors were immensely highlighted as regulators of abiotic stress responses at RNA-level studies (Jaglo-Ottosen et al., 1998; Kasuga et al., 1999; Kilian et al., 2007; Mizoi et al., 2012; Seki et al., 2001). On the other side, proteomic studies have led to the identification of various abiotic stress responsive proteins as MS-based proteomics allow isoform-specific protein identification, and hence are able to differentiate specific and shared functions within a protein family, but it is not possible in transcriptomic studies. Thus, proteome-wide identification and functional analysis of proteins provide better understanding of abiotic stress response pathways in plants at the transcriptional level as well. During drought stress, the photosynthetic electron transport chain was markedly suppressed and as a consequence the excess excitation energy was driven towards the production of reactive oxygen species or ROS. To counteract the harmful effects of these ROS, several ROS scavengers were induced during drought stress like dehydrins, dehydro ascorbate reductase, quinine reductase, γ glutamyl cysteine synthetase, and glutathione S-transferases as observed from proteomic studies pursued on soybean (Mohammadi et al., 2012), wild watermelon (Yoshimura et al., 2008) and other plant species (Mishra and Das, 2003). Additionally, increased levels of molecular chaperones, such as heat shock proteins, were detected in roots of wheat (Demirevska et al., 2008) and sugarcane (Jangpromma et al., 2010) under drought treatment. Proteins involved in signal perception were found to be higher in abundance at the early stage of salt stress (Zhao et al., 2013). These include (a) receptors in the plasma membrane (PM) or in the cytoplasm, (b) G protein, (c) Ca++ signaling protein or Ca++ binding protein, (d) phosphor proteins involving activation of kinase cascade, and (e) ethylene receptors. As adaptive responses to salt stress, root triggers several cellular and molecular events such as alteration in carbohydrate and energy metabolism (Du et al., 2010; Manaa et al., 2011; Yang et al., 2012), changes in ion homeostasis and membrane trafficking (Cavalcanti et al., 2007; Munns and Tester, 2008), ROS scavenging, (Miller et al., 2010), and dynamic reorganization of cytoskeleton and redistribution of cell wall components which are commonly altered to maintain cell turgor by adjusting cell size (Li et al., 2011). Basic cytoskeleton components such as actin (Xu et al., 2010), tubulin (Pang et al., 2010; Peng et al., 2009), and other cytoskeleton-related proteins such as some actin-binding proteins (ABPs) (Yan et al., 2005), kinesin motor (Sobhanian et al., 2010), myosin (Cheng et al., 2009), and xyloglucan endotrans-glycosylase (XET) hydrolase (Zörb et al., 2010) during salt stress.

The study of proteomics of different *Saccharum* sp. can be used to produce second-generation (2G) ethanol which may help to reduce waste and increase the yield without expanding the crop area, contributing to a cleaner, more efficient and more sustainable production. Plant cell walls are mainly composed of polysaccharides and cell wall proteins (CWPs)

(Carpita and Gibeaut, 1993). Proteomics studies have revealed the large diversity of CWPs (Albenne et al., 2013, 2014; Rose and Lee, 2010). They have been grouped in different functional classes according to predicted functional domains and experimental data: polysaccharide-modifying proteins, oxido-reductases, proteases and structural proteins such as hydroxyl proline-rich glycoproteins, namely extensins, arabinogalactan proteins and hydroxyproline/proline rich proteins, account for about 10% of the cell wall mass in dicots (Cassab and Varner, 1988) and approximately 1% in monocots (Vogel, 2008). But proteomics studies have found only few of them. CWPs are involved in growth and development, stress signaling and defence (Carpita et al., 2001; Cassab and Varner, 1988; Jamet et al., 2006; Von Groll et al., 2002). Cell wall proteomics require challenging strategies comprising several steps, from the extraction to the identification of the proteins, compared to other sub-cellular proteomics works. However, a lot of studies have been successfully done (Albenne et al., 2013, 2014). Several aerial organs have been studied in different plant species, such as alfalfa (Verdonk et al., 2012), *Linum usitatissimum* (Day et al., 2013), *Solanum tuberosum* (Lim et al., 2012) and *Arabidopsis thaliana* (Minic et al. 2007). In *Brachypodium distachyon* leaves and stems, different classes of proteins have been identified and it was possible to address some of them to the mechanism of 2G biofuel production (Douché et al., 2013). Recently, 69 CWPs have been described from isolated cells obtained from cell suspension cultures of sugarcane (Calderan-Rodrigues et al., 2014), but the description of the cell wall proteome from a differentiated organ is still under research. To achieve the 2G ethanol, two different strategies were developed to extract the CWPs of 2-month-old stems: the destructive method (DT method) and non-destructive method (ND method), that is, vacuum infiltration (Boudart et al., 2005). Quantitative MS data were used to identify the most abundant CWPs in sugarcane culms. The DT method relied on the grinding of the material and its centrifugation in solutions of increasing sucrose concentration. On the contrary, in the ND method, cell structures were intact with each other while performing the extraction of CWPs by vacuum infiltration of the tissues. Thus, it was expected that the DT method would be able to extract more wall-bound proteins than the ND one. In both protocols, protein extraction from cell walls was performed using 0.2 M CaCl2 and 2 M LiCl. The efficiency of CaCl2 to release CWPs could rely on the fact that demethylesterified homogalacturonans strongly chelate calcium (Angyal et al., 1989), solubilizing weakly bound proteins by a competition mechanism (Jamet et al., 2008). Proteins were analysed by shotgun LC-MS/MS, after tryptic digestion. The identification of proteins was performed using the translated-SUCEST database containing ESTs (Vettore et al., 2003). This work has contributed to three main aspects: (a) characterizing CWPs from sugarcane young stems; (b) comparing the CWPs found, regarding type and amount, using two different methods of extraction; and (c) candidate CWPs to be used in future research to enhance 2G ethanol production.

Metabolomics

Understanding of many physiological plant processes can be done efficiently by evaluating and monitoring the various metabolites involved in it to improve the productivity and quality. This systematic study is defined as metabolomics. In other words, metabolomics is the study of primary and secondary metabolites of an organism. Besides their use as a breeding or selection tool, metabolomics techniques have also been used to evaluate stress responses in barley (Widodo et al., 2009), citrus (Djoukeng et al., 2008), *Medicago truncatula* (Broeckling et al., 2005) and *Arabidopsis thaliana* (Fukushima et al., 2011). Metabolomics can be used to precisely measure the concentration of a limited number of known metabolites, by using either gas chromatography (GC) or liquid chromatography (LC) coupled to mass spectrometry (MS) or nuclear magnetic resonance spectroscopy (NMR).

Metabolite profiling attempts from the last few years have helped to identify and quantify a specific class or classes of chemically related metabolites that often share chemical properties as well (Seger and Sturm, 2007). The metabolome represents the downstream result of gene expression and is closer to the phenotype than transcript expression or proteins. Extensive knowledge on metabolic flows could allow assessment of genotypic or phenotypic differences between plant species or among genotypes exhibiting different tolerance to some biotic or abiotic stresses. Current strategies for metabolite characterization still face significant obstacles caused by the high degree of chemical diversity among metabolite pools as well as the complexity of spatial and temporal distribution within living tissues. Plant metabolomics methodology and instrumentation are being developed at a rapid pace to address these analytical challenges (Hegeman, 2010). Like other functional genomics research, metabolomics generates large amounts of data. Further developments in this area require improvements in both analytical science and bioinformatics (Shulaev, 2006). Metabolomics is still in the works to be applied to understand the various pathways and functions in non-ideal plant sugarcane (Bosch et al., 2003). Metabolites are the ultimate products derived from differences in DNA sequence, control of expression, and translation into protein and enzyme activity, and, therefore, are potentially useful phenotypic markers and can assist in understanding the basis of agronomically important traits.

Sucrose is a major sugar component of sugarcane which is produced in the leaf and translocated to the stem where it is stored (Moore, 1995). Immature internodes at the top of the plant have a low sucrose concentration; however, as the internodes develop, those lower down the plant have accumulated sucrose, reaching concentrations of 50% of the total dry weight (Botha and Black, 2000). Increasing sucrose content is a major research goal since an increase in sucrose yield is more valuable through increased content than increased biomass of the same sucrose content (Jackson, 2005; Jackson et al., 2000). Osmotic stress could be expected at such high concentrations of sucrose, and increased expression of many genes whose products are involved in responding to stress has been found in maturing sugarcane internodes (Casu et al., 2004). Consequently, identifying metabolites that increase in abundance with increasing sucrose concentration may assist in understanding how sugarcane is able to accumulate large amounts of sucrose. Metabolite profiles at four different points along the sugarcane stem, representing different stages of development, were acquired after methanol/water extracts were separated and analysed by GC-MS after derivatization in either N-methyl-N-(trimethylsilyl) trifluoracetamide (TMS) or N-methyl N-(tert-butyldimethylsilyl) trifluoroacetamide (TBS). AMDIS (Automated Mass Spectral Deconvolution and Identification System) was used to analyse the chromatograms obtained and identified 121 and 71 metabolites from the TMS and TBS derivatization, respectively (Donna et al., 2007). Fifty-five metabolites were identified using commercial and publicly available libraries, from which four metabolites showing the greatest correlations with sucrose – trehalose, tartaric acid, maltose and raffinose – was correlated. This approach will be employed to see if the same metabolites are correlated with sucrose concentration across a range of genotypes differing in sucrose content and if they can be used to determine the changes arising from metabolic engineering in sugarcane in an attempt to increase sucrose content or the production of novel metabolites.

References

Akhtar, S., Wahid, A., and Rasul, E. (2003). Emergence, growth and nutrient composition of sugarcane sprouts under NaCl salinity. *Biol Plants* 46: 113–117.

Albenne, C., Canut, H., Hoffmann, L., and Jamet, E. (2014). Plant cell wall proteins: A large body of data, but what about runaways? *Proteomes* 2: 224–42.

Albenne, C., Canut, H., and Jamet, E. (2013). Plant cell wall proteomics: The leadership of *Arabidopsis thaliana*. *Front Plant Sci* 4: 111.

Angyal, S.J., Craig, D.C., and Kuszmann, J. (1989). The composition and conformation of D-Threo-3,4-Hexodiulose in solution, and the X-ray crystal-structure of its beta-anomer. *Carbohydrate Res* 194: 21–29.

Anonymous. (2008). *Cooperative sugar* 40(1).

Bosch, S., Rohwer, J.M., and Botha, F.C. (2003). The sugarcane metabolome. *Proc S Afr Sugar Technol Assoc* 7: 129–133.

Botha, F.C., and Black, K.G. (2000). Sucrose phosphate synthase and sucrose synthase activity during maturation of internodal tissue in sugarcane. *Aust J Plant Physiol* 27: 81–85.

Boudart, G., Jamet, E., Rossignol, M., Lafitte, C., Borderies, G., and Jauneau, A. (2005). Cell wall proteins in apoplastic fluids of *Arabidopsis thaliana* rosettes: Identification by mass spectrometry and bioinformatics. *Proteomics* 5: 212–221.

Bray, E.A., Bailey-Serres, J., and Weretilnyk, E. (2000). Responses to abiotic stresses. In: *Biochemistry and molecular biology of plants*, B.B. Buchanan, W. Gruissem, and R.L. Jones (Eds.), 1158–1203. Rockville, MD: American Society of Plant Physiologists.

Broeckling, C.D., Huhman, D.V., and Farag, M.A. (2005). Metabolic profiling of *Medicago truncatula* cell cultures reveals the effects of biotic and abiotic elicitors on metabolism. *J Exper Bot.* 56(410): 323–336.

Calderan-Rodrigues, M.J., Jamet, E., Bonassi, M.B.C.R., Guidetti-Gonzalez, S., Begossi, A.C., and Setem, L.V. (2014). Cell wall proteomics of sugarcane cell suspension culture. *Proteomics* 14: 738–749.

Carpita, N., Tierney, M., and Campbell, M. (2001). Molecular biology of the plant cell wall: Searching for the genes that define structure, architecture and dynamics. *Plant Mol Biol* 47: 1–5.

Carpita, N.C., and Gibeaut, D.M. (1993). Structural models of primary-cell walls in flowering plants: Consistency of molecular-structure with the physical properties of the walls during growth. *Plant J.* 3: 1–30.

Cassab, G., and Varner, J.E. (1988). Cell wall proteins. *Ann Rev Plant Phys* 39: 321–53.

Casu, R.E., Dimmock, C.M., Chapman, S.C., Grof, C.P.L., McIntyre, C.L., Bonnett, G.D., and Manners, J.M. (2004). Identification of differentially expressed transcripts from maturing stem of sugarcane by in silico analysis of stem expressed sequence tags and gene expression profiling. *Plant Mol Biol* 54: 503–517.

Casu, R.E., Manners, J.M., Bonnett, G.D., Jackson, P.A., McIntyre, C.L., Dunne, R., Chapman, S.C., Rae, A.L., and Grof, C.P.L. (2005). Genomics approaches for the identification of genes determining important traits in sugarcane. *Field Crops Res* 92: 137–147.

Cavalcanti, F.R., Santos Lima, J.P.M., Ferreira-Silva, S.L., Viégas, R.A., and Silveira, J.A.G. (2007). Roots and leaves display contrasting oxidative response during salt stress and recovery in cowpea. *J Plant Physiol* 164: 591–600. doi: 10.1016/j.jplph.2006.03.004.

Chain, P.S.G, Grafiam, D.V., and Fulton, R.S. (2009). Genome project standards in a new era of sequencing. *Science* 326(5950): 236–237.

Chandran, K. (2010). *In vitro* multiplication and conservation of Saccharum germplasm. *Indian J. Plant Genetic Resour* 23(1): 65–68.

Cheng, Y., Qi, Y., Zhu, Q., Chen, X., Wang, N., and Zhao, X. (2009). New changes in the plasma-membrane-associated proteome of rice roots under salt stress. *Proteomics* 9: 3100–3114. doi:10.1002/pmic.200800340.

Day, A., Fénart, S., Neutelings, G., Hawkins, S., Rolando, C., and Tokarski, C. (2013). Identification of cell wall proteins in the flax (*Linum usitatissimum*) stem. *Proteomics* 13: 812–25.

Demirevska, K., Simova-Stoilova, L., Vassileva, V., Vaseva, I., Grigorova, B., and Feller, U. (2008). Drought-induced leaf protein alterations in sensitive and tolerant wheat varieties. *Gen Appl Plant Physiology* 34: 79–102.

Desai, N.S., Suprasanna, P., and Bapat, V.A. (2004a). A simple and reproducible method for direct somatic embryogenesis from immature inflorescence segments of sugarcane. *Curr Sci* 87(6): 764–768.

Desai, N.S., Suprasanna, P., and Bapat, V.A. (2004b). Partial desiccation of embryogenic callus improves plant regeneration frequency in sugarcane (*Saccharum* spp.). *J Plant Biotechnol* 6: 229–233.

Djoukeng J.D., Arbona, V., Argamasilla, R., and Gomez-Cadenas, A. (2008). Flavonoid profiling in leaves of citrus genotypes under different environmental situations. *J Agri Food Chem* 56(23): 11087–11097.

Donna, G., Ute, R., Anthony, B., and Graham, D.B. (2007). Changes in the sugarcane metabolome with stem development: Are they related to sucrose accumulation? *Plant Cell Physiol* 48(4): 573–584.

Douché, T., San Clemente, H., Burlat, V., Roujol, D., Valot, B., and Zivy, M. (2013). Brachypodium distachyon as a model plant toward improved biofuel crops: Search for secreted proteins involved in biogenesis and disassembly of cell wall polymers. *Proteomics* 13: 2438–2454.

Du, C.X., Fan, H.F., Guo, S.R., Tezuka, T., and Li, J. (2010). Proteomic analysis of cucumber seedling roots subjected to salt stress. *Phytochemistry* 71: 1450–1459. doi:10.1016/j.phytochem.2010.05.020.

Food and Agriculture Organization of the United Nations (FAO). (2003). Food and Agriculture Organization of the United Nations Statistical Databases (FAOSTAT).

Feuillet, C., Leach, J.E., Rogers, J., Schnable, P.S., and Eversole, K. (2010). Crop genome sequencing: Lessons and rationales. *Trends Plant Sci* 16(2): 77–88.

Fukushima, A., Kusano, M., Redestig, H., Arita, M., and Saito, K. (2011). Metabolomic correlation-network modules in *Arabidopsis* based on a graph-clustering approach. *BMC Systems Biology* 5(1).

Gomathi, R., and Thandapani, T.V. (2005). Salt stress in relation to nutrient accumulation and quality of sugarcane genotypes. *Sugar Tech* 1: 39–47.

Gonzales-Arnao, M.T., and Engelmann, F. (2006). Cryopreservation of plant germplasm using the encapsulation-dehydration technique: Review and the case study on sugarcane. *CryoLetters* 27: 155–168.

Grivet, L., Glaszmann, J.C., and Arruda, P. (2001). Sequence polymorphism from EST data in sugarcane: A fine analysis of 6-phosphogluconate dehydrogenase genes. *Genet Mol Bio* 24: 161–167.

Gupta, V., Raghuvanshi, S., Gupta, A., Saini, N., Gaur, A., Khan, M.S. et al. (2010). The water-deficit stress- and red-rot-related genes in sugarcane. *Funct Integr Genomics* 10: 207–214.

Haggag, M., and Mohamed H.A.A. (2007). Biotechnological aspects of microorganisms used in plant biological control. *Am-Eurasian J Sustainable Agri* 1(1): 7–12.

Hakeem, K.R., Chandna, R., Ahmad, P., Iqbal, M., and Ozturk, M. (2012). Relevance of proteomic investigations in plant abiotic stress physiology. *OMICS* 16, 621–635. doi:10.1089/omi.2012.0041.

Hasegawa P.M., Bressan R.A., Zhu J.K., and Bohnert H.J. (2000). Plant cellular and molecular responses to high salinity. *Annu Rev Plant Mol Plant Physiol* 51: 463–499.

Hegeman, A.D. (2010). Plant metabolomics-meeting the analytical challenges of comprehensive metabolite analysis. *Brief Funct Genomics Proteomics* 9(2): 139–148.

Hendre, R.R., Iyer, R.S., Kotwal, M., Khuspe, S.S., and Mascarenhas, A.F. (1983). Rapid multiplication of sugarcane by tissue culture. *Sugarcane* 3: 5–8.

Indian Sugar. (2008). Sugar Statistics Vii(11): 57, 70.

Jackson, P., Bonnett, G., Chudleigh, P., Hogarth, M., and Wood, A. (2000). The relative importance of cane yield and traits affecting CCS in sugarcane varieties. *Proc Aust Soc Sugar Cane Technol* 22: 23–29.

Jackson, P.A. (2005). Breeding for improved sugar content in sugarcane. *Field Crop Res* 92: 277–290.

Jaglo-Ottosen, K.R., Gilmour, S.J., Zarka, D.G., Schabenberger, O., and Thomashow, M.F. (1998). *Arabidopsis* CBF1 overexpression induces COR genes and enhances freezing tolerance. *Science* 280, 104–106. doi:10.1126/science.280 .5360.104.

Jamet, E., Canut, H., Albenne, C., Boudart, G., and Pont-Lezica, R. (2008). Cell wall. In: G.K. Agrawal and R. Rakwal, eds., *Plant proteomics: Technologies, strategies, and applications*, 293–307. London: Wiley.

Jamet, E., Canut, H., Boudart, G., and Pont-Lezica, R.F. (2006). Cell wall proteins: A new insight through proteomics. *Trends Plant Sci* 11: 33–39.

Jangpromma, N., Kitthaisong, S., Lomthaisong, K., Daduang, S., Jaisil, P., and Thammasirirak, S. (2010). A proteomics analysis of drought stress-responsive proteins as bio marker for drought-tolerant sugarcane cultivars. *Am J Biochem Biotechnol* 6, 89–102. doi:10.3844/ajbbsp.2010.89.102.

Kasuga, M., Liu, Q., Miura, S., Yamaguchi-Shinozaki, K., and Shinozaki, K. (1999). Improving plant drought, salt, and freezing tolerance by gene transfer of a single stress-inducible transcription factor. *Nat Biotechnol* 17, 287–291. doi: 10.1038/7036.

Kaur, A., Gill, M., Ruma, D., and Gosal, S. (2008). Shoot multiplication and elongation in sugarcane using cefotaxime. *Sugar Tech* 10(1): 60–64.

Kilian, J., Whitehead, D., Horak, J., Wanke, D., Weinl, S., and Batistic, O. (2007). The AtGenExpress global stress expression data set: protocols, evaluation and model data analysis of UV-B light, drought and cold stress responses. *Plant J* 50: 347–363. doi:10.1111/j.1365-313X.2007.03052.x.

Li, W., Zhang, C., Lu, Q., Wen, X., and Lu, C. (2011). The combined effect of salt stress and heat shock on proteome profiling in *Suaedasalsa*. *J Plant Physiol* 168: 1743–1752. doi:10.1016/j.jplph.2011.03.018.

Lim, S., Crisholm, K., Coffin, R.H., Peters, R.D., Al-Mughrabi, K.I., and Wang-Pruski, G. (2012). Protein profiling in potato (*Solanum tuberosum* L.) leaf tissues by differential centrifugation. *J Proteome Res* 11: 2594–2601.

Manaa, A., Ben Ahmed, H., Valot, B., Bouchet, J.P., Aschi-Smiti, S., and Causse, M. (2011). Salt and genotype impact on plant physiology and root proteome variations in tomato. *J Exp Bot* 62: 2797–2813. doi:10.1093/jxb/erq460.

Marques, M.C., Alonso-Cantabrana, H., and Forment, J. (2009). A new set of ESTs and cDNA clones from full-length and normalized libraries for gene discovery and functional characterization in citrus. *BMC Genomics* 10(1): 428.

Miller, G., Suzuki, N., Ciftci-Yilmaz, S., and Mittler, R. (2010). Reactive oxygen species homeostasis and signaling during drought and salinity stresses. *Plant Cell Environ.* 33: 453–467. doi:10.1111/j.1365-3040.2009.02041.x.

Minic, Z., Jamet, E., Negroni, L., and der Garabedian, P.A. (2007). A sub proteome of *Arabidopsis thaliana* trapped on Concanavalian A is enriched in cell wall glycoside hydrolases. *J Exp Bot* 58: 2503–2512.

Mishra, S., and Das, A.B. (2003). Effect of NaCl on leaf salt secretion and antioxidative enzyme level in roots of a mangrove, *Aegicer ascorniculatum*. *Indian J Exp Biol* 41: 160–166.

Mizoi, J., Shinozaki, K., and Yamaguchi-Shinozaki, K. (2012). AP2/ERFfamily transcription factors in plant abiotic stress responses. *Biochim Biophys Acta* 1819: 86–96. doi:10.1016/j.bbagrm.2011.08.004.

Mohammadi, P.P., Moieni, A., Hiraga, S., and Komatsu, S. (2012). Organ-specific proteomic analysis of drought-stressed soybean seedlings. *J Proteomics* 75, 1906–1923. doi:10.1016/j.jprot.2011.12.041.

Moore, P.H. (1995). Temporal and spatial regulation of sucrose accumulation in the sugarcane stem. *Aust J Plant Physiol* 22: 661–679.

Munns, R., and Tester, M. (2008). Mechanisms of salinity tolerance. *Annu Rev Plant Biol* 59: 651–681. doi:10.1146/annurev.arplant.59.032607.092911.

Pagariya, M.C., Harikrishnan, M., Kulkarni, P.A., Devarumath, R.M., and Kawar, P.G. (2010). Physio-biochemical analysis and transcript profiling of *Saccharum officinarum* L. submitted to salt stress. *Acta Physiologiae Plantarum* 33: 1411–1424.

Pang, Q., Chen, S., Dai, S., Chen, Y., Wang, Y., and Yan, X. (2010). Comparative proteomics of salt tolerance in *Arabidopsis thaliana* and *Thellungiellahalophila*. *J Proteome Res* 9: 2584–2599. doi:10.1021/pr100034f.

Patade, V.Y. (2009). Studies on salt stress responses of sugarcane (*Saccharum officinarum* L.) using physiological and molecular approaches. PhD thesis, University of Pune, India.

Patade, V.Y., Bhargava, S., and Suprasanna, P. (2011a). Salt and drought tolerance of sugarcane under iso-osmotic salt and water stress: Growth, osmolytes accumulation and antioxidant defense. *J Plant Interact*. doi:10.1080/17429145.2011 .557513.

Patade, V.Y., Bhargava, S., and Suprasanna, P. (2011b). Transcript expression profiling of stress responsive genes in response to short-term salt or PEG Stress in sugarcane leaves. *Mol Bio Rep*. doi:10.1007/s11033-011-1100-z.

Patade, V.Y., Rai, A.N., and Suprasanna, P. (2011c). Expression analysis of sugarcane shaggy-like kinase (SuSK) gene identified through cDNA subtractive hybridization in sugarcane (*Saccharum officinarum* L.). *Protoplasma* 248(3): 613–621.

Patade, V.Y., and Suprasanna, P. (2010). Short-term salt and PEG stresses regulate expression of MicroRNA, miR159 in sugarcane leaves. *J Crop Sci Biotech* 13(3): 177–182.

Patel, B.M., Suprasanna, P., Patade, V.Y., and Bapat, V.A. (2007). Simple and novel method of partial desiccation for improving plant regeneration in gamma-irradiated sugarcane tissue culture. National Conference on Biotechnology, 2–3 Feb, Mumbai.

Peng, Z., Wang, M., Li, F., Lv, H., Li, C., and Xia, G. (2009). A proteomic study of the response to salinity and drought stress in an Introgression strain of bread wheat. *Mol Cell Proteomics* 8: 2676–2686. doi:10.1074/mcp.M900052-MCP200.

Pérez-Alfocea, F., Ghanem, M.E., Gómez-Cadenas, A., and Dodd, I. (2011). Omics of root-to-shoot signaling under salt stress and water deficit. *OMICs J Integr Bio* 15(12): 893–901.

Prabu, G., Kawar P.G., Pagariya M.C., and Theertha Prasad, D. (2010). Identification of water deficit stress upregulated genes in sugarcane. *Plant Mol Bio Rep* 29(2): 291–304.

Rose, J.K.C., and Lee, S.J. (2010). Straying off the highway: Trafficking of secreted plant proteins and complexity in the plant cell wall proteome. *Plant Physiol* 153: 433–436.

Seger, C., and Sturm, S. (2007). Analytical aspects of plant metabolite profiling platforms: Current standings and future aims. *J Proteome Res* 6(2): 480–497.

Seki, M., Narusaka, M., Abe, H., Kasuga, M., Yamaguchi-Shinozaki, K., and Carninci, P. (2001). Monitoring the expression pattern of 1300 *Arabidopsis* genes under drought and cold stresses by using a full-length cDNA micro-array. *Plant Cell* 13: 61–72. doi:10.1105/tpc.13.1.61.

Shulaev, V. (2006). Metabolomics technology and bioinformatics. *Brief Bioinform* 7(2): 128–139.

Snyman, S.J., Meyer, G.M., Koch, A.C., Banasiak, M., and Watt, M.P. (2011). Applications of in vitro culture systems for commercial sugarcane production and improvement. *In Vitro Cellular Dev Bio Plant* 47: 234–249.

Sobhanian, H., Razavizadeh, R., Nanjo, Y., Ehsanpour, A.A., Jazii, F.R., and Motamed, N. (2010). Proteome analysis of soybean leaves, hypocotyls sand roots under salt stress. *Proteome Sci* 8: 19. doi:10.1186/1477-5956-8-19.

Sunkar, R., and Zhu, J.K. (2004). Novel and stress-regulated microRNAs and other small RNA from *Arabidopsis*. *Plant Cell* 16: 2001–2019.

Suprasanna, P., and Bapat, V.A. (2006). Advances in the development of in vitro culture systems and transgenics in sugarcane. International Symposium on Technologies to Improve Sugar Productivity in Developing Countries, 5–8 December, Guilin, China.

Suprasanna, P., Desai, N.S., Sapna, G., and Bapat, V.A. (2006a). Monitoring genetic fidelity in plants derived through direct somatic embryogenesis in sugarcane by RAPD analysis. *J New Seeds* 8(3): 1–9.

Suprasanna, P., Ganapathi, T.R., and Bapat, V.A. (2006b). Synthetic seeds technology. In A.S. Basra, ed., *Handbook of Seed Science and technology*, 227–267. New York: CRC Press.

Suprasanna, P., Patade, Y.V., and Bapat, V.A. (2007). Sugarcane biotechnology: A perspective on recent developments and emerging opportunities. In *Advances in plant biotechnology*, 313–342. Hauppauge, NY: Science Publishers.

Suprasanna, P., Patade, V.Y., Desai, N.S., Devarumath, R.M., Kawar, P.G., Pagariya, M.C., Ganapathi, A., Manickavasagam, M., and Babu, K.H. (2011). Biotechnological developments in sugarcane improvement: An overview. *Sugar Tech*. doi:10.1007/s12355-011-0103-3.

Ulian, E. (2000). *Functional genomics for sugar accumulation gene discovery in sugarcane*. Brisbane, Australia: Sugarcane Genomics Workshop.

Verdonk J.C., Hatfield, R.D., and Sullivan, M.L. (2012). Proteomic analysis of cell walls of two developmental stages of alfafa stems. *Front Plant Sci* 3: 279.

Vettore, A.L. (2003). Analysis and functional annotation of an expressed sequence tag collection for tropical crop sugarcane. *Genome Res* 13: 2725–2735.

Vogel, J. (2008). Unique aspects of the grasses cell wall. *Curr Opin Plant Biol* 11(3): 301–307.

Von Groll, U., Berger, D., and Altmann, T. (2002). The subtilisin-like serine protease SDD1 mediates cell-to-cell signaling during *Arabidopsis* stomatal development. *Plant Cell* 14: 1527–1539.

Wang, W., Vinocur, B., and Altman, A. (2003). Plant responses to drought, salinity and extreme temperatures: Towards genetic engineering for stress tolerance. *Planta* 218: 1–14.

Widodo, Patterson, J.H., Newbigin, E., Tester, M., Bacic, A., and Roessner, U. (2009). Metabolic responses to salt stress of barley (*Hordeum vulgare* L.) cultivars, Sahara and Clipper, which differ in salinity tolerance. *J Exp Bot* 60(14): 4089–4103.

Wiedenfeld, B. (2008). Effects of irrigation water salinity and electrostatic water treatment for sugarcane production. *Agri Water Manag* 95: 85–88.

Xiong, L., Schumaker, K.S., and Zhu, J.K. (2002). Cell signaling during cold, drought, and salt stress. *Plant Cell* 14: 165–183.

Xu, C., Sibicky, T., and Huang, B. (2010). Protein profile analysis of salt-responsive proteins in leaves and roots in two cultivars of creeping bent grass differing in salinity tolerance. *Plant Cell Rep* 29: 595–615. doi:10.1007/s00299-010-0847-3.

Yan, S., Tang, Z., Su, W., and Sun, W. (2005). Proteomic analysis of salt stress-responsive proteins in rice root. *Proteomics* 5, 235–244. doi:10.1002/pmic .200400853.

Yang, L., Ma, C., Wang, L., Chen, S., and Li, H. (2012). Salt stress induced proteome and transcriptome changes in sugarbeet monosomic addition lineM14. *J Plant Physiol* 169: 839–850. doi:10.1016/j.jplph.2012.01.023.

Yoshimura, K., Masuda, A., Kuwano, M., Yokota, A., and Akashi, K. (2008). Programmed proteome response for drought avoidance/tolerance in the root of a C3 xerophyte (wild watermelon) under water deficits. *Plant Cell Physiol* 49: 226–241. doi:10.1093/pcp/pcm180.

Zhao, Q., Zhang, H., Wang, T., Chen, S., and Dai, S. (2013). Proteomics-based investigation of salt-responsive mechanisms in plant roots. *J Proteomics* 82: 230–253. doi:10.1016/j.jprot.2013.01.024.

Zörb, C., Schmitt, S., and Mühling, K.H. (2010). Proteomic changes in maize roots after short term adjustment to saline growth conditions. *Proteomics* 10: 4441–4449. doi:10.1002/pmic.201000231.

chapter seven

Insights into biotechnological interventions for sugarcane improvement

Pooja Dhansu, Ashwani Kumar, Anita Mann,
Ravinder Kumar, B.L. Meena, Parvender Sheoran,
B. Parameswari and Neeraj Kulshreshtha

Contents

Plant adaptation to abiotic stresses is controlled by cascades of events at the molecular level. As a result, several defence mechanisms are triggered to re-establish homeostasis and protection of proteins and membranes. On the molecular level, several gene families are responsible for the induction of stress-related defence pathways. Current efforts to improve abiotic stress tolerance of plants with genes working in the stress response pathways have resulted in significant achievements. However, due to the complex nature of abiotic stress tolerance, the present technologies have to overcome several limitations. In the current situation, where dramatic changes in the environmental conditions are very common, finding alternate options for increasing crop productivity is main area that needs to be emphasized to feed the overwhelming populations and ensure their food security. Sugarcane is one of the most important field crops grown in the tropics and subtropics. Conventional and biotechnological research inputs have contributed in solving some of the constraints limiting crop productivity. However, limitations such as complex genome, narrow genetic base, poor fertility, susceptibility to biotic and abiotic stresses, and long duration to breed elite cultivars, hinder crop improvement programmes in sugarcane. Sugarcane, thus, is a suitable candidate for application of plant biotechnology and genetic engineering tools. This chapter reviews the research progress in use of biotechnological tools for sugarcane improvement related to abiotic stress.

Introduction

Plant biotechnology has made significant strides in the past two decades or so, encompassing within its folds the spectacular developments in the plant genetic engineering. In the current situation, where dramatic changes in the environmental conditions are common, finding alternate options for increasing crop productivity is the main area that needs to be emphasized to feed the overwhelming population and ensure their food security. Among environmental conditions, various biotic and abiotic stresses are playing major roles in decreasing crop productivity. Abiotic stress includes high and low temperature, salinity, drought, flooding and heavy metals. It has also been practically established that the global temperature has been on the increase with associated fluctuations in annual rainfall regimes, the resultant drought and flood events, and increasing soil and water salinization.

These stresses reduce the yield of crops, depending on the type of crop and stress period. In many semi-arid and arid regions of the world, crop yield is limited due to the increased rate of soil salinity. Salinity and drought are the two most complex stress tolerances to breed for. Timing in relation to plant growth stage and intensity of stress can all vary considerably, which have severely affected plant growth and biomass production since long. These challenges would be met with the introduction and utilization of new technologies coupled with conventional approaches. In this direction, efforts are being made to breed tolerant varieties using conventional breeding and contemporary biotechnological tools. Recent advances in this area include unraveling the physiological, biochemical and molecular mechanism of abiotic stress tolerance and corresponding development of tolerant cultivars through transgenic technology or molecular breeding (Ashraf, 2010; Niu et al., 2012; Patade et al., 2011a, 2011b, 2011c).

Biotechnology paves the way for changing the agricultural and plant scene in three major areas (Haggag et al., 2007):

1. Growth and development control (vegetative and reproduction/ propagation)
2. Protecting plants against the ever-increasing threats of abiotic and biotic stress
3. Expanding the horizons by producing specialty foods, biochemicals and pharmaceuticals

Multiple biotic and abiotic environmental stress factors negatively affect various aspects of plant growth, development and crop productivity. Plants, as sessile organisms, have developed, in the course of their evolution, efficient strategies of response to avoid, tolerate or adapt to different types of stress situations. The diverse stress factors that plants have to face often activate similar cell signaling pathways and cellular responses, such as the production of stress proteins, upregulation of the antioxidant machinery and accumulation of compatible solutes. Over the last few decades, advances in plant physiology, genetics and molecular biology have greatly improved our understanding of plant responses to abiotic stress conditions. Plant biotechnology has made significant strides in the past two decades or so, encompassing within its folds the spectacular developments in the plant genetic engineering. Nowadays, genetically engineered crops appear as the most recent technological advances to help boost food production, mainly by addressing the production constraints with minimum costs and environmental pollution. Transgenic crops offer significant production advantages such as decreased and easier herbicide use and reduced pesticide use (Baker, 2003). This has a double advantage: First, it reduces the cost of production and, second, it escapes environmental pollution due to the indiscriminate use of pesticides and herbicides.

Sugarcane (*Saccharum* ssp.) belongs to the family Poaceae, ranking among the 10 most planted crops in the world, and is widely distributed in tropical and subtropical regions of the world (Clayton and Renvoize, 1982; Sánchez-Ken and Clark, 2010), also including regions where water availability is limited or highly inconsistent (Azevedo et al., 2011; Inman-Bamber and Smith, 2005; Zhang et al., 2006). On a global perspective, sugarcane (*Saccharum* spp. hybrid complex) is an important cash crop cultivated to produce sugar (Imman-Bamber and Smith, 2005) and ethanol in tropical and subtropical regions. It was introduced in Brazil during the colonization period, and today represents one of the main cultures of the economy. India is the second largest producer as well as consumer of the sugar in the world. Sugarcane is cultivated over an area of 5.0 million hectares with an average productivity of 68 tons ha^{-1} in India. It has been estimated that by 2030, India will require 33.0 million tons of sugar with average sugar recovery of 10.75%. This will entail a productivity of 110 tons ha^{-1}, as area under sugarcane may stabilize 5.0 million hectares. The sugarcane hybrids are highly polyploid and aneuploid, and on average contain 100 to 120 chromosomes with an estimated somatic cell size of 10,000 Mbp (D'Hont, 2005). The number of chromosomes can vary in commercial cultivars. The basic genome size ranges from 760 to 926 Mbp, which is twice the size of the rice genome (389 Mbp) and similar to sorghum's (760 Mbp) (D'Hont and Glaszmann, 2001). Even in the face of the economic importance, the complexity of the sugarcane genome inhibited large efforts and investments in the development of biotechnology and genetic tools for this crop. Cultivar improvement has been achieved over the years using traditional breeding, which can take up to 15 years of selections. Nevertheless, sugarcane transgenics are still lagging behind since the complete genome sequence of sugarcane is not yet available. Large genome size, high polyaneuploidy, low fertility, complex environmental interactions, slow breeding advances, and mobilizations hinder the breeding for this crop with resistance/tolerance to drought, salinity, insect pests, fungal diseases and herbicides. Consequently, a lack of suitable multiplication procedures has long been a serious problem in sugarcane breeding programmes (Tiwari et al., 2010). Gene discovery and identification is essential for breeding programs and a significant progress has been noted with the development of expressed sequence tags (ESTs). Large collections of ESTs have led to the exploration of the large polyploid sugarcane genome and consequently renewed the interest in sugarcane genetics (Butterfield et al., 2001; Grivet and Arruda, 2002; Ming et al., 2006). Genome-based technology such as cDNA microarray data indicates genes associated with sugar content that may be used to develop new varieties improved for sucrose content or for traits that restrict the expansion of the cultivated land (Menossi et al., 2008). The genes can also be used as molecular markers of agronomic traits in traditional breeding programs, because EST collection provides information to determine an organism's genome content and also it can directly point to genes which may contribute to agronomical trait development (e.g. tolerance to abiotic and biotic stresses, mineral nutrition and sugar content).

Biotechnological innovations for sugarcane improvement

Growing and expanding populations have created a huge pressure on global agriculture to provide increased food, feed and fibre. The recent phenomenon of climate change has further added fuel to the fire. It has been practically established now that the global temperature has been on the increase with associated fluctuations in annual rainfall regimes, and the resultant drought and flood events and increasing soil and water salinization. These challenges would be compensated through the introduction and utilization of new technologies coupled with conventional approaches. The modern functional and omics era of biotechnology can provide a solution to the burning issues of global agriculture. Until recently, the models used by breeders in statistical genetics approaches have been developed for diploid organisms, which are not ideal for a polyploid genome such as that of sugarcane. Although conventional breeding programs in sugarcane are relatively not so old as compared to other major crops, interspecific hybridization within the genus Saccharum supported significant improvements in yield, ratooning ability, sugar content and disease resistance, while maintaining acceptable fibre levels for milling (Jackson, 2005; Lakshmanan et al., 2005; Ming et al., 2006). Most modern sugarcane cultivars originate from crosses between a relatively small numbers of original progenitor clones compared with the large number of basic clones that exist in the Saccharum genus, resulting in a narrow gene pool (Jackson, 2005). It remains challenging to exploit the large genetic variation existing among clones of different Saccharum species (Ming et al., 2006). Molecular markers may assist breeders in incorporating useful genes from sexually compatible sources into the gene pool of the advanced cultivars. Modern cultivars contain between $2n \times 100$ and $2n \times 130$ chromosomes, with 5% to 10% consisting of the wild *S. spontaneum* contribution and less than 5% of these being recombinant or translocated chromosomes (Ming et al. 2006). The high ploidy (5× to 14×; Burner and Legendre, 1994) and the complex genome structure of sugarcane create challenges for marker development and genome characterization (Cuadrado et al., 2004; D'Hont et al., 1996).

Tolerance to biotic and abiotic stresses, nutrient management and improved sugar recovery are some of the concerns to sustain sugarcane production and to improve productivity. Both the conventional and biotechnological methods have greatly contributed in solving some of these constraints. Use of modern molecular biology tools for elucidating the control mechanisms of stress tolerance and for engineering stress-tolerant plants are based on the expression of specific stress-related genes. To date, successes in genetic improvement of environmental stress resistance have involved manipulation of a single or a few genes involved in signaling/regulatory pathways or that encode enzymes involved in these pathways. This chapter briefly describes the biotechnological approaches for sugarcane improvement.

In vitro *culture technology*

During the 19th century, the idea of development of callus (a disorganized proliferated mass of actively dividing cells) from isolated stem fragments and root apices came into existence. Callus could also be developed from buds, and root and shoot fragments of about 1.5 mm in size without using nutrient medium. For the first time, Hamberlandt (1902) originated the concept of cell culture in Berlin. He attempted to cultivate the isolated plant cells *in vitro* on an artificial medium (Knop's solution, peptone, asparagin and sucrose). The term *tissue culture* can be applied to any multicellular culture growing on a solid medium (or attached to substratum and nurtured with a liquid medium) that consists of many cells in protoplasmic continuity. But in organ culture (e.g. excised roots) the cultured plant material maintains its morphological identity, more or less, with the same anatomy and physiology as *in vitro* of the parent plants (Doods and

Roberts, 1985). *In vitro* culture technology is a tool for obtaining rapid, mass multiplication of disease-free, true-to-type planting materials. The major advantages of *in vitro* screening are a controlled environment, large populations can be handled in a lesser space within a short span of time and the plant material can be kept disease free. Successful protocols for shoot tip culture, callus culture, embryo culture, virus-free plant production and somatic embryogenesis have already been established. Thus, *in vitro* technology can be used to enhance productivity of sugarcane. Sugarcane is highly heterogeneous and generally multiplied vegetatively by stem cuttings in many countries. However, due to a low seed multiplication rate (1:6 to 1:8) the spread of newly released varieties is slow, taking over 10 years to scale up a newly released variety to the commercial level (Cheema and Hussain, 2004; Sengar, 2010). Moreover, the method requires large nursery space: one hectare nursery for 10 to 15 hectares field planting (Sundara, 2000). According to Dookun (1998) and Lakshmanan et al. (2005), sugarcane is a suitable candidate for the application of biotechnological and genetic engineering tools in increasing its productivity. Sugarcane *in vitro* culture began in the 1960s with culture of mature parenchyma of internodal tissues for some physiological studies (Nickell, 1964). Later, after demonstration of totipotency in callus cultures of sugarcane (Barba and Nickell, 1969; Heinz and Mee, 1969), rapid progress was made in cell and tissue culture of this crop, and it was found that cultures could be raised from any part of the plant. Plant regeneration in sugarcane can occur through two main routes: direct and indirect morphogenesis. In direct morphogenesis, plants are regenerated directly from tissues such as immature leaf roll discs and also from shoot tip culture, by which sugarcane is propagated commercially. Indirect morphogenesis involves initial culturing of leaf roll sections or inflorescences on an auxin-containing medium to produce an undifferentiated mass of cells, or callus. Studies have been conducted to employ *in vitro* culture combined with radiation/chemical-induced mutagenesis for mutant isolation (Synman et al., 2011). Jalaja et al. (2008) reported that within 9 months, callus culture of apical meristem-produced planting materials from a single spindle which was sufficient to plant one hectare of land. Ramgareeb et al. (2010) in a study in sugarcane reported propagation of approximately 1300 shoots from a single 2 mm meristem in 11 weeks.

Despite several advantages of applying the micro-propagation technique in sugarcane (such as quick multiplication of newly released varieties, rejuvenation of old deteriorated varieties, production of disease-free seed, easy transportation of seed material, elimination of viruses, and high cane productivity and sugar yield, etc.), this technique is not gaining popularity to the desired extent (Wekesa, 2015). Contamination of cultures of microbes is a severe problem that not only reduces the frequency of shoot culture initiation from the source explants but also the total number of shoots produced at various cycles due to loss of cultures (Lal et al., 2014).

Somatic embryogenesis

Research on sugarcane tissue and cell culture was first started by Nickell (1964) in Hawaii. The first successful plant regeneration system in sugarcane was established about 40 years ago (Barba and Nickell, 1969), however, a persuasive evidence through somatic embryogenesis was reported by Ahloowalia and Maretzki (1983). Moreover, successful somatic embryogenesis and regeneration was further studied in sugarcane using different explants and medium composition (Asad et al., 2009; Falco et al., 1996). Development of somatic embryogenesis was a milestone in transgenesis in sugarcane (Lakshmanan et al., 2005: Ming et al., 2006). Somatic embryogenesis is an important aspect of plant tissue culture where somatic embryos do arise in culture usually from single cells and ontogeny of somatic embryogenesis is comparable with zygotic embryogenesis. Somatic embryos are uniparental and hence the plants regenerated from somatic embryos are true to type. Thus, somatic embryogenesis is being looked upon as an attractive alternative for mass cloning of plants and as an important tool for genetic transformation (Gopitha et al., 2010). Moreover, developments of the direct somatic embryogenesis system reinforce sugarcane biotechnology because embryogenic calli is the most suitable target tissue for genetic transformation (Snyman et al., 1996).

The recognition of somatic embryogenesis was a turning point in sugarcane biotechnology (Lakshmanan, 2006). Although, originally developed as an alternative system to regeneration, somatic embryogenesis has achieved prominence as an integral part of the genetic transformation system (Bower and Birch, 1992). Somatic cells are theoretically totipotent. Ho and Vasil (1983) reported the evidence of embryogenic callus development in monocots. Somatic embryogenesis has been reported from a large number of commercial sugarcane clones and can be obtained directly or indirectly from the leaf tissues (Guiderdoni et al., 1995; Raza et al., 2012; Manickavasagam and Ganapathi, 1998; Wekesa et al., 2014). Embryogenic callus can be maintained for several months without losing its regeneration potential to a significant level (Fitch and Moore, 1993). Somatic embryogenesis offers an efficient and high volume regeneration system for the production of a large number of plants within a short period (Shah et al., 2009). The system may be useful for developing transgenic plants through the Agrobacterium-mediated method in the future. These transgenic plants with desirable genes may be useful in sugarcane improvement programs. Somatic embryogenesis techniques have two main goals: (1) the development of a highly efficient method for propagating a large number of uniform plants in less time and possibly at lesser cost than the conventional propagation methods, and (2) a cell culture-based regeneration system useful for genetic transformation (Suprasanna 2010).

Use of somaclonal variation that results from either *in vitro* culture or mutagenic treatments is one of the ways of diversifying the genetic pool and potentially introducing desirable traits (Snyman et al., 2011). Somaclonal variation is a random event, so the identification of desirable somaclones is critical. Selection should be performed either *in vitro*, by the addition of a selective agent (e.g. incorporation of a fungal culture filtrate), through field-based screening of plantlets or both (Snyman et al., 2011). To further capitalize on *in vitro* somaclonal variation and to increase the frequency at which it occurs, physical and chemical mutagens may be applied to callus cultures (Snyman et al., 2011). Such induced mutagenesis has the potential to elicit beneficial modifications in cultivars (Patade and Suprasanna, 2008).

Molecular marker-assisted improvement of sugarcane

A comprehensive functional map of the sugarcane genome has been described with an enhanced resolution, creating the means for developing 'perfect markers' associated with key quantitative trait loci (QTL) (Oliveira et al., 2007). Molecular marker-assisted selection (MAS) is desirable, if visual selection is difficult and cost/time ineffective. MAS is a strategy for accelerating the crop breeding for biotic and abiotic stress tolerance (Mantri et al., 2010b; Ribaut and Ragot, 2007; Wei et al., 2009). Identification of molecular markers linked to the desired traits has made it possible to examine their usefulness in crop improvement (Ashraf, 2010; Delannay et al., 2012). Many efforts have been attempted to develop molecular markers such as restriction fragment length polymorphism (RFLPs), random amplified polymorphic DNA (RAPDs), amplified fragment length polymorphisms (AFLPs), and simple sequence repeats (SSRs) for efficient MAS in breeding programs. DNA markers have greatly contributed to fingerprinting of elite genetic stocks, assessing of genetic diversity, increasing the efficiency of trait selection and diagnostics. Sugarcane QTL mapping is mostly based on single marker analysis or (composite) interval mapping (Pastina et al., 2010). In order to provide useful results for genetic studies and breeding purposes, new models need to be developed, taking into consideration QTL versus environment interaction and epistasis. Although no MAS has been reported in sugarcane, the Bru-1 and Bru-2 haplotypes have potential use in the identification of a durable rust-resistance gene in sugarcane germplasm (Cunff et al., 2008). Various molecular marker systems including restriction fragment length polymorphism (Lu et al., 1994), ribosomal RNA (Glaszmann et al., 1990), mitochondria and chloroplast genes (D'Hont et al., 1993), RAPDs (Kawar et al., 2009; Nair et al., 1999), and simple sequence repeats (SSR) (Cordeiro et al., 2003; Selvi et al., 2003) have shown usefulness for differentiating the genera and to assess germplasm diversity within the genus Saccharum. The involvement of miRNAs in abiotic stress has been studied in plants in response to dehydration or NaCl by using expression analysis, suggesting stress-specific regulation of expression of miRNA (Patade and Suprasanna, 2010) in sugarcane. RAPD markers have been used to assess the genetic diversity in elite and exotic sugarcane germplasm (Kawar et al., 2009; Nair et al., 1999, 2002; Srivastava and Gupta, 2006;) and construct genetic maps (Mudge et al., 1996). The potential of ISSR markers for molecular profiling was assessed in sugarcane using 42 varieties from subtropical India (Srivastava and Gupta, 2008). Virupakshi and Naik (2008) used organellar genome inter-simple sequence repeat markers (cplSSR and mtISSR) to analyse red rot disease resistant/moderately resistant and susceptible elite sugarcane (*Saccharum* spp. hybrid) genotypes. The results indicated that these markers may be used as a new tool for the identification of the disease-resistant varieties. In sugarcane, AFLP markers have been used to study diversity existing among tropical and subtropical Indian sugarcane cultivars (Selvi et al., 2005), and Saccharum complex and Erianthus (Besse et al., 1998; Selvi et al., 2006). One of the major ways sugarcane industries have already benefited from molecular markers is the use of SSRs for cultivar identification. Target region amplification polymorphism (TRAP) has also been used to characterize the germplasm from the genera Saccharum, Miscanthus, and Erianthus with the help of six primers designed using sucrose- and cold tolerance-related EST sequences (Alwala et al., 2006). Development of new high-throughput marker systems like single nucleotide polymorphisms (SNPs) and diversity array technology (DArT) markers are expected to have a major impact in the future for sugarcane improvement. Pagariya et al. (2010) carried out cDNA RAPD-based gene expression at an early growth stage in tolerant sugarcane variety Co 62175 to understand the molecular basis of salt (NaCl 2%) stress response. Among 335 differentially expressed transcript-derived fragments (TDFs), 156 up regulated and 85 downregulated were sequenced. The 17% TDFs representing potential transcripts involved in stress tolerance and plant defence in sugarcane were reported. Further, they have identified 137 salinity-tolerant candidate cDNAs from sugarcane, 20% of which were novel sugarcane genes. These unique sequences that have so far not been reported to be stress related might provide further understanding on the perception, response and adaptations mechanisms of non-model plant-like sugarcane to salinity stress.

Transgenic plants with abiotic stress tolerance

A special issue of *Nature* on GM crops (Anon. 2013) reported that the GM market is dominated by herbicide- and insect-resistant varieties. There is no mention of drought-resistant GM crops. The annual report by the International Service for the Acquisition of Agri-Biotech Applications (2013) on biotech GM crop traits on the market does not mention drought resistance as a commercialized GM trait. Three GM wheats (*Triticum aestivum* L.) and one GM sugarcane (*Saccharum officinarum* L.) were subjected for evaluation under the regulatory system in Australia, but none has yet been released for commercial use between 2007 and 2013. Several important papers recognised the problem, directly or indirectly, and examined possible ways to enhance the impact of drought-resistance gene discovery work on plant production in the field (Blum, 2014).

The abundance of abiotic stress-related transcriptome analysis data generated in several plant species has revealed the importance of a number of genes catalysing biosynthetic pathways of osmoprotectants. In many ways then, these osmoprotectants protect the plants against the damaging effects of secondary stresses such as osmotic and ionic stresses. Various plant species have been engineered, using these different genes to enhance their abiotic stress tolerance. Wang et al. (2003) introduced the trehalose synthase (maltose alpha-D-glucosyltransferase) gene from *Grifola frondosa* into calli of a sugarcane hybrid using *A. tumefaciens* EHA105 strain, which also contained the bar gene. In case of sugarcane, around 50% of losses by borers have been employed (Singh et al., 2013). Insecticidal crystal protein (ICP) genes encoded from *Bacillus thuringiensis* proved highly toxic to the larvae of sugarcane borers. As of now, more than 30 species of crop plants have been transformed with the Bt (Cry genes) (Schuler, 1998). Transgenic sugarcane plants expressing transgenes against the borer, mainly Cry1Ab (Arvinth et al., 2009), aprotinin gene (Christy et al., 2009) and Cry 1Aa3 (Kalunke et al., 2009), were evaluated and expression was checked through serological and molecular techniques. A number of transgenic sugarcane lines have been developed with gene-expressing Cry protein, proteinase inhibitor or lectin resistance to borers, sucking insects or grubs (Srikant et al., 2011).

The range of potential applications through the transfer of genes from a wide variety of plant and non-plant sources (i.e. creating transgenics) is increasing rapidly in sugarcane. Suprasanna (2010) analysed the expression of different stress related genes *NHX* and *SUT1*, *P5CS* and *PDH*, involved in direct protection of plant cells (osmoprotectants, antioxidant enzymes, ion transporters) or regulatory functions (signaling genes and transcription factors). On long-term exposure to salt or PEG (polyethylene glycol) stress, the steady-state levels of both *P5CS*, a gene coding for an important enzyme of the proline biosynthesis pathway, and *PDH*, which codes for an enzyme that plays a role in proline catabolism, increased, which also correlated to proline accumulation under these stress conditions. Plant *GSK3/Shaggy*-like kinases are involved in hormone signaling, development and stress response.

Raza et al. (2016) found integration of *Arabidopsis* vacuolar H^+-pyrophosphatase (H^+-PPase) (AVP1) transgene in transgenic sugarcane plants for drought tolerance. The gene integration was confirmed by polymerase chain reaction (PCR) and Southern blotting. Transgene expression was estimated by Western blotting. When sugarcane plants were grown on soil under 50% reduced water supply, the transgenic sugarcane overexpressing AVP1 produced higher shoot biomass based on cane height, number of millable canes and Brix (%) compared with the wild type. The overexpression of AVP1 in transgenic sugarcane plants increased tolerance to drought stress, as demonstrated by increased relative water content (RWC) and leaf water (Ψw), osmotic (Ψs) and turgor potential (Ψp). Physiological parameters such as photosynthetic rate (Pn), stomatal conductance (C) and transpiration rate (E) were less affected by water-deficit stress in transgenic AVP1 plants compared with wild-type plants, which indicated that AVP1 conferred tolerance to drought or water-deficit stress, highlighting potential use of this gene for crop improvement through biotechnological applications. Zhangsun et al. (2007) reported the nptII gene (selectable marker) as most useful for selection of sugarcane callus transformation by *A. tumefaciens* system. During 1994, the U.S. Environmental Protection Agency declared that nptII is safe to use in commercial transgenic crops, including cotton, tomato and rapeseed. The NPTII protein has no deleterious effects on human beings and it is easily degradable in gastrointestinal tract. Embryonic callus of Australian sugarcane variety Q117 has been transformed availing an easy and reproducible protocol by *A. tumefaciens* system using the nptII gene as the selectable marker.

'Omics' technologies for sugarcane

Expression profiling has become an important tool to investigate how an organism responds to environmental changes. Plants have the ability to alter their gene expression patterns in response to environmental changes such as temperature, water availability or the presence of deleterious levels of ions (Rodriguez et al., 2005). High throughput detection of differential expression of genes is an efficient means of identifying genes and pathways that may play a role in biological systems under certain experimental conditions. There exists a variety of approaches that could be used to identify groups of genes that change in expression in response to a particular stimulus or environment. Almeida et al. (2013) prepared cDNA clones corresponding defence genes related abiotic stress in sugarcane, and constructed a cDNA library through SSH technique, with genes induced by salicylic acid (SA library). Four gene groups presenting different functions were identified in the SA library. Of these, 40 clones showed no similarity with data bank sequences, 37 represented genes involved in defence mechanisms, 24 were similar to putative proteins with unknown roles and 21 clones were associated to cell maintenance and plant development. Plant genes with known roles, involved in biotic and abiotic stress response and associated to cell maintenance and plant development, were identified in SA library. Water stress-related proteins were identified using a proteomics approach based on two-dimensional polyacrylamide gel electrophoresis (2-DE) and matrix-assisted laser desorption/ionization time of flight mass spectrometry (MALDI-TOF-MS). The expression of some proteins changed after the water deficit treatment in RB 72910, upregulated and downregulated. While in RB 943365 the expression of all proteins decreased after treatment. An analysis demonstrated that the functions of these proteins were associated with such functions as photosynthesis, signal transduction and regulation process (Almeida et al., 2013). These results provide insight into part of the regulatory mechanism of adaptation to water deficit through differential expression of specific proteins and highlight the potential of the variety RB 72910 for the genetic improvement of the sugarcane crop.

Molecular approaches concerning drought and salinity performance in sugarcane were carried out using techniques based on molecular hybridization such as suppression subtractive hybridization (SSH) (Patade et al., 2010) and micro-/macroarrays (Rodrigues et al., 2009). Four SuperSAGE libraries have been generated, using bulked root tissues from four drought-tolerant accessions as compared with four bulked-sensitive genotypes, aiming to generate a panel of differentially expressed stress-responsive genes (Akio Kido et al., 2011). The SuperSAGE libraries produced 8,787,315 tags (26 bp) that, after exclusion of singlets, allowed the identification of 205,975 unitags. GO categorization of the tag-related ESTs allowed *in silico* identification of 213 upregulated unitags responding basically to abiotic stresses, from which 145 presented no hits after BlastN analysis, probably concerning new genes still uncovered in previous studies.

EST sequencing has significantly contributed to gene discovery and expression studies used to associate function with sugarcane genes. A significant amount of data exists on regulatory events controlling responses to herbivory, drought and phosphate deficiency, which cause important constraints on yield and on endophytic bacteria, which are highly beneficial. The means to reduce drought, phosphate deficiency and herbivory by the sugarcane borer have a negative impact on the environment. Improved tolerance for these constraints is being sought. Sugarcane's ability to accumulate sucrose up to 16% of its culm dry weight is a challenge for genetic manipulation. Genome-based technology such as cDNA microarray data indicates genes associated with sugar content that may be used to develop new varieties improved for sucrose content or for traits that restrict the expansion of the cultivated land. Trujillo et al. (2009) identified an EST encoding a putative protein with a DNA-binding domain that is typically found in EREBP/AP2-type transcription factors through single-pass sequencing of randomly selected clones from an ë ZAP-cDNA library generated from ethephon-treated young sugarcane leaves. The full-length cDNA clone, named SodERF3 (EMBL accession number AM493723), was further isolated from the excised library. SodERF3 encodes a 240 amino acid DNA-binding protein that acts as a transcriptional regulator of the ethylene responsive factor (ERF) superfamily. SodERF3 is induced in sugar cane leaves by ethylene, abscisic acid, salt stress and wounding as judged by Northern and Western blots assays. Accordingly, SodERF3 can be a valuable tool to assist the manipulation of plants to improve their stress tolerance. Several sugarcane ESTs collections have been developed (Bower et al., 2005; Carson and Botha, 2002; Casu et al., 2003; Ma et al., 2004; Vettore et al., 2003). The publicly available sugarcane ESTs were assembled into tentative consensus sequences (virtual transcripts), singletons and mature transcripts, referred to as the Sugarcane Gene Index (SGI; http://compbio .dfci.harvard.edu/tgi/cgi-bin/tgi/gimain.pl?gudb=S.officinarum). The Brazilian sugarcane EST project collection (SUCEST, http://sucestfun.org) generated 237,954 ESTs, which were organized into 43,141 putative unique sugarcane transcripts (26,803 contigs and 16,338 singletons) referred to as sugarcane assembled sequences (SASs). SASs that presented significant similarity with *Arabidopsis* proteins (60% of the SASs) were detected. A detailed organization of sugarcane genes into functional categories (e.g. signal transduction components, regulation of gene expression, development, biotic and abiotic stresses, transposable elements, metabolism) (Vettore et al. 2003) was completed and represents the basis to develop functional genomic approaches. To increase the knowledge on the sugarcane responses to drought, cDNA microarrays were used to evaluate gene expression in plants submitted to 24, 72 and 120 hours of water deprivation (Rocha et al., 2007). Drought stress caused dramatic changes in the gene expression profile of sugarcane plants, with 93 genes being up- or down-regulated. Among the genes differentially expressed, transcription factor orthologs of the Myb, WRKY, NAC and DREB proteins, which are known as role players in the drought responses of other systems (Yamaguchi-Shinozaki and Shinozaki, 2006), were upregulated. Sugarcane plants also selectively activated proteases in response to hydric stress, since a homologue to the cysteine proteinase RD19A precursor was induced. This gene is also induced by water stress in *Arabidopsis* (Koizumi et al., 1993).

Sequencing the sugarcane genome

A sugarcane modern cultivar is a hybrid of *Saccharum officinarum* and *Saccharum spontaneum*. Sequencing the sugarcane genome poses new challenges due to its highly polyploid and aneuploid structure with a complete set of homeologous genes predicted to range from 10 to 12 copies (alleles). The monoploid genome is estimated to be around 1 Gb, but the high level of polymorphism requires new assembly algorithms that can take into account allelic variation and a high content of repetitive regions. Obtaining a reference-assembled monoploid genome for this crop is one of the greatest challenges in genomics at this time. There are 1585 nucleotide sequences (including 491 mRNA sequences), 283,158 ESTs and 10,728 genome survey sequences (GSSs) of Saccharum species at NCBI (Souza et al., 2011). Efforts underway include BAC-by-BAC and whole genome shot-gun sequencing (WGS) (Souza et al., 2011). The most comprehensive effort so far is devoted to sequencing BACs corresponding to regions of interest of the cultivar R570. A BAC library of 103,296 clones representing 14× the monoploid genome and 1.3× the total genome and 3D pools of BAC clones are available. Moreover, a total of 6021 over go probes were analysed on the library to provide links with sorghum and there is ongoing effort to obtain R570 BAC-end sequences (Souza et al. 2011). Sequencing of R570 using the BAC library is being pursued by groups in Australia, France, South Africa, the United States and Brazil (http://sugarcanegenome.org). It is also worth mentioning that BAC and WGS sequencing are underway for SP80-3280, the Brazilian cultivar that most contributed to the available ESTs, and *S. officinarum* and *S. spontaneum* genotypes (LA Purple and SES208) (G. Souza and Ray Ming, personal communication).

Conclusion

The availability of a cellular and molecular toolbox has opened up a plethora of prospects. With the advent of functional genomics, gene expression profiling is leading to the definition of gene networks. The sequencing of the sugarcane genome, which is underway, will greatly contribute to numerous aspects of research on grasses. The transgenic and the marker-assisted route for sugarcane improvement will contribute to increased sugar, stress tolerance and higher yield, and that the industry for years to come will be able to rely on sugarcane as the most productive energy crop. Genetic engineering of sugarcane cultivars that can produce economically important compounds such as medicinal proteins, sweetener, nutraceuticals, biopolymers, biopigments, precursors and various enzymes in concrete ways will launch sugarcane as a biofactory in coming years. The classical breeding of sugarcane takes 15 years of crosses and agronomical evaluation before a new cultivar is released for commercialization. Gene discovery through the SUCEST sequencing program has been a major breakthrough for the breeding programs throughout the world, and functional studies based on cDNA arrays are uncovering pathways of plant adaptation and responses to the environment. EST-SSRs have been successfully used for genetic relationship analysis, extending the knowledge of the genetic diversity of sugarcane to a functional level. Development of new markers based on ESTs and their integration in genetic maps will renew breeding programs and help Molecular marker Assistated Breeding (MAB) technology speed up the breeding programs. Profiling of gene expression under conditions that affect crop yield can aid in building an 'expression panel' for sugarcane cultivars which should become invaluable in target gene selection. The advances in sugarcane biotechnology could become remarkable in the coming years, both in terms of improving productivity as well as substantially increasing the value and utility of this crop.

References

Ahloowalia BS and Maretzki A (1983) Plant regeneration via somatic embryogenesis in sugarcane. *Plant Cell Reports,* 2: 21–25.

Akio Kido Ederson, Neto JRCF, de Oliveira RL, Pandolfi V, Guimaraes ACR, Veiga DT, Chabregas SM, Crovella S and Benko-Iseppon AM (2012) New insights in the sugarcane transcriptome responding to drought stress as revealed by Supersage. *The Scientific World Journal* 2012: 821062, 14 pages. doi:10.1100/2012/821062.

Almeida CMA, Donato VMTS, Amaral DOJ, Lima GSA, Brito GG, Lima MMA, Correia MTS and Silva MV (2013) Differential gene expression in sugarcane induced by salicylic acid and under water deficit conditions. *Agricultural Science Research Journals* 3(1): 38–44.

Almeida CMA, Túlio DS, Carolina BM, Amaral DOJ, Arruda IRS, Brito GG, Donato VMTS, Silva MV and Correia MTS (2012) Proteomic and physiological analysis of response to water deficit in sugarcane. *Wudpecker Journal of Agricultural Research* 2(1): 1–7.

Alwala S, Suman A, Arro JA, Veremis JC and Kimbeng CA (2006) Target region amplification polymorphism (TRAP) for assessing genetic diversity in sugarcane germplasm collections. *Crop Science* 46: 448–455.

Anon. (2013) GM crops: A story in numbers. *Nature* 497: 22–23. doi:10.10 38/497022a.

Arvinth S, Selvakesavan RK, Subramonian N and Premachandran MN (2009) Transmission and expression of transgenes in progeny of sugarcane clones with cry1Ab and aprotinin genes. *Sugar Tech* 11: 292–295.

Asad S, Arshad M, Mansoor S, and Zafar Y (2009) Effect of various amino acids on shoot regeneration of sugarcane (*Saccharumofficinarum* L.). *African Journal of Biotechnology* 8: 1214–1218.

Ashraf M (2010) Inducing drought tolerance in plants: Recent advances. *Biotechnology Advances* 28: 169–183.

Azevedo RA, Carvalho RF, Cia MC and Gratão PL (2011) Sugarcane under pressure: An overview of biochemical and physiological studies of abiotic stress. *Tropical Plant Biology* 4: 42–51.

Barba R and Nickell LG (1969) Nutrition and organ differentiation in tissue culture of sugarcane, a monocotyledon. *Planta* 89: 299–302.

Besse P, Taylor G, Carrol B, Berding N, Burner D and McIntyre ML (1998) Assessment of genetic diversity in a sugarcane germplasm collection using an automated AFLP analysis. *Genetica* 104: 143–153.

Blum A (2014) Genomics for drought resistance – Getting down to earth. *Functional Plant Biology* 41(11): 1191–1198.

Bower NI, Casu RE, Maclean DJ, Reverter A, Chapman SC and Manners JM (2005) Transcriptional response of sugarcane roots to methyl jasmonate. *Plant Science* 168(3): 761–772.

Bower R and Birch RG (1992) Transgenic sugarcane plants via microprojectile bombardment. *The Plant Journal* 2(3): 409–416.

Burner DM and Legendre BL (1994) Cytogenetic and fertility characteristics of elite sugarcane clones. *Sugar Cane* 1: 6–10.

Butterfield MK, D'Hont A and Berding N (2001) The sugarcane genome: A synthesis of current understanding, and lessons for breeding and biotechnology. *Proceedings of the South African Sugar Technologists' Association* 75: 1–5.

Carson D and Botha F (2002) Genes expressed in sugarcane maturing internodal tissue. *Plant Cell Reports* 20(11): 1075–1081.

Casu RE, Grof CPL, Rae AL, McIntyre CL, Dimmock CM and Manners JM (2003) Identification of a novel sugar transporter homologue strongly expressed in maturing stem vascular tissues of sugarcane by expressed sequence tag and microarray analysis. *Plant Molecular Biology* 52(2): 371–386.

Cheema KL and Hussain M (2004) Micropropataion of sugarcane through apical bud and axillary bud. *International Journal of Agriculture and Biology* 6(2): 257–259.

Christy LA, Aravith S, Saravanakumar M, Kanchana M, Mukhunthan N, Srikanth J, Thomas G and Subramonian N (2009) Engineering sugarcane cultivars with bovine pacncreatic trypsin inhibitor (aprotinin) gene for protection against to borer (*Scripophagaexcerptalis* Walker). *Plant Cell Reports* 28: 175–184.

Clayton WD and Renvoize SA (1982) Gramineae. In *Flora of tropical East Africa*, RM Polhill, editor, Part 3, 700–767. Rotterdam: Balkema.

Cordeiro GM, Pan YB and Henry RJ (2003) Sugarcane microsatellites for the assessment of genetic diversity in sugarcane germplasm. *Plant Science* 165: 181–189.

Cuadrado A, Acevedo R, Moreno Diaz de la Espina S, Jouve N and de la Torre C (2004) Genome remodeling in three modern *S. officinarum* × *S. spontaneum* sugarcane cultivars. *Journal of Experimenal Botany* 55: 847–854.

Cunff Le L, Garsmeur O, Raboin LM, Pauquet J, Telismart H, Selvi A et al. (2008) Diploid/polyploidsyntenic shuttle mapping and haplotype-specific chromosome walking toward a rust resistance gene (Bru1) in highly polyploid sugarcane (2n approximately 12× approximately 115). *Genetics* 180: 649–660.

D'Hont A (2005) Unraveling the genome structure of polyploids using FISH and GISH; examples of sugarcane and banana. *Cytogenetic and Genome Research* 109(1–3): 27–33.

D'Hont A and Glaszman JC (2001) Sugarcane genome analysis with molecular markers, a firstdecade of research. *Proceedings of the International Society for Sugar Cane Technology* 24: 556–559.

D'Hont A, Grivet L, Feldmann P, Rao S, Berding N and Glaszmann JC (1996) Characterisation of the double genome structure of modern sugarcane cultivars (*Saccharum* spp.) by molecular cytogenetics. *Molecular Genomics and Genetics* 250: 405–413.

D'Hont A, Lu YH, Feldmann P and Glaszmann JC (1993) Cytoplasmic diversity in sugar cane revealed by heterologous probes. *Sugar Cane* 1: 12–15.

Delannay X, McLaren G and Ribaut JM (2012) Fostering molecular breeding in developing countries. *Molecular Breeding* 29(4): 857–873.

Dookun A (1998). Biotechnology for sugarcane. *Agriculture Biotechnology News Information* 10: 75–80.

Falco MC, Mendes BMJ, Tulmann NA and Gloria BD (1996) Histological characterization of *in vitro* regenerated *Saccharum* spp. *Revista Brasileira de Fisiologia Vegetal* 8: 193–197.

Fitch MMM and Moore PH (1993) Long term culture of embryogenic sugarcane callus. *Plant Cell Tissue Organ and Culture* 32: 335–343.

Glaszmann JC, Lu YH and Lanaud C (1990) Variation of nuclear ribosomal DNA in sugarcane. *Journal of Genetics and Breeding* 44: 191–198.

Gopitha K, Bhavani AL and Senthilmanickam J (2010) Effect of the different auxins and cytokinins in callus induction, shoot, root regeneration in sugarcane. *International Journal of Pharma and Bio Sciences* 1: 1–7.

Grivet L and Arruda P (2002) Sugarcane genomics: Depicting the complex genome of an important tropical crop. *Current Opinion in Plant Biology* 5(2): 122–127.

Guiderdoni E, Merot B, Eksomtramage T, Paulet F, Fredman R and Gaszmann JC (1995) Somatic embryogenesis in sugarcane (Saccharum species). In *Biotechnology in agriculture and forestry*, 13th ed., YPS Bajaj, editor, 92–113. Berlin: Springer Verlag.

Haberlandt G (1902) Kulturversuchemitisolierten Pflanzenzellen Sitzungsber. *Akademie der Wissenschaften in Wien. Sitzungsberichte. Mathematisch-naturwissenschaftliche Klasse. Abteilung* 111, 69–92.

Haggag M and Wafaa Mohamed HAA (2007) Biotechnological aspects of microorganisms used in plant biological control. *American-Eurasian Journal of Sustainable Agriculture* 1(1): 7–12.

Heinz DJ and Mee G (1969). Plant differentiation from callus tissue of *Saccharum* species. *Crop Science* 9(3): 346–348.

Ho W and Vasil IK (1983). Somatic embryogenesis in sugarcane (*Sacchurumofficinarum*). I. The morphology and physiology of callus formation and the ontogeny of somatic embryos. *Protoplasm* 118: 169–180.

Inman-Bamber NG and Smith DM (2005) Water relations in sugarcane and response to water deficits. *Field Crops Research* 89(92): 185–202.

Jackson PA (2005) Breeding for improved sugar content in sugarcane. *Field Crop Research* 92: 277–290.

Jalaja NC, Neelamathi D and Sreenivasan TV (2008). *Micropropagation for quality seed production in sugarcane in Asia and the Pacific*. Rome: Food and Agriculture Organization of the United Nations (FAO).

Kalunke RM, Kolge AM, Babu KH and Prasad DT (2009) Agrobacterium mediated transformation of sugarcane for borer resistance using cry 1Aa3 gene and one step regeneration of transgenic plants. *Sugar Tech* 11: 355–359.

Kawar PG, Devarumath RM and Nerkar YS (2009) Use of RAPD markers for assessment of genetic diversity in sugarcane cultivars. *Indian Journal of Biotechnology* 8: 67–71.

Koizumi M, Yamaguchi-Shinozaki K, Tsuji H and Shinozaki K (1993) Structure and expression of two genes that encode distinct drought-inducible cysteine proteinases in *Arabidopsis thaliana*. *Gene* 129(2): 175–182.

Lakshmanan P (2006) Somatic embryogenesis in sugarcane – An addendum to the invited review 'Sugarcane biotechnology: The challenges and opportunities.' *In Vitro Cellular and Developmental Biology – Plant* 42: 201–205.

Lakshmanan P, Geijskes R, Aitken KS, Grof CLP, Bonnett GD and Smith GR (2005) Sugarcane biotechnology: The challenges and opportunities. *In Vitro Cellular and Developmental Biology – Plant* 41: 345–363.

Lal M, Tiwari AK and Gupta GN (2014) Commercial scale micropropagation of sugarcane: Constraints and remedies. *Sugar Tech* 17(4): 339–347.

Lu YH, D'Hont A, Walker DIT, Rao PS, Feldmann P and Glaszmann JC (1994) Relationships among ancestral species of sugarcane revealed with RFLP using single copy maize nuclear probes. *Euphytica* 78: 7–18.

Ma HM, Schulze S, Lee S, Yang M, Mirkov E, Irvine J, Moore P and Paterson A (2004) An EST survey of the sugarcane transcriptome. *Theoretical and Applied Genetics* 108(5): 851–863.

Manickavasagam M and Ganapathi A (1998) Direct somatic embryogenesis and plant regeneration from leaf explants of sugarcane. *Indian Journal of Experimental Biology* 832–835.

Mantri N, Pang ECK and Ford R (2010) Molecular biology for stress management. In *Climate change and management of cool season grain legume crops*, S Yadav, D McNeil, R Redden and S Patil, editors, 377–408. New York: Springer, Science + Business Media.

Menossi M, Silva-Filho MC, Vincentz M, Van-Sluys MA and Souza GM (2008) Sugarcane functional genomics: Gene discovery for agronomic trait development. *International Journal of Plant Genomics*. doi:10.1155/2008/458732.

Ming R, Moore PH, Wu KK, D'Hont A, Glaszmann JC, Tew TL et al. (2006) Sugarcane improvement through breeding and biotechnology. *Plant Breeding Reviews* 27: 15–118.

Mudge J, Anderson WR, Kehrer RL and Fairbanks DJ (1996) A RAPD genetic map of Saccharumofficinarum. *Crop Science* 36: 1362–1366.

Nair NV, Nair S, Sreenivasan TV and Mohan M (1999) Analysis of genetic diversity and phylogeny in Saccharum and related genera using RAPD markers. *Genetic Resources and Crop Evolution* 46: 73–79.

Nair NV, Selvi A, Sreenivasan TV and Pushphalatha KN (2002) Molecular diversity in Indian sugarcane varieties as revealed by randomly amplified DNA polymorphisms. *Euphytica* 127: 219–225.

Nickell LG (1964) Tissue and cell cultures of sugarcane: Another research tool. *Hawaii Planters' Record* 57: 223–229.

Niu CF, Wei W, Zhou QY, Tian AG, Hao YJ, Zhang WK et al. (2012) Wheat WRKY genes TaWRKY2 and TaWRKY19 regulate abiotic stress tolerance in transgenic *Arabidopsis* plants. *Plant, Cell and Environment* 35(6): 1156–1170.

Oliveira KM, Marconi TG, Margarido GRA et al. (2007) Functional integrated genetic linkage map based on EST markers for a sugarcane (*Saccharum* spp.) commercial cross. *Molecular Breeding* 20(3): 189–208.

Pagariya MC, Harikrishnan M, Kulkarni PA, Devarumath RM and Kawar PG (2010) Physio-biochemical analysis and transcript profiling of *Saccharum officinarum* L. submitted to salt stress. *Acta Physiologiae Plantarum* 33: 1411–1424.

Pastina MM, Pinto LR, Oliveira KM, Souza KM, Garcia AAF (2010) Molecular mapping of complex traits. In *Genetics, genomics and breeding of sugarcane*, RJ Henry and C Kole, editors, 117–148. Boca Raton, FL: CRC Press.

Patade VY and Suprasanna P (2008) Radiation induced *in vitro* mutagenesis for sugarcane improvement. *Sugar Tech* 10(1): 14–19.

Patade VY and Suprasanna P (2010) Short-term salt and PEG stresses regulate expression of MicroRNA, miR159 in sugarcane leaves. *Journal of Crop Science and Biotechnology* 13(3): 177–182.

Patade VY, Bhargava S and Suprasanna P (2011a) Transcript expression profiling of stress responsive genes in response to short-term salt or PEG stress in sugarcane leaves. *Molecular Biology Reports.* doi:10.1007/s11033-011-1100-z.

Patade VY, Bhargava S and Suprasanna P (2011b) Effects of NaCl and iso-osmotic PEG stress on growth, osmolytes accumulation and antioxidant defense in cultured sugarcane cells. *Plant Cell, Tissue and Organ Culture.* doi:10.1007/s11240-011-0041-5.

Patade VY, Bhargava S and Suprasanna P (2011c) Salt and drought tolerance of sugarcane under iso-osmotic salt and water stress: Growth, osmolytes accumulation and antioxidant defense. *J Plant Interact.* doi:10.1080/17429145.2011.557513.

Patade VY, Rai AN and Suprasanna P (2010) Expression analysis of sugarcane shaggy-like kinase (SuSK) gene identified through cDNA subtractive hybridization in sugarcane (*Saccharum officinarum* L.). *Protoplasma* 248(3): 613–621.

Ramgareeb S, Snyman S and van Antwerpen T (2010) Elimination of virus and rapid propagation of disease free sugarcane (*Saccharum* spp cultivar NCO 376) using apical meristem culture. *Plant Cell, Tissue and Organ Culture* 100: 175–181.

Raza A, Ali K, Ashraf MY, Mansoor S, Javid M and Asad S (2016) Overexpression of an H+-PPase gene from *Arabidopsis* in sugarcane improves drought tolerance, plant growth, and photosynthetic responses. *Turkish Journal of Biology* 40: 109–119.

Raza S, Qamarunisa S, Hussain M, Jamil I, Anjum S, Azhar A and Qureshi JA (2012) Regeneration in sugarcane via somatic embryogenesis and genomic instability in regenerated plants. *Journal of Crop Science and Biotechnology* 15: 131–136.

Ribaut JM and Ragot M (2007) Marker-assisted selection to improve drought adaptation in maize: The backcross approach, perspectives, limitations and alternatives. *Journal of Experimental Botany* 58: 351–360.

Rocha FR, Papini-Terzi FS, Nishiyama Jr MY, Vencio RZN, Vicentini R, Duarte RDC et al. (2007) Signal transduction-related responses to phytohormones and environmental challenges in sugarcane. *BMC Genomics* 8: 71.

Rodrigues FA, de Laia ML and Zingaretti SM (2009) Analysis of gene expression profiles under water stress in tolerant and sensitive sugarcane plants. *Plant Science* 176(2): 286–302.

Rodríguez M, Canales E and Borrás-Hidalgo O (2005) Molecular aspects of abiotic stress in plants. *Biotecnología Aplicada* 22: 1–10.

Sánchez-Ken JG and Clark LG (2010) Phylogeny and a new tribal classification of the Panicoideaes. l. (Poaceae) based on plastid and nuclear sequence data and structural data. *American Journal of Botany* 97: 1732–1748.

Schuler TH, Poppy GM, Kerry BR and Denholm I (1998) Insect resistant transgenic plants. *Trends in Biotechnology* 16: 168–175.

Selvi A, Nair NV, Balasundaram N and Mohapatra T (2003) Evaluation of maize microsatellite markers for genetic diversity analysis and fingerprinting in sugarcane. *Genome* 46(3): 394–403.

Selvi A, Nair NV, Mohapatra T, Kartikprabhu T and Sunderavelpandian K (2006) Identification of molecular markers for red rot resistance in a complex polyploid, sugarcane. *Proceedings of Second National Plant Breeding Congress,* 269–270, March 1–3. Coimbatore: TNAU.

Selvi A, Nair NV, Noyer JL, Singh NK, Balasundaram N, Bansal KC, Koundal KR and Mohapatra T (2005) Genomic constitution and genetic relationship among the tropical and sub tropical Indian sugarcane cultivars revealed by AFLP. *Crop Science* 45: 1750–1757.

Sengar K (2010) Developing an efficient protocol through tissue culture technique for sugarcane micropropagation. *BioInfoBank Library Acta* 01/2010.

Shah HA, Rashid N, Haider MS, Saleem F, Tahir M and Iqbal J (2009) An efficient, short and cost-effective regeneration system for transformation studies of sugarcane (*Saccharum officinarum* L.). *Pakistan Journal of Botany* 41: 609–614.

Singh RK, Kumar P, Tiwari NN, Rastogi J and Singh SP (2013) Current status of sugarcane transgenic: An overview. *Advancements in Genetic Engineering* 2: 112.

Snyman SJ, Meyer GM, Carson DL and Botha FC (1996) Establishment of embryogenic callus and transient gene expression in selected sugarcane varieties. *South African Journal of Botany* 62: 151–154.

Snyman SJ, Meyer GM, Koch AC, Banasiak M and Watt MP (2011) Applications of *in vitro* culture systems for commercial sugarcane production and improvement. *In Vitro Cellular and Developmental Biology – Plant* 47: 234–249.

Souza GM, Berges H, Bocs SS, Casu R, D'Hont A, Ferreira JE, Sluys MA et al. (2011) The sugarcane genome challenge: Strategies for sequencing a highly complex genome. *Tropical Plant Biology*. doi: 10.1007/s12042-011-9079-0.

Srikant J, Subramonian N and Premachandran MN (2011) Advances in transgenic research for insect resistance in sugarcane. *Tropical Plant Biology* 4: 52–61.

Srivastava S and Gupta PS (2006) Low level of genetic diversity in sugarcane varieties of India as assessed by RAPD markers. In *Proceedings of the International Symposium on 'Technologies to Improve Sugar Productivity in Developing Countries,' Session 7: Molecular Biology, Biotechnology and Tissue Culture in Sugar Crops*, 574–578. Guilin, P. R. China.

Srivastava S and Gupta PS (2008) Inter simple sequence repeat profile as a genetic marker system in sugar. *Sugar Tech* 10(1): 48–52.

Sundara B (2000) *Sugarcane cultivation*. New Delhi, India: Vikas Publications Pvt.

Suprasanna P (2010) Biotechnological interventions in sugarcane improvement: Strategies, methods and progress. *BARC News Letter* 47–53.

Tiwari AK, Bharti YP, Tripathi S, Mishra N, Lal M, Sharma PK, Roa, Gp and Sharma ML (2010) Biotechnological approaches to improve sugarcane crop with special reference to disease resistance. Acta *Phytopath et Entom Hung* 45: 235–249.

Trujillo LE, Sotolongo M, Menéndez C, Ochogavía ME, Coll Y, Hernández I, Borrás-Hidalgo O, Thomma BPHJ, Vera P and Hernández L (2008) SodERF3, a novel sugarcane ethylene responsive factor (ERF), enhances salt and drought tolerance when overexpressed in tobacco plants. *Plant and Cell Physiology* 49: 512–525.

Vettore AL, da Silva FR, Kemper EL, Souza GM, da Silva AM, Ferro MI et al. (2003) Analysis and functional annotation of an expressed sequence tag collection for tropical crop sugarcane. *Genome Research* 13(12): 2725–2735.

Virupakshi S and Naik GS (2008) ISSR analysis of chloroplast and mitochondrial genome can indicate the diversity in sugarcane genotypes for red rot resistance. *Sugar Tech* 10: 65–70.

Wang ZZ, Zhang SZ, Yang BP and Li YR (2003) Trehalose synthase gene transfer mediated by *Agrobacterium tumefaciens* enhances resistance to osmotic stress in sugarcane. *Scientia Agricultura Sinica* 36: 140–146.

Wei B, Jing R, Wang Ch, Chen J, Mao X, Chang X and Jia J (2009) Dreb1 genes in wheat (*Triticum aestivum* L.): Development of functional markers and gene mapping based on SNPs. *Molecular Breeding* 23: 13–22.

Wekesa R, Onguso JM, Nyende BA and Wamocho LS (2015) Sugarcane *in vitro* culture technology: Opportunities for Kenya's sugar industry. *African Journal of Biotechnology* 14(47): 3170–3178.

Wekesa RK, Onguso JM and Wamocho LS (2014) Effect of 2, 4- dichlorophenoxy acetic acid and naphthelene acetic acid concentrations on callogenesis for *in vitro* regeneration in selected sugarcane varieties. *Proceedings of JKUAT Scientific, Technological and Industrialization Conference*, 101–113. 13–14 November, Jomo Kenyatta University of Agriculture and Technology, Nairobi.

Yamaguchi-Shinozaki K and Shinozaki K (2006) Transcriptional regulatory networks in cellular responses and tolerance to dehydration and cold stresses. *Annual Review of Plant Biology* 57: 781–803.

Zhang SZ, Yang BP, Feng CL, Chen RK, Luo JP, Cai WW and Liu FH (2006) Expression of the Grifolafrondosa trehalose synthase gene and improvement of drought-tolerance in sugarcane (*Saccharum officinarum* L.). *Journal of Integrative Plant Biology* 48: 453–459.

Zhangsun D, Luo S, Chen R and Tang K (2007) Improved Agrobacterium-mediated genetic transformation of GNA transgenic sugarcane. *Biologia* 62: 386–393.

chapter eight

Potential applications of molecular markers for genetic diversity and DNA fingerprinting in sugarcane

Manoj Kumar Sharma, Kalpana Sengar,
Ashu Singh and R.S. Sengar

Contents

Introduction

Sugarcane (*Saccharum officinarum* L.) is one of the most important vegetatively propagated commercial cash crops in the world that is cultivated at large scales for producing sugar and biofuels (Dillon et al., 2007; Singh et al., 2014a). It is the world's largest crop by production quantity with the contribution of approximately 80% of total sugar production in the world (the rest is from sugar beet) (Commodity Research Bureau, 2015; Que et al., 2014). Sugarcane is a unique crop regarding the ability to accumulate sucrose that can reach levels up to 50% of dry weight in its stalks (Botha and Black, 2000; Bull and Glasziou, 1963). Sugarcane is commercially grown in approximately 110 countries over an area of about 27.2 mha producing about 1877 million metric tons of sugarcane with an average productivity of 70.9 tons/ha (FAOSTAT, 2015). Brazil stands first in the world with respect to area (10.2 million hectares) and production (768.09 million tons) followed by India, China, Thailand and Pakistan. India is the second largest producer of sugar (24.55 million tons) including traditional cane sugar sweeteners *khandsari* and *gur*, with a sugarcane cropping area of about 5.06 million hectares and cane production of 352.14 million tons (FAOSTAT, 2015). Economically, sugarcane is an important industrial raw material for sugar and allied industries producing acetic acid, paper, plywood, industrial enzymes and animal feeds (Arencibia et al., 1998). Apart from the traditional use as a source of sugar, sugarcane is also fast becoming and promising source for butanol, ethanol and biomass production as an alternative energy source (Lam et al., 2009; Waclawovsky et al., 2010). The sugar industry is the second largest industry in India after textiles and as an agro-based industry plays a vital role in the socioeconomic transformation of the country (FAO, 2013).

Sugarcane is a complex polyaneuploid and highly heterozygous crop plant that taxonomically belongs to the grass family *Poaceae* (or *Gramineae*), sub-family *Panicoideae*, tribe *Andropogoneae*, sub-tribe *Sacharinae*, under the group *Saccharastrae* (Amalraj and Balasundaram, 2006). The 'Saccharum complex' constitutes a closely related inbreeding group of five genera namely, *Saccharum*, *Erianthus* (sect. Rimpidium), *Miscanthus* (sect. Diantra), *Sclerostachya* and *Narenga* (Daniels et al., 1975; Mukherjee, 1957). Typically, the genus 'Saccharum' is classified into six different species, namely, *Saccharum officinarum* (Noble clones, $2n = 80$), *S. edule* Hassk. ($2n = 60$–80), *S. robustum* E.W. Brandes and Jeswiet ex Grassl ($2n = 60$, 80), *S. spontaneum* L. ($2n = 40$–180), *S. sinense* Roxb. ($2n = 81$–124) and *S. barberi* Jeswiet ($2n = 111 - 120$), in which *S. spontaneum* and *S. robustum* are wild species; *S. officinarum*, *S. barberi*, and *S. sinense* are early cultivars; and *S. edule* is a marginal specialty cultivar (Amalraj and Balasundaram, 2006; Daniels and Roach, 1987; D'Hont et al., 2008; Sreenivasan et al., 1987). These genera represent the genetic variability pool available for sugarcane breeding programmes. The genus *Saccharum* is reported to be highly polyploid (the ploidy level ranges from 5× to 16×), and therefore considered among the most complex plant genomes (Manners et al., 2004). The previous studies reported that the basic chromosome number for *S. spontaneum* is 8; whereas, it is 10 for all other species including *S. officinarum*, *S. robustum*, *S. sinense*, *S. barberi* and *S. edule* (D'Hont et al., 1998; Ming et al., 1998; Panje and Babu, 1960; Piperidis et al., 2010). The size of the sugarcane genome is about 10 GB, while its genome complexity is due to the mixture of euploid and aneuploid chromosome sets with homologous genes present in from 8 to 12 copies (Souza et al., 2011). The estimated monoploid genome size ranges from 760 to 926 Mbp, which is twice that of rice genome (389 Mbp) and similar to sorghum genome (760 Mbp) (D'Hont and Glaszmann, 2001).

Sugarcane is originated in Southeast Asia and New Guinea (Lebot, 1999). Modern commercial cultivars of sugarcane are complex hybrids developed by the interspecific hybridization of *Saccharum officinarum* (noble clones with higher sugar and low fibre content) as the female parent and *S. spontaneum* (a wild species with no sugar, thin culms and resistant to biotic and abiotic stresses) as the male parent with minor contribution from *S. sinense* Roxb (Chinese clones) and *S. barberi* Jesw (North Indian clones). The segregating progenies were further repeatedly backcrossed with *S. officinarum* clones to recover the favourable alleles for sugar content and to transfer disease resistance genes from *S. spontaneum*. This process is referred to as the 'nobilization of canes' (Edme et al., 2005; Roach, 1972, 1989). It has been determined through the use of the genome *in situ* hybridization (GISH) technique that modern commercial sugarcane cultivars constitute approximately 80% chromosomes from *S. officinarum*, 10% to 15% chromosomes from *S. spontaneum* and approximately 10% recombinant chromosomes (D'Hont et al., 1996; Piperidis et al., 2010).

From time to time, several earlier workers suggested the utilization of possible crossing combinations among sugarcane and its related genera for the improvement of this bioenergy crop through breeding procedures (Daniels and Roach, 1987; Dutt and Rao, 1951). In case of sugarcane, the majority of breeding programmes have focused on intercrossing between the hybrids, though in recent decades the larger increases in genetic gain have been made by incorporating more diverse germplasm into the cultivated backgrounds (Edme et al., 2005). The better understanding of interrelationships among the members of a species and its related genera is a primary and promising step towards the improvement of a crop plant by utilization of breeding programmes (Swapna and Srivastava, 2012). The utilization of traditional breeding with molecular techniques may be an effective way for the improvement and sustainable use of existing sugarcane diverse germplasm for developing genotypes with improved qualitative and quantitative characters.

Molecular markers and their applications in sugarcane

The study of genetic variations/polymorphism has become an active area of research in all important crops and model plant species. The study of polymorphism involves the development and selective use of molecular markers, which have proved highly useful not only for marker-assisted plant breeding, but also in understanding the crop domestication and evolutionary processes (Gupta et al., 2008). Due to the rapid developments in the field of molecular genetics, several varieties of different molecular or genetic marker techniques have emerged to analyse genetic variations in the last few decades. To date, a lot of molecular marker techniques have become extensively available in plant systems for basic and applied studies (Gupta et al., 1999). Increasingly, these techniques are being developed to more precisely, quickly and cheaply assess genetic variations. In the current scenario of plant research, the use of DNA-based molecular or genetic marker techniques has become the routine choice of breeders instead of morpho-physio-biochemical markers for the study of genetic variations among plant species because they are stable, more in number and are not influenced by environmental factors or the developmental stages of the plant (Winter and Kahl, 1995).

Molecular markers have an identifiable DNA sequence that is readily detected and whose inheritance can be easily monitored. These markers are found at specific locations in the genome that are transmitted in the fashion of standard laws of inheritance from one generation to the next, and thus provide valuable information for improvement practices in plant breeding programmes (Agarwal et al., 2008; Collard et al., 2005; Davey et al., 2011; Li and Quiros, 2001). These marker techniques may differ with respect to their important features, such as their principle, detection method, level of polymorphism, genomic abundance, inheritance pattern, reproducibility, locus specificity, genome coverage, technical requirements and developmental cost. To date, there is no single marker that is superior to all others for a wide range of applications in plant systems. An appropriate molecular or genetic marker system with wide applicability has to be dependent on some ideal properties, including polymorphic in nature, co-dominant in inheritance, high reproducibility, discriminating, neutral, no tissue specificity, easily amenable in automation, easy, fast and cheap in detection, and evenly and frequently distributed throughout the genome (Kumar et al., 2009). Molecular markers can be broadly categorized into following three main categories:

1. Hybridization-based DNA markers such as restriction fragment length polymorphisms (RFLPs) and oligonucleotide fingerprinting
2. Polymerase chain reaction (PCR)-based DNA markers such as random amplified polymorphic DNAs (RAPDs), sequence characterized amplified regions (SCARs), microsatellites, sequence-tagged sites (STSs), inter-simple sequence repeats (ISSRs), DNA amplification fingerprinting (DAF), amplified fragment length polymorphic DNAs (AFLPs), cleaved amplified polymorphic sequence (CAPS), single-strand conformation polymorphism (SSCP), sequence-related amplified polymorphism (SRAP), target region amplification polymorphism (TRAP), conserved region amplification polymorphism (CoRAP), start codon targeted (SCoT) polymorphism, conserved DNA-derived polymorphism (CDDP) and CAAT box-derived polymorphism (CBDP)
3. DNA chips or sequencing-based DNA markers such as expressed sequence tags (ESTs), internal transcribed spacer regions of nuclear ribosomal genes (ITS), single nucleotide polymorphisms (SNPs) and diversity array technology (DArT)

Due to the wide applicability among plant systems, molecular markers are used for several applications including germplasm characterization, genetic diagnostics, characterization of transformants, study of genome organization, genetic stability assessment, analysis of phylogenetic relationships, DNA fingerprinting, genetic and linkage mapping, quantitative trait loci (QTL) mapping, gene tagging, association studies, and marker-assisted selection (Archak et al., 2003; Azhaguvel et al., 2006; Gupta et al., 1999; Lateef, 2015; Varshney et al., 2005). Molecular markers seem to be the best candidates for efficient evaluation, selection and characterization of plant materials at a commercial scale. The uses of molecular markers are based on the naturally occurring DNA polymorphisms, which form the basis for designing strategies to exploit for applied purposes (Kumar et al., 2009). The polygenic characters, which were previously very difficult to analyse using traditional plant breeding methods, can now be easily tagged by using molecular markers. It is also possible to establish genetic relationships between sexually incompatible crop plants (Mohan et al., 1997).

In the case of sugarcane, several novel and trait-targeted molecular marker techniques have been extensively used for genetic diversity analysis, DNA fingerprinting, gene expression studies, gene tagging, genetic mapping, QTL mapping, association mapping, linkage mapping, map based cloning, marker assisted selection, segregation analysis, genetic stability analysis and for comparative and functional genomics studies (Aitken et al., 2014; Aitken et al., 2005, 2006, 2007, 2008; Alwala et al., 2006a, 2006b, 2008, 2009; Da Silva et al., 1993; D'Hont et al., 1993, 1994; Diola et al., 2015; Hemaprabha et al., 2013; Khaled et al., 2011; Le Cunff et al., 2008; Lu et al., 1994; Lu et al., 2015; Ming et al., 1998, 2001, 2002; Nair et al., 1999; Parida et al., 2009, 2010; Pinto et al., 2004; Singh et al., 2008, 2011b; Singh et al., 2012; Singh et al., 2013c; Srivastava et al., 2005; Swapna et al., 2011a, 2011b; Wu et al., 2013). The comparisons of characteristic features of important molecular marker technologies which have been widely applied in sugarcane for various purposes during more than the last two decades are summarized in Table 8.1.

Molecular markers for genetic diversity/phylogenetic analysis

Genetic diversity refers to the variations present among nucleotides, genes, chromosomes or whole genomes of a population of a species or genus (Kurane et al., 2009). The genetic diversity describes the sum total of genetic characteristics within a species or genus, while genetic variability shows the presence of total variations within or among these genetic characteristics (Rao and Hodgkin, 2002). In a species, the assessment of

Table 8.1 Comparison of different major molecular marker techniques used for genetic diversity analysis and DNA fingerprinting in sugarcane

Feature	RFLP	RAPD	AFLP	SSR	EST-SSR	ISSR	SSCP	SCAR	SNP	SRAP	TRAP	SCoT
Principle involved	Restriction digestion	DNA amplification	Restriction digestion and DNA amplification	DNA amplification	DNA amplification	DNA amplification	DNA amplification	DNA amplification	DNA amplification	DNA amplification	DNA amplification	DNA amplification
Detection method	Southern blotting	DNA staining	DNA staining	DNA staining	DNA staining	DNA staining	DNA staining	DNA staining	Pyro-phosphate sequencing	DNA staining	DNA staining	DNA staining
DNA quantity	High	Low	Medium	Low	Low	Low	Low	Low	Low	Low	Low	Low
DNA quality	Relatively pure	Crude	Relatively pure	Relatively pure	Relatively pure	Relatively pure	Relatively pure	Relatively pure	Pure	Relatively pure	Relatively pure	Relatively pure
Genomic abundance	High	High	High	Medium	Medium	Low	Medium	Low	High	High	Medium	High
Genomic coverage	Low copy coding regions	Whole genome	Whole genome	Whole genome	Whole genome	Whole genome	Whole genome	Whole genome	Whole genome	Whole genome	Whole genome	Whole genome
Inheritance pattern	Co-dominant	Dominant	Dominant/co-dominant	Co-dominant	Co-dominant	Dominant	Co-dominant	Dominant/co-dominant	Co-dominant	Co-dominant	Co-dominant	Dominant/co-dominant
Reproducibility	High	Low to medium	High	High	Medium	Medium	Medium to high	High	High	Medium	High	Medium
Degree of polymorphism	Medium	Medium	Medium	Medium	Medium	Low	High	Medium	High	Medium	High	Medium
Allelic	Bi-allelic	Multi-allelic	Bi-allelic	Multi-allelic	Multi-allelic	Multi-allelic	Bi-allelic	Bi-allelic	Bi-allelic	Bi-allelic	Bi-allelic	Multi-allelic
Prior sequence information	No	No	No	Yes	Yes	No	Yes	Yes	Yes	No	Yes	No
Locus specificity	Yes	No	No	No	Yes	Yes	No	Yes	Yes	Yes	Yes	Yes

(Continued)

Table 8.1 (Continued) Comparison of different major molecular marker techniques used for genetic diversity analysis and DNA fingerprinting in sugarcane

Feature	RFLP	RAPD	AFLP	SSR	EST-SSR	ISSR	SSCP	SCAR	SNP	SRAP	TRAP	SCoT
Technical requirement	High	High	Medium	Medium	Medium	Medium	Medium	Medium	High	Medium	Medium	Medium
Ease of use	Labour intensive	Easy	Difficult initially	Easy	Easy	Easy	Labour intensive	Easy	Easy	Medium	Easy	Easy
Amenable to automation	Low	Medium	Medium	High	High	Medium	Can not Automated	Medium	High	Medium	Medium	Medium
Developmental cost	High	Low	Medium	High	High	High	High	Medium	Very high	Medium	Medium	Medium
Major applications	Physical mapping	Gene tagging	Gene tagging	Genetic diversity and gene mapping	Genetic diversity	Genetic diversity and gene mapping	SNP mapping	Gene tagging and physical mapping	Genetic diversity	Plant genotyping	Plant genotyping	QTL mapping and genetic diversity
Discovered by	Botstein et al., 1980	Williams et al., 1990	Vos et al., 1995	Litt and Luty, 1989	Kantety et al., 2002	Zietkiewicz, 1994	Orita et al., 1989	Paran and Michelmore, 1993	Paterson et al., 1991	Li and Quiros, 2001	Hu and Vick, 2003	Collard and Mackill, 2009

Source: Adapted with modifications from Singh RB, Srivastava S, Rastogi J, Gupta GN, Tiwari NN, Singh B, Singh RK, 2014, *Research in Environment and Life Sciences*, 7(4): 223–232.

genetic diversity can be accomplished at intraspecific, interspecific or intergeneric levels (Mittal and Dubey, 2009). The genetic diversity can be conventionally estimated by using various biometrical techniques such as D^2, divergence analysis, meteroglyph and principal component analysis (PCA), phenotypic diversity index (H) or coefficient of parentage at morphological, agronomical and biochemical levels (Ahmad et al., 2008; Jaradat et al., 2004; Matus and Hayes, 2002; Mohammadi and Prassana, 2003). However, this type of evaluation is laborious and took several years to make a fruitful conclusion. The diversity among plant species based on morphological, physiological and biochemical characters usually depends upon environmental conditions and the trait evaluation needs maturity stage of plants prior to identification. Nowadays, the detection of genetic variations within or among plant species has become a promising and potential tool in crop improvement (Jonah et al., 2011). The increasing levels of genetic diversity among crop plant species establish important basis for sustainable crop production through enhancing better plant survival under biotic and abiotic stresses in their natural habitats (Hajjar et al., 2008; Trethowan and Kazi, 2008).

Phylogenetic relationships reflect the genetic relatedness of a group of species based on the calculated genetic distance in their evolutionary history (Wang et al., 2009). Molecular markers exhibit powerful systems for assessing phylogenetic relationships within or among species or closely related species. The assessment of genetic diversity and phylogenetic relationships among available germplasm provides valuable information to realize the overall genetic potential of plant species to exploit them in developing superior productive germplasm for crop improvement practices (Hoshino et al., 2012). Molecular phylogenetics studies would also be helpful to clarify the taxonomic identity and evolutionary relationships among the crop species and its wild relatives (Poczai et al., 2012).

One of the major contributions of molecular markers in plant breeding is the better understanding of genetic diversity, phylogenetic relationships and genetic relationships within or among species, germplasm collections, breeding lines, crops and their wild relatives for their immediate use in crop improvement (Hemaprabha and Simon, 2012). The development of molecular marker techniques allows easy identification and characterization of large numbers of genotypes of a species in a short period of time. Molecular makers have become the most powerful tools for the assessment of genetic variations and elucidating phylogenetic relationships among crop plant species at a large scale for identification of superior cultivars for crop breeding programmes (Chakravarthi and Naravaneni, 2006). In terms of genetic diversity, molecular markers generate high quality genetic data that may not be possible through other genetic methods. The advent of different molecular marker techniques led the foundation for rapid and accurate analysis of vast germplasm

collections and estimation of genetic diversity found to corroborate phenotypic data for breeder use. Beside these aspects, the genetic diversity
estimated through molecular markers is necessary for mapping genes/
QTLs, gene tagging and marker-assisted selection (Lapitan et al., 2007).

Modern sugarcane cultivars exhibit a high level of polyploidy and
aneuploidy in their genomes with the presence of multiple alleles at each
locus. These cultivars have been developed over decades in earlier breeding programmes by interspecific hybridization among several species of
Saccharum genus including *Saccharum officinarum, Saccharum spontaneum,
Saccharum barberi, Saccharum robustum* and *Saccharum sinense* (Besse et
al., 1998). Cross breeding is the most promising and effective way for the
development of new sugarcane genotypes with desirable characteristics
(Jackson, 2005). In earlier breeding programmes, the number of original
parental lines used in sugarcane crosses was limited. Then, maximum
sugarcane crosses were made among superior clones of modern sugarcane; this led to the foundation of a narrow genetic base. The high genome
complexity and narrow genetic base of sugarcane genotypes has made
cane breeding slow and cumbersome (Arruda, 2012; Nayak et al., 2014;
Singh et al., 2011b).

As per the heterosis theory, it is concluded that the utilization of
parental lines having a larger genetic distance is a promising requirement for crossing during the development of cultivars in a breeding programme (Loomis and Williams, 1963; Ming et al., 2006; Stevenson, 1965).
Therefore, a better understanding and evaluation of genetic diversity
among domestic and wild sugarcane genetic resources, especially for
those used as progenitors, is an essential step towards the identification of
potential germplasm collections for the optimization of hybridization and
selection procedures for improvement of this important cash crop (Singh
et al., 2013a; You et al., 2016). Germplasm collections having high levels
of genetic diversity would enable breeders to broaden the genetic base of
parental genotypes and thereby facilitate genetic gains of sugarcane cultivars (Cooper et al., 2001; Ming et al., 2006).

In the case of sugarcane, several methods mainly based on pedigree
records and morpho-physio-biochemical traits have been used for estimating genetic diversity at large scales (Cordeiro et al., 2003). However,
these traits are directly or indirectly influenced by environmental factors;
therefore they do not reflect true levels of genetic diversity in sugarcane
(Creste et al., 2010a; Singh et al., 2010). Therefore, the utilization of molecular markers in estimating genetic diversity is the best option to get efficient
and more reliable data. A lot of hybridization, PCR or sequencing-based
molecular markers such as ITS, RFLP, RAPD, AFLP, SSR, EST-SSR, STMS,
ISSR, CISP, SSCP, SNP, DArT, SRAP, TRAP, and SCoT have been extensively used for determining genetic diversity and phylogenetic relationships among vast germplasm collections of sugarcane (Table 8.2).

Table 8.2 An overview of different molecular marker techniques applied in sugarcane for genetic diversity/phylogenetic analysis and DNA fingerprinting

Marker used	Reference
Genetic diversity/Phylogenetic analysis	
5S nrRNA ITS	Glaszmann et al., 1990
5S nrRNA ITS	Pan et al., 2000
5S nrRNA ITS	Hodkinson et al., 2002
5.8S nrDNA ITS	Chen et al., 2003
5.8S nrDNA ITS	Liu et al., 2010
5.8S nrDNA ITS	Liu et al., 2016
5.8S nrDNA ITS	Yang et al., 2016
RFLP	D'Hont et al., 1993
RFLP	Lu et al., 1994
RFLP	Burnquist et al., 1995
RFLP	Besse et al., 1997
RFLP	Jannoo et al., 1999
RFLP	Coto et al., 2002
RAPD	Harvey et al., 1994
RAPD	Harvey and Botha, 1996
RAPD	Burner et al., 1997
RAPD	Nair et al., 1999
RAPD	Ubayasena and Perera, 1999
RAPD	Chen et al., 2001
RAPD	Hemaprabha and Rangasamy, 2001
RAPD	Leon et al., 2001
RAPD	Nair et al., 2002
RAPD	Xu et al., 2003
RAPD	Zeng et al., 2003
RAPD, ISSR and ITS	Pan et al., 2004
RAPD	Zhang et al., 2004
RAPD	Afghan et al., 2005
RAPD	Pan et al., 2005
RAPD	Mary et al., 2006
RAPD and SSR	Khan et al., 2007
RAPD	Selvi et al., 2008
RAPD	Singh et al., 2008
RAPD	Alvi et al., 2008
RAPD	Da Silva et al., 2008
RAPD	Khan et al., 2009
RAPD	Kawar et al., 2009

(Continued)

Table 8.2 (Continued) An overview of different molecular marker techniques applied in sugarcane for genetic diversity/phylogenetic analysis and DNA fingerprinting

Marker used	Reference
Genetic diversity/Phylogenetic analysis	
RAPD	Alejandre Rosas et al., 2010
RAPD	Nawaz et al., 2010b
RAPD	Zhang et al., 2010
RAPD	Govindaraj et al., 2011
RAPD	Srivastava et al., 2011
RAPD	Mumtaz et al., 2011
RAPD and SSR	Shahid et al., 2012
RAPD	Ali et al., 2013
RAPD, ISSR and AFLP	Ismail, 2013
RAPD, STS and TRAP	Khan et al., 2013
RAPD	Sarid Ullah et al., 2013
RAPD	Rao et al., 2014
RAPD and STMS	Saravanakumar et al., 2014
RAPD	Seema et al., 2014
RAPD	Mahmud et al., 2014
RAPD and ISSR	Khaled et al., 2015
AFLP	Besse et al., 1998
AFLP	Lima et al., 2002
AFLP	Selvi et al., 2005
AFLP	Cai et al., 2005a
AFLP	Zhuang et al., 2005
AFLP	Arro et al., 2006
AFLP	Aitken et al., 2006
AFLP	Selvi et al., 2006
AFLP	Jimenez et al., 2008
AFLP	Lao et al., 2008
AFLP	Lao et al., 2009a
AFLP, TRAP and SSRs	Creste et al., 2010a
AFLP and SSR	Perera et al., 2012
SSR	Cordeiro et al., 2000
SSR and RAPD	Ceron and Angel, 2001
SSR	Cordeiro et al., 2001
SSR	Cordeiro et al., 2003
SSR	Cordeiro et al., 2003
SSR	Pan et al., 2003b

(Continued)

Table 8.2 (Continued) An overview of different molecular marker techniques applied in sugarcane for genetic diversity/phylogenetic analysis and DNA fingerprinting

Marker used	Reference
Genetic diversity/Phylogenetic analysis	
SSR	Riasco et al., 2003
SSR	Selvi et al., 2003
SSR and AFLP	Cai et al., 2005b
SSR	Queme et al., 2005
SSR	Brown et al., 2007
SSR	Liang et al., 2008
SSR	Chen et al., 2009
SSR	Glynn et al., 2009
SSR	Lao et al., 2009b
SSR	Parida et al., 2009
SSR	Huang, 2009
SSR	Singh et al., 2010
SSR	Filho et al., 2010
SSR	Banumathi et al., 2010
SSR	Liang et al., 2010
SSR	Pandey et al., 2011
SSR	Singh et al., 2011b
SSR	Qi et al., 2011
SSR	Qi et al., 2012
SSR	Liu et al., 2011
SSR	Park et al., 2012
SSR	Hameed et al., 2012
SSR	Santos et al., 2012
SSR	Abbas et al., 2013
SSR	Fan et al., 2013
SSR	Singh et al., 2013a
SSR	You et al., 2013
SSR	Chandra et al., 2014
SSR	Maranho et al., 2014
SSR	Nayak et al., 2014
SSR	Sharma et al., 2014
SSR	Tena et al., 2014
SSR	Liu et al., 2016
cpSSR	Raj et al., 2016
SSR-SSCP	Srivastava et al., 2011

(Continued)

Table 8.2 (Continued) An overview of different molecular marker techniques applied in sugarcane for genetic diversity/phylogenetic analysis and DNA fingerprinting

Marker used	Reference
	Genetic diversity/Phylogenetic analysis
SSR-SSCP	Swapna et al., 2011
SSR-SSCP	Kalwade and Devarumath, 2014
EST-SSR	Da Silva, 2001
EST-SSR	Pinto et al., 2004
EST-SSR	Pinto et al., 2006
EST-SSR	Oliveira et al., 2009
EST-SSR	Marconi et al., 2011
EST-SSR	Duarte et al., 2012
EST-SSR	Silva et al., 2012a
EST-SSR	Singh et al., 2012a
EST-SSR	Ukoskit et al., 2012
EST-SSR	Diola et al., 2013
EST-SSR	Singh et al., 2013b
EST-SSR	Singh et al., 2015
EST-SSR and SSR	You et al., 2016
STMS	Hemaprabha et al., 2005
STMS	Hemaprabha et al., 2006a
STMS	Babu et al., 2010
STMS	Lavanya and Hemaprabha, 2010a
STMS	Lavanya and Hemaprabha, 2010b
STMS	Sindhu et al., 2011
STMS	Hemaprabha and Simon, 2012
STMS	Saravanakumar et al., 2014
ISSR	Yu et al., 2002
ISSR	Shrivastava and Gupta, 2008
ISSR	Virupakshi and Naik, 2008
ISSR	Alves de Almeida et al., 2009
ISSR	Da Costa et al., 2011
ISSR	Kalwade et al., 2012
ISSR and SSR	Devarumath et al., 2012
ISSR	Latorre et al., 2013
CISP	MS Khan et al., 2011
CISP	Chandra et al., 2013
SNP	Cordeiro et al., 2006
SRAP	Suman et al., 2008

(Continued)

Table 8.2 (Continued) An overview of different molecular marker techniques applied in sugarcane for genetic diversity/phylogenetic analysis and DNA fingerprinting

Marker used	Reference
Genetic diversity/Phylogenetic analysis	
SRAP and TRAP	Song, 2009
SRAP	Chang et al., 2012
TRAP	Arro, 2005
TRAP	Alwala et al., 2006a
TRAP	Alwala et al., 2006b
TRAP	Que et al., 2009
TRAP	Creste et al., 2010b
TRAP	Suman et al., 2012
TRAP and SNP	Devarumath et al., 2013
TRAP	Sosa et al., 2015
DArT	Heller-Uszynska et al., 2011
SCoT	Yachun et al., 2012
SCoT	Que et al., 2014
cpDNA sequence	Takahashi et al., 2005
cpDNA sequence	Zhu et al., 2014
DNA fingerprinting/Cultivar identification	
5.8S rDNA ITS	Piperidis et al., 2000
5.8S nrDNA ITS	Zhang et al., 2009
5S nrRNA ITS	Pan et al., 2001
cDNA probes	Schenck et al., 2004
RFLP	D'Hont et al., 1995
RAPD	Huckett and Botha, 1995
RAPD	Oropeza et al., 1995
RAPD and SSR	Abdel-Tawab et al., 2003
RAPD and ISSR	Fahmy et al., 2008
RAPD	Zhang et al., 2008a
RAPD	Zhang et al., 2008b
RAPD	Ahmed and Khaled, 2009
RAPD	Khaled, 2010
RAPD	Khaled and Fathalla, 2010
SCAR	Tabasum et al., 2010
AFLP	Srivastava et al., 2012
AFLP	Thokoane et al., 1999
	Rodriguez et al., 2005

(Continued)

Table 8.2 (Continued) An overview of different molecular marker techniques applied in sugarcane for genetic diversity/phylogenetic analysis and DNA fingerprinting

Marker used	Reference
DNA fingerprinting/Cultivar identification	
SSR	Cordeiro and Henry, 1999
SSR	Piperidis et al., 2001
SSR	Hack et al., 2002
SSR	Piperidis, 2003
SSR	Selvi et al., 2003
SSR and RAPD	Pan et al., 2003a
SSR	Manigbas and Villegas, 2004
SSR	Piperidis et al., 2004
SSR	Govindaraj et al., 2005
SSR	Tew and Pan, 2005
SSR	Tew et al., 2005
SSR	Pan, 2006
SSR	Pan et al., 2007
SSR	Gilbert et al., 2008
SSR	Comstock et al., 2009
SSR	Glaz et al., 2009
SSR	Glynn et al., 2009
SSR	Huang, 2009
SSR	Maccheroni et al., 2009
SSR	Milligan et al., 2009
SSR	Pan et al., 2009
SSR	Tew et al., 2009
SSR	Long et al., 2010
SSR	Matsuoka et al., 2010
SSR	Nawaz et al., 2010a
SSR	Pan, 2010
SSR	Tew and Pan, 2010
SSR	Liu et al., 2011
SSR	Singh et al., 2011a
SSR	Tew et al., 2011
SSR	White et al., 2011
SSR	Hameed et al., 2012
SSR	Hale et al., 2013
SSR	Joshi and Albertse, 2013
EST-SSR	Silva et al., 2012b

(Continued)

Table 8.2 (Continued) An overview of different molecular marker techniques applied in sugarcane for genetic diversity/phylogenetic analysis and DNA fingerprinting

Marker used	Reference
DNA fingerprinting/Cultivar identification	
SSR-SSCP	Singh et al., 2008
SSR-SSCP	Swapna et al., 2011
STMS	Hemaprabha et al., 2006b
STMS	Hemaprabha et al., 2010
STMS	Hemaprabha et al., 2011
STMS	Govindaraj et al., 2012
STMS	Hemaprabha et al., 2013
ISSR	Wang et al., 2007
ISSR	Markad et al., 2014
TRAP	Khan et al., 2011

Molecular markers for DNA fingerprinting/cultivar identification

Plant DNA fingerprinting is a molecular marker-based advance technology that is extensively used for paternity testing, to assess seed purity, cultivar identification, cross monitoring of a variety/germplasm and to protect intellectual property rights (Soller and Beckmann, 1983). Fingerprinting databases are most probably implemented to prevent mis-identification, effective and reliable practices for management of valuable germplasm at large scales (Poczai et al., 2012). Fingerprinting profiles are also beneficial to confirm pedigree and genetic relationships, or to distinguish closely related species/lines with incomplete genealogical records (Christiansen et al., 2002; Landjeva et al., 2006; Manifesto et al., 2001). Nowadays, plant DNA fingerprinting has become a reliable and invaluable tool in several scientific and industrial laboratories all over the world (Joshi and Albertse, 2013). Since the International Plant Variety Protection Act (1970) was signed, the protection of intellectual property rights of plant varieties has received so much attention worldwide. To date, a lot of varieties have been released and registered worldwide in sugarcane and other crop plants; therefore, it has been essential to acquire their DNA fingerprints to support the protection of these varieties (Hemaprabha et al., 2013; Zhang et al., 2015).

In sugarcane, the varieties are usually identified on the basis of their morphological characters. For clear-cut identification of plants varieties, DUS (distinctiveness, uniformity and stability) guidelines are also based on a set of morphological descriptors (Amalraj et al., 2011). This type of plant characterization is easy and economical, but is impeded by several environmental factors that alter sugarcane morphology; therefore, varietal purity assurance strictly needs constant attention (Hemaprabha et al., 2013). The accurate identification of germplasm is strictly required for the conservation and sustainable utilization of sugarcane genetic resources. The emergence of several new and advanced molecular marker technologies has changed the face of varietal identification. Nowadays, a lot of molecular marker techniques are available that can be effectively used for the precise identification of sugarcane varieties. The International Union for the Protection of New Varieties of Plants also recommended the use of molecular tools (UPOV, 2005), especially DNA-based molecular markers for the accurate examination of crop plant's distinctness, uniformity and stability (Zhang et al., 2013). Several types of molecular markers that have been earlier used by researchers for the identification/characterization or DNA fingerprinting in sugarcane are shortlisted in Table 8.2.

Conclusion and future prospects

Sugarcane improvement through traditional breeding is impeded by cytological complexities in genome and long breeding and selection cycles. The advent of novel molecular approaches in the past two decades addressed these problems and made sugarcane breeding programmes faster and more precise. Genetic diversity is the study of comparing an individual or population with another individual or population, while DNA fingerprinting is the unambiguous or accurate identification of an individual or a population. Both genetic diversity and fingerprinting can be assessed on morphological, biochemical and molecular levels. It has been estimated that morphological and biochemical character-based genetic diversity and fingerprinting is not reliable or more precise due to their interference by environmental factors. Therefore, the utilization of molecular markers for characterization and fingerprinting of crop plant species may be a reliable and effective strategy because they provide fast and accurate results in small time periods. Several molecular markers technologies including ITS, RFLP, RAPD, AFLP, SSR, EST-SSR, STMS, ISSR, CISP, SSCP, SNP, DArT, SRAP, TRAP and SCoT have been extensively used in sugarcane for germplasm characterization and DNA fingerprinting. Vegetatively propagated crops like sugarcane usually have a narrow genetic base; therefore, it is necessary to broaden the genetic base of sugarcane through germplasm characterization for developing superior clones for maintaining yields and sustaining in harshening climatic conditions. DNA fingerprinting has been a valuable strategy in sugarcane to protect plant breeders' rights and germplasm misidentification. In the near future, the discovery and implementation of high throughput platforms would save time, money and researchers' laborious efforts in the management of valuable germplasm and its sustainable utilization for breeding superior genotypes for sugarcane improvement.

Acknowledgements

The authors sincerely acknowledge the Vice Chancellor, SVPUA&T, Meerut, Uttar Pradesh for providing the support, encouragement and laboratory facilities for conducting research on sugarcane.

References

Abbas SR, Gardazi SDA, Iqbal MZ, Khan MY, Batool A, Abbas MR, Shazad S, Khan AM, Shah SH. 2013. Characterization of twenty-six genotypes of sugarcane using SSR markers. *International Journal of Scientific and Engineering Research*, 4(7): 1002–1007.

Abdel-Tawab FM, Fahmy EM, Allam AI, El-Rashidy HA, Shoaib RM. 2003. Development of RAPD and SSR marker associated with stress tolerance and some technological traits and transient transformation of sugarcane (*Saccharum* spp.). In: *Proceedings of the International Conference The Arab Region and Africa in the World Sugar Context*, 10–12 March, Aswan, Egypt, pp. 1–23.

Afghan S, Haider MS, Shah AH, Rashid N, Iqbal J, Tahir M, Akhtar M. 2005. Detection of genetic diversity among sugarcane (*Saccharum* spp.) genotypes using random amplified polymorphic DNA markers. *Sugarcane International*, 23(6): 15–19.

Agarwal M, Shrivastava N, Padh H. 2008. Advances in molecular marker techniques and their applications in plant sciences. *Plant Cell Reports*, 27: 617–631.

Ahmad Z, Ajmal SU, Munir M, Zubair M, Masood MS 2008. Genetic diversity for morpho-genetic traits in barley germplasm. *Pakistan Journal of Botany*, 40 (3): 1217–1224.

Ahmed AZ, Khaled KAM. 2009. Detection of genetic similarity of sugarcane genotypes. *Electronic Journal of Gene Conserve*, 31: 686–697.

Aitken KS, Jackson PA, McIntyre CL. 2005. A combination of AFLP and SSR markers provides extensive map coverage and identification of homoeologous linkage groups in a sugarcane cultivar. *Theoretical and Applied Genetics*, 110: 789–801.

Aitken KS, Li JC, Jackson P, Piperidis G, McIntyre CL. 2006. AFLP analysis of genetic diversity within *Saccharum officinarum* and comparison with sugarcane cultivars. *Australian Journal of Agricultural Research*, 57(11): 1167–1184.

Aitken KS, Jackson PA, McIntyre CL 2007. Construction of genetic linkage map of *Saccharum officinarum* incorporating both simplex and duplex markers to increase genome coverage. *Genome*, 50: 742–756.

Aitken KS, Hermann S, Karno K, Bonnett GD, McIntyre LC, Jackson PA. 2008. Genetic control of yield related stalk traits in sugarcane. *Theoretical and Applied Genetics*, 117: 1191–1203.

Alejandre Rosas JA, Galindo Tovar ME, Lee Espinosa HE, Alvarado Gomez OG. 2010. Genetic variability in 22 sugarcane hybrid varieties (*Saccharum* spp. Hybrid). *International Journal of Experimental Botany*, 79: 87–94.

Aitken KS, McNeil MD, Hermann S, Bundock PC, Kilian A, Heller-Uszynska K, Henry RJ, Li J. 2014. A comprehensive genetic map of sugarcane that provides enhanced map coverage and integrates high-throughput Diversity Array Technology (DArT) markers. *BMC Genomics*, 15: 152.

Ali W, Muhammad K, Nadeem MS, Inamullah, Ahmad H, Iqbal J. 2013. Use of RAPD markers to characterize commercially grown rust resistant cultivars of sugarcane. *International Journal of Biosciences*, 3(2): 115–121.

Alves de Almeida CM, Nascimento de Lima SE, de Andrade Lima GS, de Brito JZ, Tenorio VM, Donato S, da Silva MV. 2009. Molecular characterization of the sugarcane cultivars obtained by ISSR markers. *Ciencia e Agrotecnologia*, 33: 1771–1776.

Alvi AK, Iqbal J, Shah AH, Pan YB. 2008. DNA based genetic variation for red rot resistance in sugarcane. *Pakistan Journal of Botany*, 40(4): 1419–1425.

Alwala S, Suman A, Arro JA, Veremis JC, Kimbeng CA. 2006a. Target region amplification polymorphism (TRAP) for assessing genetic diversity in sugarcane germplasm collections. *Crop Science*, 46: 448–455.

Alwala S, Kimbeng CA, Gravois KA, Bischoff KP. 2006b. TRAP a new tool for sugarcane breeding: Comparison with AFLP and coefficient of parentage. *Journal of the American Society of Sugar Cane Technologists*, 26(1): 62–87.

Alwala S, Collins A, Kimbeng J, Veremis C, Gravois KA. 2008. Linkage mapping and genome analysis in a Saccharum interspecific cross using AFLP, SRAP and TRAP markers. *Euphytica*, 164: 37–51.

Alwala S, Collins A, Kimbeng J, Veremis C, Gravois KA. 2009. Identification of molecular markers associated with sugar-related traits in a *Saccharum* interspecific cross. *Euphytica*, 167: 127–142.

Amalraj VA, Balasundaram N. 2006. On the taxonomy of the members of 'Saccharum complex'. *Genetic Resources and Crop Evolution*, 53: 35–41.

Amalraj VA, Premachandran MN, Ram B, Karuppaiyan R, Devi S, Viola R, Remadevi AK. 2011. DUS characteristics of reference varieties of sugarcane (*Saccharum* L.). *Sugarcane Breeding Institute Newsletter*, p. 128.

Archak S, Ambika B, Gaikwad AB. 2003. DNA fingerprinting of Indian cashew (*Anacardium occidentale* L.) varieties using RAPD and ISSR techniques. *Euphytica*, 230: 397–404.

Arencibia AE, Carmona P, Tellez MT, Chan SM, Yu L, Trujillo, Oramas P. 1998. An efficient protocol for sugarcane (*Saccharum* spp.) transformation mediated by *Agrobacterium tumefaciens*. *Transgenic Research*, 7: 213–222.

Arro JA. 2005. Genetic diversity among sugarcane clones using target region amplification polymorphism (TRAP) markers and pedigree relationships. Master's thesis submitted to Louisiana State University, Baton Rouge, Louisiana.

Arro JA, Veremis JC, Kimbeng CA, Botanga C. 2006. Genetic diversity and relationships revealed by AFLP markers among *Saccharum spontaneum* and related species and genera. *Journal of the American Society of Sugar Cane Technologists*, 26: 101–115.

Arruda P. 2012. Genetically modified sugarcane for bioenergy generation. *Current Opinion in Biotechnology*, 23: 315–322.

Azhaguvel P, Saraswathi DV, Sharma A, Varshney RK. 2006. Methodological advancement in molecular markers to delimit the gene(s) for crop improvement. In: JA Teixeira da Silva (ed.), *Floriculture, ornamental and plant biotechnology*, vol. 1, pp. 460–469. Isleworth, UK: Global Science Books.

Babu C, Kodalingam K, Natarajan US, Govindaraj P, Shanthi RM. 2010. Assessment of genetic diversity detected by co-ancestry of crosses and STMS markers among sugarcane genotypes (*Saccharum* spp. hybrids). *Sugarcane International*, 28(5): 206–215.

Banumathi G, Krishnasamy V, Maheswaran M, Samiyappan R, Govindaraj P, Kumaravadivel N. 2010. Genetic diversity analysis of sugarcane (*Saccharum* spp.) clones using simple sequence repeat markers of sugarcane and rice. *Electronic Journal of Plant Breeding*, 1(4): 517–526.

Besse P, McIntyre CL, Berding N. 1997. Characterisation of *Erianthus* sect. *Ripidium Saccharum* germplasm (Andropogoneae-Saccharinae) using RFLP markers. *Euphytica*, 93: 283–292.

Besse P, Taylor G, Carroll B, Berding N, Burner D, McIntyre CL. 1998. Assessing genetic diversity in a sugarcane germplasm collection using an automated AFLP analysis. *Genetica*, 104: 143–153.

Botha FC, Black KG. 2000. Sucrose phosphate synthase and sucrose synthase activity during maturation of internodal tissue in sugarcane. *Functional Plant Biology*, 27: 81–85.

Botstein D, White RL, Skolnick M, Davis RW. 1980. Construction of a genetic map in man using restriction fragment length polymorphisms. *American Journal of Human Genetics*, 32: 314–331.

Brasileiro BP, Marinho CD, Costa PMA, Peternelli LA, Resende MDV, Cursi DE, Hoffmann HP, Barbosa MHP. 2014. Genetic diversity and coefficient of parentage between clones and sugarcane varieties in Brazil. *Genetics and Molecular Research*, 13(4): 9005–9018.

Brown JS, Schnell RJ, Power EJ, Douglas SL, Kuhn DN. 2007. Analysis of clonal germplasm from five *Saccharum* species: *S. barberi, S. robustum, S. officinarum, S. sinense* and *S. spontaneum*. A study of inter- and intra species relationships using microsatellite markers. *Genetic Resources and Crop Evolution*, 54: 627–648.

Bull TA, Glasziou KT. 1963. The evolutionary significance of sugar accumulation in *Saccharum*. *Australian Journal of Biological Science*, 16: 737–742.

Burnquist WL, Sorrells ME, Tanksley SD. 1995. Characterization of genetic variability in *Saccharum* germplasm by means of restriction fragment length polymorphism (RFLP) analysis. *Proceedings of the International Society of Sugarcane Technology*, 21: 355–365.

Burner DM, Pan YB, Webster RD 1997. Genetic diversity of North American and Old World *Saccharum* assessed by RAPD analysis. *Genetic Resources and Crop Evolution*, 44: 235–240.

Cai Q, Fan YH, Aitken K, Piperidis G, McIntyre CL, Jackson P. 2005a. Assessment of the phylogenetic relationships with in the '*Saccharum complex*' using AFLP markers. *Acta Agronomica Sinica*, 31(5): 551–559.

Cai Q, Aitken K, Fan YH, Piperidis G, Jackson P, McIntyre CL. 2005b. A preliminary assessment of the genetic relationship between *Erianthus rockii* and the '*Saccharum complex*' using microsatellite (SSR) and AFLP markers. *Plant Science*, 169: 976–984.

Ceron A, Angel F. 2001. Genetic diversity in sugarcane hybrids in Colombia measured using molecular markers. In: *Proceedings of the 24th ISSCT Congress, Molecular Biology Workshop*, 16–21 September, Brisbane, Australia, p. 626.

Chandra A, Jain R, Solomon S, Shrivastava S, Roy AK. 2013. Exploiting EST data-bases for the development and characterisation of 3425 gene-tagged CISP markers in biofuel crop sugarcane and their transferability in cereals and orphan tropical grasses. *BMC Research Notes*, 6: 47–57.

Chandra A, Grisham MP, Pan YB. 2014. Allelic divergence and cultivar-specific SSR alleles revealed by capillary electrophoresis using fluorescence labeled SSR markers in sugarcane. *Genome*, 57: 363–372.

Chang D, Yang FY, Yan JJ, Wu YQ, Bai SQ, Liang XZ, Zhang YW, Gan YM. 2012. SRAP analysis of genetic diversity of nine native populations of wild sug-arcane, *Saccharum spontaneum*, from Sichuan, China. *Genetics and Molecular Research*, 11: 1245–1253.

Chen H, Fan YH, Shi X, Cai Q, Zhang M, Zhang YP. 2001. Research on genetic diversity and systemic evolution in *Saccharum spontaneum* L. *Acta Agronomica Sinica*, 27(5): 645–652.

Chen H, Fan YH, Xiang YJG, Cai Q, Zhang YP. 2003. Phylogenetic relationships of *Saccharum* and related species inferred from sequence analysis of the nrDNA ITS region. *Acta Agronomica Sinica*, 29(3): 379–385.

Chen PH, Pan YB, Chen RK, Xu LP, Chen YQ. 2009. SSR marker-based analysis of genetic relatedness among sugarcane cultivars (*Saccharum* spp. hybrids) from breeding programs in China and other countries. *Sugar Tech*, 11(4): 347–354.

Chakravarthi BK, Naravaneni R. 2006. SSR marker based DNA fingerprinting and diversity study in rice (*Oryza sativa* L.). *African Journal of Biotechnology*, 5(9): 684–688.

Christiansen MJ, Andersen SB, Ortiz R. 2002. Diversity changes in an intensively bred wheat germplasm during the 20th century. *Molecular Breeding*, 9: 1–11.

Collard BCY, Jahufer MZZ, Brouwer JB, Pang ECK. 2005. An introduction to mark-ers, quantitative trait loci (QTL) mapping and marker assisted selection for crop improvement: The basic concepts. *Euphytica*, 142: 169–96.

Collard BCY, Mackill DJ. 2009. Start codon targeted (SCoT) polymorphism: A sim-ple, novel DNA marker technique for generating gene-targeted markers in plants. *Plant Molecular Biology Reporter*, 27: 86–93.

Commodity Research Bureau. 2015. *The 2015 CRB commodity yearbook*. Chicago, IL: Commodity Research Bureau.

Comstock JC, Glaz B, Edme SJ, Davidson RW, Gilbert RA, Glynn NC, Miller JD, Tai PYP. 2009. Registration of 'CP 00-1446' Sugarcane. *Journal Plant Registrations*, 3(1): 28–34.

Cooper H, Spillane C, Hodgkin T, Cooper H 2001. Broadening the genetic base of crops: An overview. In: *Broadening the genetic base of crop production*, HD Copper, C Spillane, T Hodgkin (eds.), pp. 1–23. Wallingford, UK: CABI Publishing.

Cordeiro GM, Henry RJ. 1999. Microsatellite markers as an important tool in the genetic analysis of sugarcane (*Saccharum* spp.). Paper presented to the Plant and Animal Genome VII Conference, San Diego, California, pp. 17–21.

Cordeiro GM, Taylor GO, Henry RJ. 2000. Characterisation of microsatellite mark-ers from sugarcane (*Saccharum* sp.) a highly polyploid species. *Plant Science*, 155(2): 161–168.

Cordeiro GM, Casu R, McIntyre CL, Manners JM, Henry RJ. 2001. Microsatellite markers from sugarcane (*Saccharum* spp.) ESTs cross transferable to *Erianthus* and *Sorghum*. *Plant Science*, 160: 1115–1123.

Cordeiro GM, Pan YB, Henry RJ. 2003. Sugarcane microsatellites for the assessment of genetic diversity in sugarcane germplasm. *Plant Science*, 165: 181–189.

Cordeiro GM, Eliott F, McIntyre CL, Casu RE, Henry RJ. 2006. Characterization of single nucleotide polymorphisms in sugarcane ESTs. *Theoretical and Applied Genetics*, 113(2): 331–343.

Coto O, Cornide MT, Calvo D, Canales, E, D'Hont A, de Prada F. 2002. Genetic diversity among wild sugarcane germplasm from Laos revealed with markers. *Euphytica*, 123: 121–130.

Creste S, Sansoli D, Tardiani A, Silva D, Goncalves F, Favero T et al. 2010a. Comparison of AFLP, TRAP and SSRs in the estimation of genetic relationships in sugarcane. *Sugar Tech*, 12: 150–154.

Creste S, Accoroni KAG, Pinto LR, Vencovsky R, Gimenes MA, Xavier MA, Landell MGA. 2010b. Genetic variability among sugarcane genotypes based on polymorphisms in sucrose metabolism and drought tolerance genes. *Euphytica*, 172: 435–446.

Da Costa MLM, Amorim LLB, Onofre AVC, Tavares de Melo LJO, de Oliveira MBM, de Carvalho R, Benko-Iseppon AM. 2011. Assessment of genetic diversity in contrasting sugarcane varieties using inter-simple sequence repeat (ISSR) markers. *American Journal of Plant Sciences*, 2: 425–432.

Daniels J, Daniels CA. 1975. Geographical, historical and cultural aspects of the origin of the Indian and Chinese sugarcanes *S. barberi* and *S. sinense*. *Sugarcane Breeding Newsletters*, 36: 4–23.

Daniels J, Roach BT. 1987. Taxonomy and evolution. In: *Developments in crop science*, JH Don (ed.), 7–84. Amsterdam: Elsevier.

Da Silva JAG, Sorrells ME, Burnquist WL, Tanksley SD. 1993. RFLP linkage map and genome analysis of *Saccharum spontaneum*. *Genome*, 36(4): 782–791.

Da Silva CM, Mangolin CA, Mott AS, Machado MFPS. 2008. Genetic diversity associated with *in vitro* and conventional bud propagation of *Saccharum* varieties using RAPD analysis. *Plant Breeding*, 127: 160–165.

Da Silva JAG 2001. Preliminary analysis of microsatellite markers derived from sugarcane expressed sequence tags (ESTs). *Genetics and Molecular Biology*, 24 (1–4): 155–159.

Davey JW, Hohenhole PA, Etter PD, Boone JO, Catchen JM, Blaxter ML. 2011. Genome-wide genetic marker discovery and genotyping using next generation sequencing. *Nature Reviews Genetics*, 12: 499–510.

Devarumath RM, Kalwade SB, Bundock P, Eliott FG, Henry R. 2013. Independent target region amplification polymorphism and single-nucleotide polymorphism marker utility in genetic evaluation of sugarcane genotypes. *Plant Breeding*, 132: 736–747.

Devarumath RM, Kalwade SB, Kawar PG, Sushir KV. 2012. Assessment of genetic diversity in sugarcane germplasm using ISSR and SSR markers. *Sugar Tech*, 14: 334–344.

D'Hont A, Lu YH, Feldmann P, Glaszmann, JC. 1993. Cytoplasmic diversity in sugar cane revealed by heterologous probes. *Sugar Cane*, 1: 12–15.

D'Hont A, Lu YH, Le'on DGD, Grivet L, Feldmann P, Panaud E, Glaszmann JC. 1994. A molecular approach to unraveling the genetics of sugarcane, a complex polyploid of Andropogonaea tribe. *Genome*, 37: 222–230.

D'Hont A, Rao PS, Feldmann P, Grivet L, IslamFaridi N, Taylor P, Glaszmann JC. 1995. Identification and characterization of sugarcane intergeneric hybrids, *Saccharum officinarum×Erianthus arundinaceus*, with molecular markers and DNA *in situ* hybridization. *Theoretical and Applied Genetics*, 91(2): 320–326.

D'Hont A, Grivet L, Feldmann P, Rao PS, Berding N, Glauszmann JC. 1996. Characterisation of the double genome structure of modern sugarcane cultivars (*Saccharum* spp.) by molecular cytogenetics. *Molecular Genetics and Genomics*, 250: 404–413.

D'Hont A, Ison D, Alix K, Roux C, Glaszmann JC. 1998. Determination of basic chromosome numbers in the genus *Saccharum* by physical mapping of ribosomal RNA genes. *Genome*, 41: 221–225.

Dillon SL, Shapter FM, Henry RJ, Cordeiro G, Izquierdo L, Lee LS. 2007. Domestication to crop improvement: Genetic resources for sorghum and *Saccharum* (Andropogoneae). *Annals of Botany*, 100: 975–989.

Diola V, Barbosa MHP, Veiga CFM, Fernandes EC. 2014. Molecular markers EST-SSRs for genotype-phenotype association in sugarcane. *Sugar Tech*, 16(3): 241–249.

Duarte LSC, Silva PP, Santos JM, Barbosa GVS, Ramalho CE, Soares L, Andrade JCF, Almeida C. 2010. Genetic similarity among genotypes of sugarcane estimated by SSR and coefficient of parentage. *Sugar Tech*, 12(2): 145–149.

Edme SJ, Miller JD, Glaz B, Tai PYP, Comstock JC. 2005. Genetic contribution to yield in the Florida sugarcane industry across 33 years. *Crop Science*, 45: 92–97.

Fahmy EM, Abd El-Gawad NM, El-Geddawy IH, Saleh OM, El-Azab NM. 2008. Development of RAPD and ISSR markers for drought tolerance in sugarcane (*Saccharum officinarum* L.). *Egyptian Journal of Genetics and Cytology*, 37(1): 1–15.

Fan LN, Deng HH, Luo QW, He HY, Li Y, Wang QN et al. 2013. Genetic diversity of *Saccharum spontaneum* from geographical regions of China assessed by simple sequence repeats. *Genetics and Molecular Research*, 12 (4): 5916–5925.

Food and Agricultural Organization of the United Nations (FAO). 2013. www.fao .org.

FAOSTAT. 2015. Statistical Division: Production domain – Crops. Updated on November 24, 2015; retrieved on 18 December 2015.

Filho LSCD, Silva PP, Santos JM. 2010. Genetic similarity among genotypes of sugarcane estimated by SSR and coefficient of parentage. *Sugar Tech*, 12(2): 145–149.

Gilbert RA, Comstock JC, Glaz B, Edme SJ, Davidson RW, Glynn NC, Miller JD, Tai PYP. 2008. Registration of 'CP 00-1101' Sugarcane. *Journal Plant Registrations*, 2(2): 95–101.

Glaszmann JC, Lu YH, Lanaud C. 1990. Variation of nuclear ribosomal DNA in sugarcane. *Journal of Genetics and Breeding*, 44(3): 191–198.

Glaz B, Edme SJ, Davidson RW, Gilbert RA, Comstock JC, Glynn NC, Miller JD, Tai PYP. 2009. Registration of 'CP 00-2180' Sugarcane. *Journal Plant Registrations*, 3(1): 35–41.

Glynn NC, McCorkle K, Comstock JC. 2009. Diversity among mainland USA sugarcane cultivars examined by SSR genotyping. *Journal of American Society for Sugar Cane Technology*, 29: 36–52.

Govindaraj P, Natarajan US, Balasundaram N, Premachandran MN, Sharma TR, Koundal KR, Singh NK. 2005. Development of new microsatellite markers for the identification of interspecific hybrids in sugarcane. *Sugarcane International*, 23: 30–38.

Govindaraj P, Sindhu R, Balamurugan A, Appunu C. 2011. Molecular diversity in sugarcane hybrids (*Saccharum* spp. complex) grown in peninsular and east coast zones of tropical India. *Sugar Tech*, 13(3): 206–213.

Govindaraj P, Ramesh R, Appunu C, Swapna S, Priji PJ. 2012. DNA fingerprinting of subtropical sugarcane (*Saccharum* spp.) genotypes using sequences tagged microsatellites sites (STMS) markers. *Plant Archives*, 12: 347–352.

Gupta PK, Varshney RK, Sharma PC, Ramesh B. 1999. Molecular markers and their applications in wheat breeding. *Plant Breeding*, 118: 369–390.

Gupta PK, Rustgi S, Mir RR. 2008. Array-based high-throughput DNA markers for crop improvement. *Heredity*, 101: 5–18.

Hack SM, Huckett BI, Butterfield MK. 2002. Application of microsatellite analysis to the screening of putative parents of sugarcane cross Aa40. *Proceedings of South African Sugarcane Technologists Association*, p. 76.

Hajjar R, Jarvis DI, Gemmill-Herren B. 2008. The utility of crop genetic diversity in maintaining ecosystem services. *Agriculture Ecosystems and Environment*, 123(4): 261–270.

Hale AL, Dufrene EO, Tew TL, Pan YB, Viator RP, White PM et al. 2013. Registration of 'Ho 02-113' Sugarcane. *Journal of Plant Registrations*, 7(1): 51–57.

Hameed U, Pan YB, Muhammad K, Afghan S, Iqbal J. 2012. Use of simple sequence repeat markers for DNA fingerprinting and diversity analysis of sugarcane (*Saccharum* spp) cultivars resistant and susceptible to red rot. *Genetics and Molecular Research*, 11(2): 1195–1204.

Harvey M, Huckett B, Botha FC. 1994. Use of the polymerase chain reaction (PCR) and random amplification of polymorphic DNAs (RAPDs) for the determination of genetic distances between 21 sugarcane varieties. *Proceedings of the South African Sugar Technologists' Association*, pp. 36–40.

Harvey M, Botha FC. 1996. Use of PCR-based methodologies for the determination of DNA diversity between *Saccharum* varieties. *Euphytica*, 89: 257–265.

Heller-Uszynska K, Uszynski G, Huttner E, Evers M, Carlig J, Caig V et al. 2011. Diversity arrays technology (DArT) effectively reveals DNA polymorphism in a large and complex genome of sugarcane. *Molecular Breeding*, 28(1): 37–55.

Hemaprabha G, Govindaraj P, Balasundaram N, Singh NK. 2005. Genetic diversity analysis of Indian sugarcane breeding pool based on sugarcane specific STMS markers. *Sugar Tech*, 7: 9–14.

Hemaprabha G, Govindaraj P, Singh NK. 2006b. STMS markers for fingerprinting of varieties and genotypes of sugarcane (*Saccharum* spp.) *Indian Journal of Genetics and Plant Breeding*, 66(2): 95–99.

Hemaprabha G, Krishna A, Vincy J, Priji P, Simon S, Govindaraj P. 2010. DNA fingerprinting for identification and protection of elite sugarcane (*Saccharum* spp.) varieties. *Electronic Journal of Plant Breeding*, 1(4): 420–425.

Hemaprabha G, Lavanya L, Priji PJ, Krishna A, Vincy J, Simon S, Govindaraj P. 2011. Molecular fingerprinting of commercial varieties and elite clones of sugarcane (Saccharum sp.) using STMS markers. In: *Proceedings of 4th IAPSIT International Sugar Conference*, 21–25 November, pp. 624–629.

Hemaprabha G, Natarajan US, Balasundaram N, Singh NK. 2006a. STMS based genetic divergence among commercial parents and its use in identifying productive cross combinations for varietal evolution in sugarcane (*Saccharum* spp.) *Sugar Cane International*, 22(6): 22–27.

Hemaprabha G, Priji PJ, Padmanabhan TSS. 2013. Molecular fingerprinting of recently notified sugarcane (Saccharum officinarum L.) varieties using STMS markers. *Journal of Sugarcane Research*, 3(2): 107–117.

Hemaprabha G, Rangasamy SR. 2001. Genetic similarity among five species of Saccharum based on isozyme and RAPD markers. *Indian Journal of Genetics and Plant Breeding*, 61: 341–347.

Hemaprabha G, Simon S. 2012. Genetic diversity and selection among drought tolerant genotypes of sugarcane using microsatellite markers. *Sugar Tech*, 14(4): 327–333.

Hodkinson TR, Chase MW, Lledo MD, Salamin N, Renvoize SA. 2002. Phylogenetics of *Miscanthus, Saccharum* and related genera (*Saccharinae, Andropogoneae, Poaceae*) based on DNA sequences from ITS nuclear ribosomal DNA and plastid trnL intron and trnL-F intergenic spacers. *Journal of Plant Research*, 115(5): 381–392.

Hoshino AA, Bravo JP, Nobile PM, Morelli KA. 2012. Microsatellites as Tools for Genetic Diversity Analysis. In: *Genetic diversity in microorganisms*, Mahmut Caliskan (ed.), 149–170. In Tech.

Hu J, Vick BBA. 2003. Target region amplification polymorphism: A novel marker technique for plant genotyping. *Plant Molecular Biology Reporter*, 21: 289–294.

Huang XD. 2009. DNA profiling and genetic diversity of sugarcane germplasms by SSR loci. M.S. thesis, Fujian Agriculture and Forestry University, Fuzhou, China.

Huckett BI, Botha FC. 1995. Stability and potential use of RAPD markers in a sugarcane genealogy. *Euphytica*, 86(2): 117–125.

International Union for the Protection of New Varieties of Plants (UPOV). 2005. Guidelines for molecular marker selection and database construction, BMT guidelines (Proj. 3). UPOV, Geneva.

Ismail RM. 2013. Evaluation of genetically modified sugarcane lines carrying Cry1AC gene using molecular marker techniques. *GM Crops and Food*, 4(1): 58–66.

Jackson PA. 2005. Breeding for improved sugar content in sugarcane. *Field Crops Research*, 92: 277–290.

Jannoo N, Grivet L, Seguin M, Paulet F, Domaingue R, Rao PS, Dookun A, D'Hont A, Glaszmann JC. 1999. Molecular investigation of the genetic base of sugarcane cultivars. *Theoretical and Applied Genetics*, 99(1): 171–184.

Jaradat AA, Shahid M, Al-Maskri A. 2004. Genetic diversity in the Batini barley landrace from Oman: II. Response to salinity stress. *Crop Science*, 44: 997–1007.

Jimenez HR, Cabrera FAV, Padron IS. 2008. Molecular characterization of Columbian wild cane accessions with AFLP. *Acta Agronomica (Palmira)*, 57(4): 227–231.

Jonah PM, Bello LL, Lucky O, Midau A, Moruppa SM. 2011. The importance of molecular markers in plant breeding programmes. *Global Journal of Science Frontier Research*, 11(5): 1–9.

Joshi SV, Albertse EH. 2013. Development of a DNA fingerprinting database and cultivar identification in sugarcane using a genetic analyzer. *Proceedings of South African Sugarcane Technologists Association*, 86: 200–212.

Kalwade SB, Devarumath RM, Kawar PG, Sushir KV. 2012. Genetic profiling of sugarcane genotypes using inter simple sequence repeat (ISSR) markers. *Electronic Journal of Plant Breeding*, 3(1): 621–628.

Kalwade SB, Devarumath RM. 2014. Single strand conformation polymorphism of genomic and EST-SSRs marker and its utility in genetic evaluation of sugarcane. *Physiology and Molecular Biology of Plants*, 20(3): 313–321.

Kantety RV, La Rota M, Matthews DE, Sorrells ME. 2002. Data mining for simple sequence repeats in expressed sequence tags from barley, maize, rice, sorghum and wheat. *Plant Molecular Biology*, 48: 501–10.

Kawar PG, Devarumath RM, Nerkar Y 2009. Use of RAPD markers for assessment of genetic diversity in sugarcane cutivars. *Indian Journal of Biotechnology*, 8(1): 67–71.

Khaled KAM, El-Demardash IS, Amer EAM. 2015. Genetic polymorphism among some sugarcane germplasm collections as revealed by RAPD and ISSR analyses. *Life Science Journal*, 12(3): 159–167.

Khaled KAM, Fathalla MAK. 2010. Molecular profiling and genetic diversity in sugarcane using RAPD. *Dynamic Biochemistry, Process Biotechnology and Molecular Biology*, 4(1): 104–106.

Khaled KAM. 2010. Molecular characterization and genetic similarity of three sugarcane genotypes. *Egyptian Journal of Genetics and Cytology*, 39: 229–257.

Khaled SKM, Abdel-Tawab FM, Fahmy EM, Khaled KAM. 2011. Marker-assisted selection associated with sugar content in sugarcane (*Saccharum Spp.*) In: *Proceedings of The Third International Conference of Genetic Engineering and its Applications, Sharm EL Sheikh City, South Sinai, Egypt, 5–8 October*, pp. 113–129.

Khan FA, Khan A, Azhar FM, Rauf S. 2009. Genetic diversity of *Saccharum officinarum* accessions in Pakistan as revealed by random amplified polymorphic DNA. *Genetics and Molecular Research*, 8(4): 1376–1382.

Khan IA, Bibi S, Yasmin S, Khatri A, Seema N. 2013. Phenotypic and genotypic diversity investigations in sugarcane for drought tolerance and sucrose content. *Pakistan Journal of Botany*, 45(2): 359–366.

Khan IA, Bibi S, Yasmin S, Seema N, Khatri A, Siddiqui MA, Nizamani GS, Afghan S. 2011. Identification of elite sugarcane clones through TRAP. *Pakistan Journal of Botany*, 43(1): 261–269.

Khan IA, Dahot MU, Khatri A. 2007. Study of genetic variability in sugarcane induced through mutation breeding. *Pakistan Journal of Botany*, 39(5): 1489–1501.

Khan MS, Yadav S, Srivastava S, Swapna M, Chandra A, Singh RK. 2011. Development and utilization of conserved intron scanning marker in sugarcane. *Australian Journal of Botany*, 59(1): 38–45.

Kumar LS. 1999. DNA markers in plant improvement: An overview. *Biotechnology Advances*, 17: 143–182.

Kumar P, Gupta VK, Misra AK, Modi DR, Pandey BK. 2009. Potential of molecular markers in plant biotechnology. *Plant Omics Journal*, 2(4): 141–162.

Kurane J, Shinde V, Harsulkar A. 2009. Application of ISSR marker in pharmacognosy: Current update. *Pharmacognosy Review*, 3(6): 216–228.

Lam E, Shine Jr J, Da Silva J, Lawton M, Bonos S, Calvino M et al. 2009. Improving sugarcane for biofuel: Engineering for an even better feedstock. *Global Change Biology Bioenergy*, 1: 251–255.

Landjeva S, Korzun V, Ganeva G. 2006. Evaluation of genetic diversity among Bulgarian winter wheat (*Triticum aestivum* L.) varieties during the period 1925–2003 using microsatellites. *Genetic Resources and Crop Evolution*, 53: 1605–1614.

Lao FY, Liu R, He HY, Deng HH, Chen ZH, Chen JW, Fu C, Qi YW, Zhang CM. 2009b. Analysis of genetic similarity among commonly-used sugarcane parents and between two parents of cross. *Molecular Plant Breeding*, 7: 505–512.

Lao FY, Liu R, He HY, Deng HH, Chen ZH, Chen JW, Fu C, Zhang CM, Yang YH. 2008. AFLP analysis of genetic diversity in series sugarcane parents developed at HSBS. *Molecular Plant Breeding*, 6: 517–522.

Lao FY, Liu R, He HY, Deng HH, Li QW, Chen ZH, Chen JW, Fu C, Qi YW, Zhang CM. 2009a. Genetic diversity analysis of sugarcane parents with AFLP in China. *Genomics and Genome Applied Biology*, 28(3): 503–508.

Lapitan VC, Brar DS, Abe T, Edilberto DR. 2007. Assessment of genetic diversity of Philippine rice cultivars carrying good quality traits using SSR markers. *Breeding Science*, 57(4): 263–270.

Lateef DD. 2015. DNA marker technologies in plants and applications for crop improvements. *Journal of Biosciences and Medicines*, 3: 7–18.

Latorre C, Rea R, Sosa D, Molina S, Demey J, Briceno R, De Sousa O. 2013. Genetic diversity in sugarcane germplasm *Saccharum* spp using ISSR markers. *Multiciencias*, 13(1): 7–15.

Lavanya DL, Hemaprabha G. 2010a. Analysis of genetic diversity among high sucrose genotypes of sugarcane (*Saccharum* spp.) derived from CoC 671 using sugarcane specific microsatellite markers. *Electronic Journal of Plant Breeding*, 1(4): 399–406.

Lavanya DL, Hemaprabha G. 2010b. Genetic diversity within sucrose rich parental pool of sugarcane and its application in sugarcane breeding through hybridization and selection. *Indian Journal of Genetics and Plant Breeding*, 70(2): 172–181.

Lebot V. 1999. Biomolecular evidence for plant domestication in Sahul. *Genetic Resources and Crop Evolution*, 46: 619–628.

Le Cunff L, Garsmeur O, Raboin LM, Pauquet J, Telismart H, Selvi A et al. 2008. Diploid/polyploid syntenic shuttle mapping and haplotype-specific chromosome walking toward a rust resistance gene (*Bru1*) in highly polyploid sugarcane. *Genetics*, 180: 649–660.

Leon F, Prodriguez J, Azurdia C, Amador D, Queme J, Melgar M. 2001. Use of RAPD markers to detect genetic variability among 39 sugarcane varieties. *Proceedings of the 24th ISSCT Congress, Molecular Biology Workshop, 16–21 September, Brisbane, Australia*, pp. 634–636.

Li G, Quiros CF. 2001. Sequence-related amplified polymorphism (SRAP), a new marker system based on a simple PCR reaction: Its application to mapping and gene tagging in *Brassica. Theoretical and Applied Genetics*, 103: 455–461.

Liang J, You JH, Wu KC, Fang FX, Liao JX, Mo LX, Srivastava M, Li YR. 2008. The assessment of genetic relationships amongst sugarcane cultivars using SSR markers. In: *Meeting the Challenging of Sugar Crop and Integrated Industries in Developing Countries*, Al-Arish, Egypt, pp. 278–282.

Liang J, Pan YB, Li YR, Fang FX, Wu KC, You JH. 2010. Genetic diversity assessment of *Saccharum* species and elite cultivars from China using SSR markers. *Guangxi Zhiwu/Guihaia*, 30(5): 594–600.

Lima MLA, Garcia AAF, Oliveira KM, Matsuoka S, Arizono H, Souza CL, Souza AP. 2002. Analysis of genetic similarity detected by AFLP and coeffecient of parentage among genotypes of sugar cane (*Saccharum* spp.) *Theoretical and Applied Genetics*, 104: 30–38.

Litt M, Luty JA. 1989. A hypervariable microsatellite revealed by *in vitro* amplification of a dinucleotide repeat within the cardiac muscle actin gene. *American Journal of Human Genetics*, 44: 397–401.

Liu P, Que Y, Pan YB. 2011. Highly polymorphic microsatellite DNA markers for sugarcane germplasm evaluation and variety identity testing. *Sugar Tech*, 13(2): 129–136.

Liu X, Li X, Liu H, Xu C, Lin X, Li C, Deng Z. 2016. Phylogenetic analysis of different ploidy *Saccharum spontaneum* based on rDNA-ITS sequences. *PLoS ONE*, 11(3): e0151524.

Liu XL, Li XJ, Xu CH, Lin XQ, Deng ZH. 2016. Genetic diversity of populations of *Saccharum spontaneum* ith different ploidy levels using SSR molecular markers. *Sugar Tech*, 18(4): 365–372.

Liu XL, Su HS, Ma L, Lu X, Ying XM, Cai Q, Fan, YH. 2010. Phylogenetic relationships of sugarcane related genera and species based on ITS sequences of nuclear ribosomal DNA. *Acta Agronomica Sinica*, 36(11): 1853–1863.

Loomis RS, Williams WA. 1963. Maximum crop productivity: An estimate. *Crop Science*, 3(1): 67–72.

Long LX, Li M, Kuan CX, Mei YX, Qing C, Yong LJ, Wen WC. 2010. Establishment of DNA fingerprinting identity for sugarcane cultivars in Yunnan, China. *Acta Agronomica Sinica*, 36(2): 202–210.

Lu YH, D'Hont A, Walker DIT, Rao PS, Feldman P, Glaszmann JC. 1994. Relationships among ancestral species of sugarcane revealed by RFLP using single-copy maize nuclear probes. *Euphytica*, 78: 7–18.

Lu X, Zhou H, Pan YB, Chen CY, Zhu JR, Chen PH, Li YR, Cai Q, Chen RK. 2015. Segregation analysis of microsatellite (SSR) markers in sugarcane polyploids. *Genetics and Molecular Research*, 14(4): 18384–18395.

Maccheroni W, Jordao H, De Gaspari R, De Moura GL, Matsuoka S. 2009. Development of a dependable microsatellite based fingerprinting system for sugarcane. *Sugarcane International*, 27(2): 47–52.

Mahmud S, Azad AK, Rahman MM, Mahmud S, Yasmin M, Ali R, Hossain A. 2014. Assessment of genetic variability of sugarcane varieties and somaclones based on RAPD markers. *JNSST*, 8(1): 1–9.

Manifesto MM, Schlatter AR, Hopp HE, Suarez EY, Dubcovsky J. 2001. Quantitative evaluation of genetic diversity in wheat germplasm using molecular markers.*Crop Science*, 41: 682–690.

Manigbas NL, Villegas LC. 2004. Microsatellite markers in hybridity tests to identify true hybrids of sugarcane. *Philippine Journal of Crop Science*, 29(2): 23–32.

Maranho RC, Augusto R, Mangolin CA, Machado MFPS. 2014. Use of differential levels of mean observed heterozygosity in microsatellite loci of commercial varieties of sugarcane (*Saccharum* spp.) *Genetics and Molecular Research*, 13(4): 10130–10141.

Marconi TG, Costa EA, Miranda HR, Mancini MC, Cardoso-Silva CB, Oliveira KM, Pinto LR, Mollinari M, Garcia AAF, Souza AP. 2011. Functional markers for gene mapping and genetic diversity studies in sugarcane. *BMC Research Notes,* 4: 264.

Markad NR, Kale AA, Pawar BD, Jadhav AS, Patil SC. 2014. Molecular characterization of sugarcane (*Saccharum officinarum* L.) genotypes in relation to salt tolerance. *The Bioscan,* 9(4): 1785–1788.

Mary S, Nair NV, Chaturvedi PK, Selvi A. 2006. Analysis of genetic diversity among *Saccharum spontaneum* L. from four geographical regions of India, using molecular markers. *Genetic Resources and Crop Evolution,* 53(6): 1221–1231.

Matsuoka S, Araras BR, Maccheroni WJR, Campinas BR. 2010. Microsatellite based fingerprinting system for *Saccharum* complex. *Patent application publication,* Pub. No. US 2010/0021916 AI.

Matus IA, Hayes PM. 2002. Genetic diversity in three groups of barley germplasm assessed by simple sequence repeats. *Genome,* 45: 1095–1106.

Milligan SB, Davidson RW, Edme SJ, Comstock JC, Hu CJ, Holder DG, Glaz B, Glynn NC, Gilbert RA. 2009. Registration of 'CPCL 97-2730' Sugarcane. *Journal Plant Registrations,* 3(2): 158–164.

Ming R, Liu SC, Lin YR, Da Silva J, Wilson W, Braga D. 1998. Detailed alignment of *Saccharum* and *Sorghum* chromosomes: Comparative organization of closely related diploid and polyploid genomes. *Genetics,* 150: 1663–1682.

Ming R, Liu SC, Moore PH, Irvine JE, Paterson AH. 2001. QTL analysis in a complex autopolyploid: Genetic control of sugar content in sugarcane. *Genomic Research,* 11: 2075–2084.

Ming R, Moore PH, Wu KK, D Hont A, Glaszmann JC, Tew TL. 2006. Sugarcane improvement through breeding and biotechnology. *Plant Breeding Reviews,* 27: 15–118.

Ming R, Wang YW, Draye X, Moore PH, Irvine JE, Paterson AH. 2002. Molecular dissection of complex traits in autopolyploids: Mapping QTLs affecting sugar yield and related traits in sugarcane. *Theoretical and Applied Genetics,* 105: 332–345.

Mittal N, Dubey AK. 2009. Microsatellite markers – A new practice of DNA based markers in molecular genetics. *Pharmacognosy Reviews,* 3(6): 235–246.

Mohammadi SA, Prasanna BM. 2003. Analysis of genetic diversity in crop plants – salient statistical tools and considerations. *Crop Science,* 43: 1235–1248.

Mohan M, Nair S, Bhagwat A, Krishna TG, Yano M, Bhatia CK, Sasaki T. 1997. Genome mapping, molecular markers and marker-assisted selection in crop plants. *Molecular Breeding,* 3: 87–103.

Mukherjee SK. 1957. Origin and distribution of *Saccharum. Botanical Gazette,* 119: 55–61.

Mumtaz AS, Nayab DE, Iqbal MJ, Shinwari ZK. 2011. Probing genetic diversity to characterized rot resistance in sugarcane. *Pakistan Journal of Botany,* 43(5): 2513–2517.

Nair NV, Nair S, Sreenivasan TV, Mohan M. 1999. Analysis of genetic diversity and phylogeny in *Saccharum* related genera using RAPD markers. *Genetic Resources and Crop Evolution,* 46: 73–79.

Nair NV, Selvi A, Sreenivasan TV, Pushpalatha KN. 2002. Molecular diversity in Indian sugarcane varieties as revealed by randomly amplified DNA polymorphisms. *Euphytica,* 127: 219–225.

Nawaz S, Khan FA, Tabasum S, Iqbal MZ, Saeed A. 2010a. Genetic studies of "noble cane" for identification and exploitation of genetic markers. *Genetics and Molecular Research*, 9(2): 1011–1022.

Nawaz S, Khan FA, Tabasum S, Zakria M, Saeed A, Iqbal MZ. 2010b. Phylogenetic relationships among *Saccharum* clones in Pakistan revealed by RAPD markers. *Genetics and Molecular Research*, 9(3): 1673–1682.

Nayak SN, Song J, Villa A, Pathak B, Ayala-Silva T, Yang X et al. 2014. Promoting utilization of (*Saccharum* spp.) genetic resources through genetic diversity analysis and core collection construction. *PLoS One*, 9(10): e110856.

Oliveira KM, Pinto LR, Marconi TG, Mollinari M, Ulian EC, Chabregas SM, Falco MC, Burnquist W, Garcia AAF, Souza AP. 2009. Characterization of new polymorphic functional markers for sugarcane. *Genome*, 52: 191–209.

Orita M, Suzuki Y, Sekya T, Hayashi K. 1989. Rapid and sensitive detection of point mutations and DNA polymorphisms using the polymerase chain reaction. *Genomics*, 5: 874–879.

Oropeza M, Guevara P, Garcia E, Ramirez JL. 1995. Identification of sugarcane (*Saccharum* spp.) somaclonal variants resistant to sugarcane mosaic virus *via* RAPD markers. *Plant Molecular Biology Reporter*, 13(2): 182–191.

Pan YB. 2006. Highly polymorphic microsatellite DNA markers for sugarcane germplasm evaluation and variety identity testing. *Sugar Tech*, 8(4): 246–256.

Pan YB. 2010. Databasing molecular identities of sugarcane (*Saccharum* spp.) clones constructed with microsatellite (SSR) DNA markers. *American Journal of Plant Sciences*, 1(2): 87–94.

Pan YB, Burner DM, Legendre BL, Grisham MP, White WH. 2004. An assessment of the genetic diversity within a collection of *Saccharum spontaneum* with RAPD-PCR. *Genetic Resources and Crop Evolution*, 51: 895–903.

Pan YB, Burner DM, Legendre BL, Grisham MP, White WH. 2005. An assessment of the genetic diversity within a collection of *Saccharum spontaneum* L. with RAPD-PCR. *Genetic Resources and Crop Evolution*, 51(8): 895–903.

Pan YB, Burner DM, Legendre BL. 2000. An assessment of the phylogenetic relationship among sugarcane and related taxa based on the nucleotide sequence of 5S rRNA intergenic spacers. *Genetica*, 108(3): 285–295.

Pan YB, Burner DM, Wei Q. 2001. Developing species-specific DNA markers to assist in sugarcane breeding. *Proceedings of International Society Sugar Cane Technology*, 24: 337–342.

Pan YB, Cordeiro G, Richards E, Henry R. 2003b. Molecular genotyping of sugarcane clones with microsatellite DNA markers. *Maydica*, 48: 319–329.

Pan YB, Cordeiro GM, Henry RJ, Schnell RJ. 2009. Microsatellite fingerprints of Louisiana sugarcane varieties and breeding lines. *Plant Breeding and Genetics*, Id: 47845624.

Pan YB, Miller JD, Schnell RJ, Richard Jr. EP, Wei Q. 2003a. Application of microsatellite and RAPD fingerprints in the Florida sugarcane variety program. *Sugarcane International*, 19–28.

Pan YB, Scheffler BS, Richard JE. 2007. High throughput molecular genotyping of commercial sugarcane clones with microsatellite (SSR) DNA markers. *Sugar Tech*, 9(2 and 3): 176–181.

Pandey A, Mishra RK, Mishra S, Singh YP, Pathak S. 2011. Assessment of genetic diversity among sugarcane cultivars (*Saccharum officinarum* L.) using simple sequence repeats markers. *Online Journal of Biological Sciences*, 11(4): 105–111.

Panje RR, Babu CN. 1960. Studies in saccharum spontaneum distribution and geographical association of chromosome numbers. *Cytologia*, 25: 152–72.

Paran I, Michelmore RW. 1993. Development of reliable PCR-based markers linked to downy mildew resistance genes in lettuce. *Theoretical and Applied Genetics*, 85: 985–993.

Paterson A, Tanksley S, Sorrel ME. 1991. DNA markers in plant improvement. *Advances in Agronomy*, 44: 39–90.

Perera M, Arias M, Costilla D, Luque A, Garcia MB, Diaz Romero C et al. 2012. Genetic diversity assessment and genotype identification in sugarcane based on DNA markers and morphological traits. *Euphytica*, 185: 491–510.

Parida SK, Kalia SK, Dalal KS, Hemprapha V, Selvi G, Pandit A et al. 2009. Informative genomic microsatellite markers for efficient genotyping application in sugarcane. *Theoretical and Applied Genetics*, 118(2): 327–338.

Parida SK, Pandit A, Gaikwad K, Sharma TR, Srivastava PS, Singh NK, Mohapatra T. 2010. Functionally relevant microsatellites in sugarcane unigenes. *BMC Plant Biology*, 10: 251.

Park JW, Solis-Gracia N, Trevino C, da Silva JA. 2012. Exploitation of conserved intron scanning as a tool for molecular marker development in the Saccharum complex. *Molecular Breeding*, 30: 987–999.

Pinto LR, Oliveira KM, Ulian EC, Garcia AA, de Souza AP. 2004. Survey in the sugarcane expressed sequence tag database (SUCEST) for simple sequence repeats. *Genome*, 47: 795–804.

Pinto LR, Oliveira KM, Marconi T, Garcia AAF, Ulian EC, de Souza AP. 2006. Characterization of novel sugarcane expressed sequence tag microsatellites and their comparison with genomic SSRs. *Plant Breeding*, 125: 378–384.

Piperidis G. 2003. Progress towards evaluation of SSR as a tool for sugarcane variety identification. ISSCT IV Molecular Biology Workshop, Montpellier, France.

Piperidis G, Christopher MJ, Carroll BJ, Berding N, D'Hont A. 2000. Molecular contribution to selection of intergeneric hybrids between sugarcane and the wild species *Erianthus arundinaceus*. *Genome*, 43(6): 1033–1037.

Piperidis G, Piperidis N, D'Hont A. 2010. Molecular cytogenetic investigation of chromosome composition and transmission in sugarcane. *Molecular Genetics and Genomics*, 284: 65–73.

Piperidis G, Rattey AR, Taylor GO, Cox MC. 2004. DNA markers: A tool for identifying sugarcane varieties. In: *Proceeding of Australian Society of Sugar Cane Technologists*, May, Brisbane, Australia, vol. 26, pp. 634–642.

Piperidis G, Taylor GO, Smith GR. 2001. A microsatellite marker database for fingerprinting sugarcane clones. *XXIV Proceedings of International Society of Sugar Cane Technology*, 632–633.

Poczai P, Varga I, Bell NE, Hyvonen J. 2012. Genomics meets biodiversity: Advances in molecular marker development and their applications in plant genetic diversity assessment. In: *The molecular basis of plant genetic diversity*, Mahmut Caliskan (ed.), 3–32. In Tech.

Qi YW, Lao FY, Zhang CM, Fan LN, He HY, Liu SM, Li QW, Deng HH. 2011. Comparative analysis of genetic diversity of Chinese and American sugarcane (*Saccharum* spp.) using SSR markers. *Chinese Journal of Tropical Crops*, 32(1): 99–104.

Qi Y, Pan Y, Lao F, Zhang C, Fan L, He H et al. 2012. Genetic structure and diversity of parental cultivars involved in china mainland sugarcane breeding programs as inferred from DNA microsatellites. *Journal of Integrative Agriculture*, 11(11): 1794–1803.

Que YX, Chen TS, Xu LP, Chen RK. 2009. Genetic diversity among key sugarcane clones revealed by TRAP markers. *Journal of Agricultural Biotechnology*, 17(3): 496–503.

Que YX, Pan YB, Lu YH, Yang C, Yang YT, Huang N, Xu L. 2014. Genetic analysis of diversity within a Chinese local sugarcane germplasm based on start codon targeted polymorphism (SCoT). *BioMed Research International*, 1–10.

Queme JL, Molina L, Melgar M. 2005. Analysis of genetic similarity among 48 sugarcane varieties using microsatellite DNA sequences. *Proceedings of International Society for Sugarcane Technology*, 25: 592–595.

Rao VP, Chaudhary R, Singh S, Sengar RS, Sharma V. 2014. Assessment of genetic diversity analysis in contrasting sugarcane varieties using random polymorphic DNA (RAPD) markers. *African Journal of Biotechnology*, 13(37): 3736–3741.

Raj P, Selvi A, Prathima PT, Nair NV. 2016. Analysis of genetic diversity of *Saccharum* complex using chloroplast microsatellite markers. *Sugar Tech*, 18(2): 141–148.

Riasco JJ, Victoria JI, Angel F. 2003. Genetic diversity among sugarcane (*Saccharum* spp.) varieties using molecular markers. *Revista Colombiana de Biotecnologia*, 5(1): 6–15.

Roach BT. 1972. Nobilization sugarcane. *Proceedings of the International Society of Sugarcane Technologist*, 14: 206–16.

Roach BT. 1989. Origin and improvement of the genetic base of sugarcane. *Proceedings of the Australian Society of Sugar Cane Technology*, 34–47.

Rodriguez AMH, Castillo MAC, Flores EPB 2005. Genetic diversity of the most important sugar cane cultivars in Mexico. e-*Gnosis*, 3(1): 1–10.

Santos JM, Duarte Filho LSC, Soriano ML, Silva PP. 2012. Genetic diversity of the main progenitors of sugarcane from the RIDESA germplasm bank using SSR markers. *Industrial Crops and Products*, 40: 145–150.

Saravanakumar K, Govindaraj P, Appunu C, Senthilkumar S, Kuma R. 2014. Analysis of genetic diversity in high biomass producing sugarcane hybrids (*Saccharum* spp. complex) using RAPD and STMS markers. *Indian Journal of Biotechnology*, 214–220.

Sarid Ullah SM, Amzad Hossain M, Musharaf Hossain M, Barman S, Sohag MMH, Prodhan SH. 2013. Genetic diversity analysis of chewing sugarcane (*Saccharum officinarum* L.) varieties by using RAPD markers. *Journal of Bioscience and Biotechnology*, 2(2): 145–150.

Schenck S, Crepeau MW, Wu KK, Moore PH, Yu Q, Ming R. 2004. Genetic diversity and relationships in native Hawaiian *Saccharum officinarum* sugarcane. *Journal of Heredity*, 95(4): 327–331.

Seema N, Khan IA, Raza S, Yasmeen S, Bibi S, Nizamani GS. 2014. Assessment of genetic variability in somaclonal population of sugarcane. *Pakistan Journal of Botany*, 46(6): 2107–2111.

Selvi A, Mukunthan N, Shanthi RM, Govindaraj P, Singaravelu B, Prabu, Karthik T. 2008. Assessment of genetic relationships and marker identification in sugarcane cultivars with different levels of top borer resistance. *Sugar Tech*, 10(1): 53–59.

Selvi A, Nair NV, Balasundaram N, Mohapatra T. 2003. Evolution of maize microsatellite markers for genetic diversity analysis and fingerprinting in sugarcane. *Genome*, 46: 394–403.

Selvi A, Nair NV, Noyer JL, Singh NK, Balasundaram N, Bansal KC, Koundal KR, Mohapatra T. 2005. Genomic constitution and genetic relationship among the tropical and subtropical Indian sugarcane cultivars revealed by AFLP. *Crop Science*, 45: 1750–1757.

Selvi A, Nair NV, Singh JL, Balasundaram NK, Bansal KC, Koundal KR, Mohapatra T. 2006. AFLP analysis of the phenetic organization and genetic diversity in the sugarcane complex, *Saccharum* and *Erianthus*. *Genetic Resources and Crop Evolution*, 53(4): 831–842.

Sharma MD, Dobhal U, Singh P, Kumar S, Gaur AK, Singh SP, Jeena AS, Koshy EP, Kumar S. (2014). Assessment of genetic diversity among sugarcane cultivars using novel microsatellite markers. *African Journal of Biotechnology*, 13(13): 1444–1451.

Shahid MTH, Khan FA, Saeed A, Aslam M, Rasul F. 2012. Development of somaclones in sugarcane genotype BF-162 and assessment of variability by random amplified polymorphic DNA (RAPD) and simple sequence repeats (SSR) markers in selected red rot resistant somaclones. *African Journal of Biotechnology*, 11(15): 3502–3513.

Sindhu R, Govindaraj P, Balamurugan A, Appunu C. 2011. Genetic diversity in sugarcane hybrids (*Saccharum* spp. complex) grown in tropical India based on STMS markers. *Journal of Plant Biochemistry and Biotechnology*, 20(1): 118–124.

Silva DC, de Souza MCP, Filho LSCD, dos Santos JM, Barbosa GVS, Almeida C. 2012a. New polymorphic EST-SSR markers in sugarcane. *Sugar Tech*, 14(4): 357–363.

Silva DC, Filho LSCD, dos Santos JM, Barbosa GVS, Almeida C 2012b. DNA fingerprinting based on simple sequence repeat (SSR) markers in sugarcane clones from the breeding program RIDESA. *African Journal of Biotechnology*, 11(21): 4722–4728.

Singh PK, Srivastava S, Gupta PS, Singh J, Jain R, Lal P. 2008. Identification of genetically diverse *Saccharum spontaneum* genotypes from India using SSR-SSCP and RAPD markers. In: *Meeting the Challenging of sugar crop and Integrated Industries in Developing Countries*, Al-Arish, Egypt, pp. 273–277.

Singh RB, Singh B, Singh RK. 2015. Development of microsatellite (SSRs) markers and evaluation of genetic variability within sugarcane commercial varieties (*Saccharum* spp. hybrids). *International Journal of Advanced Research*, 3(12): 700–708.

Singh RB, Srivastava S, Rastogi J, Gupta GN, Tiwari NN, Singh B, Singh RK. 2014b. Molecular markers exploited in crop improvement practices. *Research in Environment and Life Sciences*, 7(4): 223–232.

Singh RB, Srivastava S, Verma AK, Singh B, Singh RK. 2014a. Importance and progresses of microsatellite markers in Sugarcane (*Saccharum* spp. hybrids) *Indian Journal of Sugarcane Technology*, 29(1): 1–12.

Singh RK, Jena SN, Khan S, Yadav S, Banarjee N, Raghuvanshi S et al. 2013b. Development, cross-species/genera transferability of novel EST-SSR markers and their utility in revealing population structure and genetic diversity in sugarcane. *Gene*, 524: 309–329.

Singh RK, Khan MS, Singh R, Pandey DK, Kumar S, Lal S. 2011a. Analysis of genetic differentiation and phylogenetic relationships among sugarcane genotypes differing in response to red rot. *Sugar Tech*, 13(2): 137–144.

Singh RK, Singh RB, Singh SP, Mishra N, Rastogi J, Sharma ML, Kumar A. 2013a. Genetic diversity among *Saccharum spontaneum* clones and commercial hybrids through SSR markers. *Sugar Tech*, 15(2): 109–115.

Singh RK, Singh RB, Singh SP, Sharma ML. 2011b. Identification of sugarcane microsatellites associated to sugar content in sugarcane and transferability to other cereal genomes. *Euphytica*, 182: 335–354.

Singh RK, Singh RB, Singh SP, Sharma ML. 2012. Genes tagging and molecular diversity of red rot susceptible/tolerant sugarcane hybrids using c-DNA and unigene derived markers. *World Journal of Microbiology Biotechnology*, 8: 1669–79.

Singh RK, Singh SP, Tiwari DK, Srivastava S, Singh SB, Sharma ML, Singh R, Mohopatra T, Singh NK. 2013c. Genetic mapping and QTL analysis for sugar yield-related traits in sugarcane. *Euphytica*, 191: 333–353.

Singh RK, Sushil KM, Sujeet PS, Neha M, Sharma ML. 2010. Evaluation of sugarcane microsatellite markers for genetic diversity analysis among sugarcane species and commercial hybrids. *Australian Journal of Crop Science*, 4(2): 116–25.

Soller M, Beckmann JS. 1983. Genetic polymorphism in varietal identification and genetic improvement. *Theoretical and Applied Genetics*, 67: 25–33.

Song XX. 2009. Genetic diversity of sugarcane and its relatives by SRAP and TRAP markers. M.S. thesis, Fujian Agriculture and Forestry University, Fuzhou, China.

Sosa D, Rea R, Latorre C, Molina S, Demey J, Briceno R, De Sousa-Vieira O. 2015. Genetic diversity in sugarcane germplasm using TRAP markers based on sucrose-related genes. *Revista de la Faculated de Agronomia*, 32: 175–190.

Souza GM, Berges H, Bocs S, Casu R, D'Hont A, Ferreira JE et al. 2011. The sugarcane genome challenge: Strategies for sequencing a highly complex genome. *Tropical Plant Biology*, 4(3 and 4): 145–156.

Sreenivasan TV, Ahloowalia BS, Heinz DJ. 1987. Cytogenetics. In: *Sugarcane improvement through breeding*, Heinz DJ (ed.), 211–253. Amsterdam: Elsevier.

Srivastava MK, Li CN, Li YR. 2012. Development of sequence characterized amplified region (SCAR) marker for identifying drought tolerant sugarcane genotypes. *Australian Journal of Crop Science*, 6(4): 763–767.

Srivastava S, Gupta PS. 2008. Inter simple sequence repeat profile as a genetic marker system in sugarcane. *Sugar Tech*, 10(1): 48–52.

Srivastava S, Gupta PS, Singh PK, Singh J, Jain R. 2011. Genetic diversity analysis of *Saccharum spontaneum* germplasm using SSR-SSCP and RAPD markers. *Indian Journal of Agricultural Science*, 81: 914–920.

Srivastava S, Jain R, Gupta PS, Singh J. 2005. Analysis of genetic fidelity in micropropagated plants of sugarcane using SSR-SSCP assay. *Indian Journal of Genetics and Plant Breeding*, 65(4): 327–328.

Stevenson GC. 1965. *Genetic and breeding of sugarcane*. London: Longmans, Green.

Suman A, Kimbeng CA, Edme SJ, Veremis J. 2008. Sequence related amplified polymorphism (SRAP) markers for assessing genetic relationships and diversity in sugarcane germplasm collections. *Plant Genetic Resources: Characterization and Utilization*, 6(3): 222–231.

Suman A, Kazim, AJ Arro, Parco A, Kimbeng C, Baisakh N. 2012. Molecular diversity among members of the *Saccharum* complex assessed using TRAP markers based on lignin related genes. *Bioenergy Research*, 5(1): 197–205.

Swapna M, Sivaraju K, Sharma RK, Singh NK, Mohapatra T. 2011a. Single-strand conformational polymorphism of EST-SSRs: A potential tool for diversity analysis and varietal identification in sugarcane. *Plant Molecular Biology Reporter*, 29(3): 505–513.

Swapna M, Srivastava S. 2012. Molecular marker applications for improving sugar content in sugarcane. *Springer Briefs in Plant Science*, 1–49.

Swapna M, Srivastava S, Pandey DK. 2011b. Targeting genes linked to sugar content for quality improvement: A molecular marker approach in sugarcane. *Proceedings of X Agricultural Science Congress*, 8–10 February, National Bureau of Fish Genetic Resources (NBFGR), Lucknow, p. 214.

Tabasum S, Khan FA, Nawaz S, Iqbal MZ, Saeed A. 2010. DNA profiling of sugarcane genotypes using randomly amplified polymorphic DNA. *Genetics and Molecular Research*, 9(1): 471–483.

Takahashi S, Furukawa T, Asano T, Terajima Y. 2005. Very close relationship of the chloroplast genomes among *Saccharum* species. *Theoretical and Applied Genetics*, 110: 1523–1529.

Tena E, Mekbib F, Ayana A. 2014. Analysis of genetic diversity and population structure among exotic sugarcane spp. cultivars in Ethiopia using simple sequence repeats (SSR) molecular markers. *African Journal of Biotechnology*, 13(46): 4308–4319.

Tew TL, Dufrene EO, White WH, Cobill RM, Legendre BL, Grisham MP, Garrison DD, Pan YB, Richard EP, Miller JD. 2011. Registration of 'HoCP 91-552' High-Fiber Sugarcane. *Journal of Plant Registrations*, 5(2): 181–190.

Tew TL, Pan Y. 2005. Molecular assessment of the fidelity of sugarcane crosses with high throughput microsatellite genotyping. *Journal of the American Society of Sugar Cane Technologists*, 25: 119.

Tew TL, Pan YB. 2010. Microsatellite (simple sequence repeat) marker–based paternity analysis of a seven-parent sugarcane polycross. *Crop Science*, 50(4): 1401–1408.

Tew TL, White WH, Grisham MP, Dufrene EO, Garrison DD, Pan YB, Richard EP, Legendre BL, Miller JD. 2009. Registration of 'HoCP 00-950' sugarcane. *Journal Plant Registrations*, 3(1): 42–50.

Tew TL, White WH, Legendre BL, Grisham MP, Dufrene EO, Garrison DD, Veremis JC, Pan YB, Richard EP, Miller JD. 2005. Registration of 'HoCP 96-540' sugarcane. *Crop Science*, 45: 785–786.

Thokoane LN, Butterfield MK, Harvey M, Huckett B. 1999. Progress towards a fingertyping database for sugarcane varieties. *Proceedings of South African Sugarcane Technologists Association*, 73: 164–166.

Trethowan RM, Kazi AM. 2008. Novel germplasm resources for improving environmental stress tolerance of hexaploid wheat. *Crop Science*, 48: 1255–1265.

Ubayasena WLC, Perera ALT. 1999. Assessment of genetic diversity within wild sugarcane germplasm using randomly amplified polymorphic DNA techniques (RAPD). *Tropical Agricultural Research*, 11: 110–122.

Ukoskit K, Thipmongkolcharoen P, Chatwachirawong P. 2012. Novel expressed sequence tag-simple sequence repeats (EST-SSR) markers characterized by new bioinformatics criteria reveal high genetic similarity in sugarcane (*Saccharum* spp.) breeding lines. *African Journal of Biotechnology*, 11(6): 1337–1363.

Varshney RK, Graner A, Sorrells ME. 2005. Genic microsatellite markers in plants: Features and applications. *Trends in Biotechnology*, 23: 48–55.

Virupakshi S, Naik GR. 2008. ISSR Analysis of chloroplast and mitochondrial genome can indicate the diversity in sugarcane genotypes for red rot resistance. *Sugar Tech*, 10(1): 65–70.

Vos P, Hogers R, Bleeker M, Reijans M, Lee van de T, Hornes M, Frijters A, Pot J, Peleman J, Kuiper M, Zabeau M. 1995. AFLP: A new technique for DNA fingerprinting. *Nucleic Acids Research*, 23: 4407–4414.

Waclawovsky AJ, Sato PM, Lembke CG, Moore PH, Souza GM. 2010. Sugarcane for bioenergy production: An assessment of yield and regulation of sucrose content. *Plant Biotechnology Journal*, 8(3): 263–276.

Wang Y, Zhuang NS, Gao HG, Huang DY. 2007. ISSR analysis for sugarcane germplasm. *Journal of Hunan Agricultural University*, 2007–S1.

Wang ML, Barkley NA, Jenkins TM. 2009. Microsatellite markers in plants and insects. Part I: Applications of biotechnology. *Genes, Genomes and Genomics*, 3(1): 54–67.

White WH, Cobill RM, Tew TL, Burner DM, Grisham MP, Dufrene EO, Pan YB, Richard JEP, Legendre BL. 2011. Registration of 'Ho 00-961' Sugarcane. *Journal of Plant Registrations*, 5(3): 332–338.

Williams JGK, Kubelik AR, Livak KJ, Rafalski JA, Tingey SV. 1990. DNA polymorphisms amplified by arbitrary primers are useful as genetic markers. *Nucleic Acids Research*, 18: 6531–6535.

Winter P, Kahl G. 1995. Molecular marker technologies for plant improvement. *World Journal of Microbiology and Biotechnology*, 11(4): 438–448.

Wu JM, Li YR, Yang LT, Fang FX, Song HZ, Tang HQ, Wang M, Weng ML. 2013. cDNA-SCoT: A novel rapid method for analysis of gene differential expression in sugarcane and other plants. *Australian Journal of Crop Science*, 7(5): 659–664.

Xu JS, Xu LP, Zhang MQ, Chen RK. 2003. Phylogenetic relationships of *Saccharum* to its relative genus based on RAPD analysis. *Acta Agriculturae Universitatis Jiangxiensis*, 25(6): 925–928.

Yachun S, Hui L, Hengbo W, Youxiong Q, Qibin W, Shanshan C, Liping X. 2012. Optimization of SCoT-PCR reaction system, and screening and utilization of polymorphic primers in sugarcane. *Chinese Journal of Applied Environmental Biology*, 18(5): 810–818.

Yang CF, Yang LT, Li YR, Zhang GM, Zhang CY, Wang WZ. 2016. Sequence characteristics and phylogenetic implications of the nrDNA internal transcribed spacers (ITS) in protospecies and landraces of sugarcane (*Saccharum officinarum* L.). *Sugar Tech*, 18(1): 8–15.

You Q, Pan YB, Xu L, Gao S, Wang Q, Su Y, Yang Y, Wu Q, Zhou D, Que Y. 2016. Genetic diversity analysis of sugarcane germplasm based on fluorescence labeled simple sequence repeat markers and a capillary electrophoresis based genotyping platform. *Sugar Tech*, 18(4): 380–390.

You Q, Xu LP, Zheng YF, Que YX. 2013. Genetic diversity analysis of sugarcane parents in Chinese breeding programmes using gSSR markers. *The Scientific World Journal*, 11 pages. http://dx.doi.org/10.1155/2013/613062.

Yu AL, Zhang MQ, Chen RK. 2002. Applicability of inter simple sequence repeat polymorphisms in sugarcane and its related genera as DNA markers. *Journal of Fujian Agricultural and Forestry University*, 31(4): 484–489.

Zeng HZ, Zheng CM, Zhu W, Gao HQ. 2003. RAPD analysis on the relationship and parental specific markers among sugarcane germplasm. *Journal of Plant Genetic Resources,* 4(2): 99–103.

Zhang G, Li Y, He W, He H, Liu X, Song H, Liu H, Zhu R, Fang W. 2010. Analysis of the genetic diversity in *Saccharum spontaneum* L. accessions from Guangxi Province of China with RAPD-PCR. *Sugar Tech,* 12(1): 31–35.

Zhang HY, He LL, Zhong HQ, Li FS, He SC, Yang QH. 2009. Identification of inter-generic hybrids between (*Saccharum* spp.) and *Erianthus fulvus* with ITSs. *African Journal of Biotechnology,* 8(9): 1841–1845.

Zhang HY, Li FS, He IL, Zhong HQ, Yang QH. 2008a. Identification of sugar-cane interspecies hybrids with RAPDs. *African Journal of Biotechnology,* 7(18): 1072–1074.

Zhang HY, Li FS, Liu XZ, He LL, Yang QH, He SC. 2008b. Analysis of genetic variation in *Erianthus arundinacues* by random amplified polymorphic DNA markers. *African Journal of Biotechnology,* 7(19): 3414–3418.

Zhang L, Cai R, Yuan M, Tao A, Xu J, Lin L, Fang P, Qi J. 2015. Genetic diversity genetic diversity and DNA fingerprinting in jute (*Corchorus* spp.) based on SSR markers. *The Crop Journal,* 3: 416–422.

Zhang MQ, Zheng XF, Yu AL, Xu JS, Zhou H. 2004. Molecular marker application in sugarcane. *Sugar Tech,* 6(4): 251–259.

Zhang YC, Kuang M, Yang WH, Xu HX, Zhou DY, Wang YQ, Feng XA, Su C, Wang F. 2013. Construction of a primary DNA fingerprint database for cotton cul-tivars. *Genetics and Molecular Research,* 12(2): 1897–1906.

Zhuang NS, Zheng CM, Huang DY, Tang YQ, Gao HQ. 2005. AFLP analysis for sugarcane germplasms. *Acta Agronomica Sinica,* 31: 444–450.

Zhu JR, Zhou H, Pan YB, Lu X. 2014. Genetic variability among the chloroplast genomes of sugarcane (*Saccharum* spp.) and its wild progenitor species *Saccharum spontaneum* L. *Genetics and Molecular Research,* 13(2): 3037–3047.

Zietkiewicz E, Rafalski A, Labuda D. 1994. Genome fingerprinting by simple sequence repeat (SSR)-anchored polymerase chain reaction amplification. *Genomics,* 20: 176–183.

chapter nine

Prospects of non-conventional approaches for sugarcane improvement

Gulzar S. Sanghera and Rajinder Kumar

Contents

Sugarcane (*Saccharum* spp. hybrid) belongs to the family Poaceae and is a major agricultural crop. It is an important crop in the tropical and subtropical parts of the world where it is grown mainly for sugar and ethanol, playing a significant role in the world economy. Recently, it has become a major industrial cash crop, having potential to be a key crop in bio-factory evolution as it produces high yield of valuable products like sugar, biofibres, waxes, bioplastic and biofuel (Lakshmanan et al., 2005, 2006; McQualter et al., 2004b; Nonato et al., 2001). Sugar from sugarcane is produced in 58 countries around the globe. Sugarcane is highly heterogeneous and generally multiplies vegetatively by stem cutting. Large genome size, high polyaneuploidy, low fertility, complex environmental interactions, slow breeding advances and nobilization hinder the breeding for this crop. The initial breakthrough in sugarcane breeding occurred a century ago when Dutch plant breeders established an innovative breeding and selection program to incorporate the wild *Saccharum spontaneum* L. genes, controlling biotic and abiotic stresses in the sugar producing germplasm of *S. officinarum* L. (Moore, 2005). These initial hybrids paved the way for many modern cultivars (Roach, 1989) through conventional breeding which resulted in the improvement of productivity of the crop (Berding et al., 1997; Hogarth et al., 1997). Consequently, the lack of a suitable multiplication procedure has long been serious problem in sugarcane breeding programmes (Tiwari et al., 2010). Some of the most vexing problems faced in sugarcane cultivation are attributed to low cane and sugar yields, which involve development of cultivars endowed with resistance/tolerance to drought, salinity, insect pests, fungal diseases and herbicides as major constraints (Khaliq et al., 2005). With the rapidly growing demand for sugar and ethanol worldwide, the productivity of sugarcane needs considerable enhancement. The major objectives to improve sugarcane crops using conventional and biotechnological techniques include (1) high cane yield potential, (2) early maturity, (3) resistance to lodging, (4) resistance to stress environments, (5) disease resistance, (6) insect resistance and (7) juice quality components.

Plant biotechnology has made significant strides in the past two decades or so, encompassing within its folds the spectacular developments in the plant genetic engineering. Nowadays, genetically engineered crops appear as the most recent technological advances to help boost food production, mainly by addressing the production constraints with minimum costs and environmental pollution (due to the indiscriminate use of pesticides and herbicides) (Baker and Preston, 2003). The high level of genetic complexity in sugarcane creates challenges in the application of both conventional and molecular breeding to the genetic improvement of sugarcane as a sugar and energy crop. Recent technology developments indicate the potential to greatly increase our understanding of the sugarcane plant by application of emerging genomic technologies. This chapter outlines some of the biotechnological developments that are in place and tailored to address important issues related to sugarcane improvement.

In vitro culture and genetic transformation in sugarcane

For the last two decades, plant breeders have been using genetic engineering techniques for transmission of noble genes into crops to improve plant characteristics. Genetic engineering deals with introduction of foreign genes into plant genome through cells, protoplasts or tissues for the production of transgenic plants that exhibit normal physiological and biological functions (Jenes et al., 1993). In vitro culture methods in sugarcane have had a great impact both on basic research and applied commercial interest. These include micropropagation of elite clones, production of disease-free planting material, generation of agronomically superior somaclones, screening methods for biotic and abiotic stress tolerance, and conservation of novel and useful germplasm. *In vitro* techniques for the mass propagation of healthy sugarcane plantlets via direct and indirect regeneration pathways are well established and are critical in numerous ongoing efforts to improve sugarcane germplasm through genetic engineering (Snyman et al., 2011).

Somatic embryogenesis has been useful for propagating large numbers of uniform plants in less time, for obtaining virus-resistant plants through somaclonal variation, mutagenesis and *in vitro* selection, and developing transgenic plants (Suprasanna and Bapat, 2006; Suprasanna et al., 2007). The *in vitro* conservation methods can be useful in order to obviate problems of viability during storage for extended periods, manpower requirements and need for large growth facilities. These methods can also facilitate maintenance of elite lines, transgenic material and mutant effect analysis still in their field establishment and/or approval.

Sugarcane is the most suitable candidate for genetic engineering because of its complex polyploidy nature, variable fertility and genotype versus environment interactions. The availability of a high-frequency *in vitro* regeneration system from various explants makes this crop a suitable candidate for genetic manipulation. Several genes (for disease/pest resistance, salt and drought tolerance, and sugar accumulation) targeted towards sugarcane improvement have been introduced into sugarcane (Altpeter and Oraby, 2010; Hotta et al., 2011; Table 9.1). The success of transgenic sugarcane plant production depends on the method used for transformation, the target tissue/explants and tissue culture regeneration system used. Various explant types (e.g. axillary buds, apical meristems, immature inflorescences, leaf segments) have been successfully used to regenerate full plants in sugarcane indicating that a wide range of totipotent target tissues are available for genetic transformation.

Transgenic plant production requires selectable marker genes that enable the selection of transformed cells, tissue and plants. The most routinely practiced are those that exhibit resistance to antibiotics or herbicides. Since there are perceived risks in the deployment of transgenic plants containing these markers, alternate selection systems referred to as positive selection and marker-free systems have become useful (Suprasanna et al. 2010). In sugarcane, Jain et al. (2007) used mannose for the selection of embryogenic callus and found that increased mannose improved the overall transformation efficiency by reducing the number of selection escapes.

Sugarcane cultivars differ in their capacity to accumulate sucrose, and breeding programmes routinely perform crosses to identify genotypes able to produce more sucrose. In this regard, transgenic approaches to manipulate native genes that influence metabolism may have significant application. Most of the field trials of transgenic sugarcane are related to

Table 9.1 Genetic engineering of sugarcane for different traits

Trait	Gene	Transformation method	Reference
Reporter and selection system			
Neopmycin phosphotransferase	npt-II	Microprojectile	Bower and Birch, 1992
B-Glucuronidase	uid-A	Microprojectile	Bower and Birch, 1992
Hygromycin phosphotransferase	hpt	Agrobacterium	Arencibia et al., 1998
Green fluorescent protein	gfp	Agrobacterium	Elliott et al., 1998
Phosphinothricin acetyl transferase	bar	Agrobacterium	Manickavasagam et al., 2004
Phosphomannose isomerase	manA	Microprojectile	Jain et al., 2007
Herbicide resistance			
Bialophos	*bar*	Microprojectile	Gallo-Meagher and Irvin, 1996
Phosphinothricine	*bar*	Agrobacterium	Enriquez-Obregon et al., 1998
Glufosinate ammonium	*pat*	Microprojectile	Leibbrandt and Snyman, 2003
Disease resistance			
SCMV	SCMV-CP	Microprojectile	Joyce et al., 1998
Sugarcane leaf scald	*albD*	Microprojectile	Zhang and Birch, 1999
SrMV	SrMV-CP	Microprojectile	Ingelbrecht et al., 1999
Puccinia melanocephala Glucanase	*Chitanase* and *ap24*	Agrobacterium	Enriquez et al., 2000
SCYLV	SCYLV-CP	Microprojectile	Gilbert et al., 2005
Fiji leaf gall	FDVS9 ORF 1	Microprojectile	McQualter et al., 2004a
Pest resistance			
Sugarcane stem borer	*cry1A*	Microprojectile	Arencibia et al., 1999
Sugarcane stem borer	*cry1Ab*	Microprojectile	Braga et al., 2003
Sugarcane stem borer	*cry1Ab*	Microprojectile	Arvinth et al., 2010

(Continued)

Table 9.1 (Continued) Genetic engineering of sugarcane for different traits

Trait	Gene	Transformation method	Reference
Sugarcane stem borer	cry1Aa3	Agrobacterium	Kalunke et al., 2009
Proceras venosatus	Modified cry1Ac	Microprojectile	Weng et al., 2010
Sugarcane canegrub	gna	Microprojectile	Legaspi and Mirkov, 2000
Mexican rice borer	gna	Microprojectile	Setamou et al., 2002
Ceratovacuna lanigera	gna	Agrobacterium	Zhangsun et al., 2007
Scirpophaga excerptalis	Aprotinin	Microprojectile	Christy et al., 2009
Metabolic engineering/Alternative products			
Sucrose accumulation	Antisense soluble acid invertase	Microprojectile	Ma et al., 2004
Fructo oligosaccharide	IsdA	Agrobacterium	Enriquez et al., 2000
Polyphenol oxidase	ppo	Microprojectile	Vickers et al., 2005
Polyhydroxybutyrate	phaA, phaB, phaC	Microprojectile	Brumbley et al., 2007
p-Hydroxybenzoic acid	hch1 and cp1	Microprojectile	McQualter et al., 2004b
Mannose	manA	Microprojectile	Jain et al., 2007
Isomaltulose	SI	Microprojectile	Wu and Birch, 2007
Proline overproduction	P5CS	Microprojectile	Molinari et al., 2008

the transgene expression for agronomic traits and are being undertaken in Brazil and Australia. The agronomic traits like height, diameter and the number of stalks, fibre content, disease resistance, and yield of transgenic clones were not significantly different from that of untransformed sugarcane plants. However, the field trials of insect-resistant transgenic sugarcane revealed some morphological, physiological and phytopathological variations (Arencibia et al., 1999).

Commercial sugarcane cultivars rarely flower or produce seed in the field, and exposure to non-GM sugarcane has not been associated with any reports of allergic responses in Australia (Office of the Gene Technology Regulator, 2011). The report suggests that some measures can be imposed to minimize exposure include harvesting the GM sugarcane plants before flowering or removing flower heads before anthesis.

Variability enhancement through in vitro culture and mutagenesis

In vitro techniques for the mass propagation of healthy sugarcane plantlets via direct and indirect regeneration pathways are well established and are critical in numerous ongoing efforts to improve sugarcane germplasm through genetic engineering (Snyman et al., 2011). In direct morphogenesis, plants are regenerated directly from tissues such as immature leaf roll discs and also from shoot tip culture, by which sugarcane is propagated commercially (Hendre et al., 1983). Indirect morphogenesis involves initial culturing of leaf roll sections or inflorescences on an auxin-containing medium to produce an undifferentiated mass of cells, or callus.

Mutation induction of *in vitro* cultures will require that meristematic cells or tissues, and mitotically active cells are cultured to prepare sufficient material for mutagenic treatment. Intrasomatic competition discriminating mutagen affected cells and causing a loss of their cell progenies may be controlled by modifying *in vitro* conditions (medium composition or some other culture treatments like dessication) resulting in a better competitiveness of mutant cells. In sugarcane, partial desiccation has been used successfully to stimulate and enhance somatic embryo differentiation (Desai et al., 2004) and enhance regeneration response of gamma-irradiated embryogenic callus cultures (Suprasanna et al., 2008b). Partial desiccation treatment can offer as a simple and novel method in stimulating a regeneration response of higher-dose gamma-irradiated cultures.

Somaclonal variation has emerged as an important *in vitro* culture tool for crop improvement. This system has been adopted for improving the quality and production of sugarcane and somaclones for yield, sugar recovery, disease resistance, drought tolerance, and maturity have been isolated. Sugarcane was amongst the first plants where somaclonal variation was reported (Heinz and Mee, 1969; Larkin and Scowcroft, 1981). Physical and chemical mutagens have been applied to *in vitro* cultures so as to enhance the frequency of genetic variation and obtain beneficial modifications in cultivars (Patade and Suprasanna, 2008; Suprasanna 2010). Physical (gamma rays, ion beams) and chemical (ethyl methanesulfonate [EMS], sodium azide and sodium nitrite) mutagens have been used successfully, and their optimum mutagenic treatments have been devised (Ali et al., 2008; Kenganal et al., 2008; Koch et al., 2010; Patade et al., 2008). *In vitro* selection at the cellular level has been successful in isolation of mutants for desirable traits by imposing *in vitro* selection pressure either by incorporating fungal pathotoxins or fungal culture filtrates for selecting disease resistance (Rai et al., 2011) or incorporation of sodium chloride, polyethylene glycol, or mannitol for selecting salt or drought tolerance (Suprasanna et al., 2008a). In sugarcane, somaclonal variant lines resistant to eyespot disease caused by *Helminthosporium sacchari* were selected

(Larkin and Scowcroft, 1983). Various researchers have used mutagenesis and selection to isolate embryogenic cells and plants tolerant to the causal agent of red rot (Ali et al., 2008; Sengar et al., 2009; Singh et al., 2008).

Engineering sugarcane for pest management

Sugarcane is attacked by a range of insects including tissue borers, sucking pests and canegrubs. Losses due to these pests are estimated to be around 10%. Sugarcane pests exhibit a wide variation in species composition and importance in the diverse agroclimatic conditions of tropics and subtropics where the crop is cultivated. Some important tissue borers in sugarcane-growing countries of the world include the sugarcane stem borer *Diatraeasaccharalis* (F.) (Lepidoptera: Crambidae) (Rossato et al., 2010) and the sugarcane giant borer *Telchin licus licus* (Drury) (Lepidoptera: Castniidae) (Craveiro et al., 2010) in the Americas including Brazil; the Mexican rice borer *Eoreuma loftini* (Dyar) (Lepidoptera: Pyralidae) in south Texas (Tomov and Bernal, 2003); the African stem borer *Eldana saccharina* Walker (Lepidoptera: Pyralidae) in South Africa (Way and Goebel, 2007); the Asian spotted stem borer *Chilo sacchariphagus* (Bojer) (Lepidoptera: Pyralidae) in Mauritius, Réunion, Madagascar and Mozambique (Goebel and Way, 2007); *Proceras venosatus* Wlk. (Lepidoptera: Pyralidae) in China (Weng et al., 2006); and the early shoot borer *Chilo infuscatellus* (Snellen) (Lepidoptera: Crambidae) (Srikanth et al., 2009), the internode borer *Chilo sacchariphagus indicus* (Kapur) (Lepidoptera: Crambidae) (Srikanth and Kurup, 2011) and the top borer *Scirpophaga excerptalis* Walker (Lepidoptera: Pyralidae) (Mukunthan and Singaravelu, 2005) in India. The sugar industry throughout the world suffers considerable economic loss due to various insects and diseases, although precise quantification has not been reported for many of the pests.

Although chemical control and integrated pest management are regularly practiced for the control of insect pests, success is often limited due to practical difficulties. The sugarcane canopy and internal habitat pose serious limitations to deployment of chemical control measures. Since larvae of borers enter the plants soon after eclosion, then typically remain on the inner side of the leaf sheath or inside the stem, burrow tunnels and feed on the stem tissues, they are inaccessible to insecticides. Systemic insecticides are also largely ineffective due to poor translocation within the plant. Alternative control measures, including mechanical methods, cultural practices and biological agents, followed under specific situations, are limited by costs and only moderate efficacies. The genetic complexity of sugarcane coupled with the non-availability of resistance genes in the germplasm has made conventional breeding for insect resistance difficult. Advances in genetic transformation technology and knowledge of gene expression have led to rapid progress in genetic engineering of crop plants for protection against insect pests (Romeis et al., 2006). Advantageous use of this technology to produce plants for pest control using different molecules, such as proteinase inhibitors (PI), plant lectins, ribosome inactivating proteins, secondary plant metabolites, delta endotoxins and vegetative insecticidal protein from *Bacillus thuringiensis* (Bt) and related species, either alone or in combination with the Bt genes (Bates et al., 2005), has now been widely recognized. Therefore, engineering insect resistance in the sugarcane plant through expression of such molecules appears to be a realistic approach to mitigate potential damage caused by pests. Engineering crop plants for enhanced resistance to insect pests has been one of the successes of transgenic technology. As a trait, insect resistance, either alone or stacked with herbicide resistance, currently ranks second in terms of the global area occupied by biotech crops during 1996 to 2009 (James, 2009).

Significant progress has been made towards the development of transgenic sugarcane resistant to particularly lepidopteran borers such as *D. saccharalis, E. loftini, C. infuscatellus* and *S. excerptalis* using various cry genes over the last 15 years (Table 9.2). Arencibia et al. (1997) reported the first sugarcane transgenics expressing a truncated *Cry1A(b)* gene coding the active region of Bt δ-endotoxin for resistance to *D. saccharalis*. Screening of the transgenic lines resulted in the selection of five events exhibiting significant protection against the borer in spite of very low expression of the toxin (0.59 to 1.35 ng/mg of soluble leaf protein). The low-level expression was attributed to the use of native truncated *Cry1A(b)* gene that was driven by CaMV 35S promoter, known for its weak expression in monocots. However, a low to medium level of internode infestation was noticed even in the transgenic line expressing the highest amount of *Cry1A(b)* (Arencibia et al., 1997, 1999). In field trials for borer resistance, these selected transgenic lines showed little variation for the majority of the agronomic and industrial traits (Arencibia et al., 1999).

Engineering sugarcane for disease management

Different kinds of plant pathogens, namely fungi, bacteria, virus and phytoplasmas, infect sugarcane. At least 150 diseases were recorded in sugarcane in different countries. In India more than 50 diseases were recorded (Alexander and Viswanathan, 2002). Among them red rot, smut, wilt, sett rot, grassy shoot, ratoon stunting, leaf scald and mosaic are the major diseases seriously affecting sugarcane production. For the past 100 years, new varieties were developed with higher yield with high sugar through breeding approaches. However, combining higher sugar yield with disease resistance and favourable agronomic traits is difficult. Many sugarcane varieties can be cited here. The best example is CoC 671, an early maturing, high sugar variety that revolutionized the sugar industry

Table 9.2 List of sugarcane transgenics with insect resistance

Candidate gene	Target pest	Promoter	Method of gene transfer	Reference
Transgenics with Cry toxins				
cry1A(b)	Diatraea saccharalis	CaMV35S	Electroporation	Arencibia et al., 1997
cry1A(b)	Diatraea saccharalis	Maize PEPC, maize pith	Particle bombardment	Braga et al., 2001, 2003
Synthetic cry1Ac	Proceras venosatus	Maize Ubi-1	Particle bombardment	Weng et al., 2006
cry1Aa3	Scirpophaga excerptalis	CaMV35S	Agrobacterium	Kalunke et al., 2009
cry1Ab	Chilo infuscatellus	Maize Ubi-1	Particle bombardment	Arvinth et al., 2010
Agrobacterium modified cry1Ac	Proceras venosatus	Maize Ubi-1	Particle bombardment	Weng et al., 2010
Transgenics with proteinase inhibitors and lectins				
Potato proteinase inhibitor II, Nicotiana alata proteinase inhibitor, Snowdrop lectin	Antitrogus consanguineous, Dermolepida albohirtum	Maize Ubi-1	Particle bombardment	Nutt et al, 1999, 2001
Snowdrop lectin	Eoreuma loftini, Diatraea saccharalis	Maize Ubi-1	Paint-sprayer delivery	Setamou et al., 2002
Soybean Kunitz trypsin inhibitor (SKTI), soybean Bowman–Birk inhibitor (SBBI)	Diatraea saccharalis	Maize Ubi-1	Particle bombardment	Falco et al., 2003
Snowdrop lectin	Ceratovacuna lanigera	RSs-1, Maize Ubi-1	Agrobacterium	Zhangsun et al., 2007
Fusion Amaranthus viridis agglutinin, and SKTI genes Aprotinin	Diatraea saccharalis, Ceratovacuna lanigera, Scirpophaga excerptalis	Maize Ubi-1, Maize Ubi-1	Agrobacterium, particle bombardment	Deng et al., 2008; Christy et al., 2009

in Peninsular India. However, it could not be sustained due to its suscep-tibility to red rot. Other examples include CoJ 64 and CoC 92061, which were susceptible to red rot; and CoSi 95071, which was resistant to red rot but highly susceptible to smut. Since the genome is complex polyploidy, backcrossing in these varieties may not be useful for the introduction of disease-resistant genes. Disease control via genetic engineering strate-gies may have particular value where resistant germplasm is not avail-able in the target species for breeding, where pathogen adaptation can quickly overcome resistance genes and where there is little information on sources of resistance and bioassays are difficult or very time consum-ing to carry out. Several genes for resistance to diseases have already been isolated and shown to be effective in field trials with transgenic plants of other crop species.

Chitin is a significant component in the cell walls of large groups of fungi except members of oomycetes. It is made up of molecules of N-acetylglucosamine, which are the building blocks linked together by 1,4-β-glycosidic bonds. Chitinases are enzymes that cleave the bond between the C1 and C4 of two consecutive N-acetylglucosamines of chi-tin. Chitinases and other microbial hydrolytics have been shown to be involved in a variety of functions such as cell wall digestion, germina-tion of spores, assimilation of chitin and mycoparasitism. Thus chitin-ases have the potential as effective antifungal agents. Many *Trichoderma* and *Gliocladium* spp. isolates used in biocontrol kill the host by direct hyphal contact, causing the affected cells to collapse or disintegrate; veg-etative hyphae of all species have been found susceptible. Cloning of genes encoding for lytic enzymes, characterizing their products and elu-cidating their individual roles in the mycoparasitic activity has opened the way to improving the biocontrol capacity of these fungi. Also the genes coding for these lytic enzymes were expressed in plants to resist invading pathogens. Cloning of genes encoding for endochitinase and N-acetylglucanase from *Trichoderma* spp. was reported by previous work-ers (Limon et al., 1999).

Lytic enzymes of fluorescent *Pseudomonas* strains and *Trichoderma har-zianum* strain T5 were inhibitory to conidial germination, germ tube elon-gation and hyphal growth red rot pathogen. Electrolytic leakage studies indicated that the lytic enzymes produced by these strains caused cell wall degradation and brought about loss of electrolytes from the myce-lium. The lytic enzymes produced by *T. harzianum* T5 and *P. fluorescens* FP7 showed more antifungal activity (Viswanathan et al., 2003).

Detailed studies were conducted on inactivation of *C. falcatum* toxin by bacterial and fungal antagonists. The results clearly proved that *P. fluorescens* VPT4 and *T. harzianum* T5 have completely arrested bio-logical activity of the toxin and altered spectral properties of the toxin

(Malathi et al., 2002). Further characterization of the inactivating protein revealed it as a 97 kDa protein (Malathi et al., 2011).

In Australia, Smith et al. (1992) initially demonstrated that coat protein constructs SC were expressed in a sugarcane protoplast cell system, and these constructs were subsequently used to transform sugarcane callus. The lines generated from the transformed and selected callus resisted mechanical challenge inoculation. In the United States, Mirkov (2001) reported that sugarcane plants containing a coat protein construct derived from SrMV-H resisted subsequent infection by SrMV strains H, I and M, but not SCMV strains A, D or E. The mechanism virus resistance in these cases did not appear to be protein mediated but appeared to be RNA mediated, which corresponds with the observations recorded in other transgenic plant potyvirus interactions (Smith and Gambley, 1995). The transgenic sugarcane plants did not contain virions, and extracts prepared from these plants were not infectious when back inoculated onto maize (Joyce et al., 1998). Other PDR genes considered for developing SCMV resistance are nuclear protein b (Nib), known as the 'replicase' or RNA-dependent RNA polymerase (Smith and Harding, 2001). Zhu et al. (2008) reported that transgenic sugarcane containing an untranslatable SCYLV coat protein gene had increased resistance to SCYLV. Gene silencing seems to be a universal mechanism for plant resistance to viral infection as suggested by Baulcombe (1996). Posttranscriptional gene silencing has already been achieved in sugarcane by transformation with an untranslatable piece of sorghum mosaic virus. Albicidins, a family of phytotoxins produced by the LSD pathogen X. albilineans, are known to cause white pencil-like symptoms in sugarcane by inhibition of DNA replication in the plastids of young leaves. Several potentially useful albicidin resistance genes have been cloned from different bacteria, including *albA* gene from K. oxytoca, which encodes an albicidin binding protein, and *albD* gene from enzyme detoxification of albicidin in *Pantoea dispersa*. The *albD* protein irreversibly inactivates albicidin toxin and is likely to be an albicidin hydrolase. These two genes have been introduced separately into leaf scald-susceptible sugarcane cultivar Q63 and Q87 by co-bombardment with a gene. Both *albA* and *AlbD* genes can confer resistance to LSD, but *albD* is more effective than *albA* (Zhang and Birch, 1999). Analysis of transgenic sugarcane lines expressing the *albD* gene for hydrolysis of albicidin toxins show a clear correlation between the level of the resistance gene product and resistance to leaf scald disease following challenge by X. albilineans. A peptide from fish has been found to possess antibacterial activity against the bacterial pathogens X. albilineans and X. campestris pv vasculorum. The gene construct with the peptide gene has been introduced to sugarcane through biolistic methods (Office of the Gene Technology Regulator, 1999).

Abiotic stress tolerance in sugarcane

Salinity and drought are important environmental factors that limit crop productivity. Sugarcane, being a typical glycophyte, exhibits stunted growth or no growth under salinity, with its yield falling to 50% or even more as compared to its true potential (Akhtar and Rasul, 2003; Wiedenfeld, 2008). A high salt environment adversely affects plant growth due to alterations in water relations, ionic and metabolic perturbations, generation of reactive oxygen species (ROS) and tissue damage (Patade et al., 2011a), and enzymes involved in sugar metabolism (Gomathi and Thandapani, 2005), and respond with an altered expression of stress responsive genes, which may ameliorate the detrimental effects of salinity. Therefore construction of cDNA libraries enriched for differentially expressed transcripts is an important first step in attempting to study stress responsive genes. Patade et al. (2011b) constructed a forward subtracted cDNA library from sugarcane plants stressed with NaCl (200 mM) for 0.5 to 18 h to find mRNA species that are differentially expressed in sugarcane in response to salinity stress. Sequencing the differentially expressed few cDNAs clones led to the identification of salinity induced shaggy-like kinase (designated as sugarcane shaggy-like protein kinase [SuSk]). The expression was induced by salt as well as polyethylene glycol (PEG) stress indicating that the induction of this gene probably occurred in response to the osmotic component of salt stress rather than the ionic component.

In the study of various tissue-specific EST (expressed sequence tag) libraries sequence data of Indian subtropical sugarcane variety (CoS 767), 25 water-deficit stress-related clusters showed greater than twofold relative expression during 9 h dehydration stress (Gupta et al., 2010). Further, Prabu et al. (2010), based on sqRT-PCR analysis, showed higher transcript expression of WRKY, 22-kDa drought-induced protein, MIPS and ornithine-oxo-acid amino transferase at initial stages of stress induction with a gradual decrease in advanced stages. Analysis of the expression of these stress responsive genes in sugarcane plants under water deficit stress revealed a different transcriptional profile compared with sucrose accumulation. Prabu et al. (2010) identified differentially expressed transcripts in response to water deficiency stress in sugarcane cv. Co740 using a PCR-based cDNA suppression subtractive hybridization technique. Of the sequenced 158 cDNA clones based on Dot blot, 62% showed similarity with known functional genes and 12% with hypothetical proteins of plant origin, while 26% represented new unknown sequences. Annotation of these differentially ESTs indicated their possible function in cellular organization, protein metabolism, signal transduction and transcription.

RNAi technology for abiotic stress in sugarcane

MicroRNAs (miRNAs) are small, single-stranded, non-coding, naturally occurring, highly conserved families of transcripts (18–25 nt in length). Several miRNAs are either upregulated or downregulated by abiotic stresses, suggesting that they may be involved in stress-responsive gene expression and stress adaptation (Sunkar and Zhu, 2004). The involvement of miRNAs in abiotic stress has been studied in plants in response to dehydration or NaCl by using expression analysis, suggesting stress-specific regulation of expression of miRNA (Patade and Suprasanna, 2010) in sugarcane. In response to long-term (15 days) isoosmotic (−0.7 MPa) NaCl or PEG stress, no change in mature transcript level of miR159 over the control was detected. However, under short-term (up to 24 h) salt stress, the transcript level of the mature miRNA increased to 112% of the control at 16 h treatment. The mature transcript level of miR159 was higher under all the PEG-induced osmotic stress treatments as compared to the control, and it progressively increased with the stress exposure period (1.3 fold at 8 h treatment). This indicated that expression of the miR159 gene was more responsive to osmotic stress than ionic stress. The authors studied expression of one of the predicted target MYB under the same stress (NaCl or PEG) conditions to study the changes in target gene expression in response to over- or underexpression of miR159. The results on the expression of specific miR159 and its targets could be useful in developing appropriate markers for selection of tolerant cultivars in sugarcane.

Case study: Drought-tolerant sugarcane in Indonesia

In May 2013, Indonesia, the second largest (2.4 million tonnes, valued at US$1.6 billion) raw sugar importing country in the world, issued food and environmental safety certificates for the country's first home-grown genetically modified drought-tolerant sugarcane. The biotech sugarcane variety Cane PRG Drought Tolerant NX1-4T was developed under a public–private partnership between the Indonesian state-owned sugar company, PT Perkebunan Nusantara XI, and the Ajinomoto Company, Japan, in collaboration with Jember University in East Java, Indonesia. The drought-tolerant sugarcane varieties can withstand water stress up to 36 days, and under drought stress can yield substantially higher than the control variety BL-19. Yield increases from 2% to 75% in the first planting, 14% to 57% in the first ratoon, and from 11% to 44% in the second ratoon. It is expected that the first home-grown drought-tolerant sugarcane will be officially planted in Indonesia in 2015, pending approval of the product for feed.

Future prospects and conclusion

Sugarcane is a source of food and fuel, and biotechnology can contribute to substantially increase the utility of this crop. The successful application of biotechnological tools will require reliable, high levels of transgene expression and their stability over generations. The availability of a cellular and molecular toolbox has opened a plethora of prospects. Innovative *in vitro* culture systems have become available with the potential for rapid propagation and generating novel germplasm with desirable traits. A greater understanding of the crop using functional genomics and cellular methods will accelerate understanding responses to biotic and abiotic stresses and their management. Gene silencing is being used in transgenic research aimed at downregulation of endogenous genes in sugarcane. Some of the important challenges include gene discovery, transgenics and controlled transgene expression, sucrose metabolism and photosynthesis. In sugarcane, biotechnological applications are yielding positive results in the areas of genome characterization, mapping of specific traits, molecular variability of pathogens, marker-aided selection for insect/disease resistance, transformation and precise detection of plant pathogens. Technologies for the efficient production of useful transgenic sugarcane lines should be improved to become an integral part of varietal development programmes. The manipulation of plant DNA is rapidly becoming the world's third technological revolution, having effects comparable to those of the earlier Industrial Revolution and the present information technology. The advances in sugarcane biotechnology could become remarkable in the coming years, both in terms of improving productivity as well as substantially increasing the value and utility of this crop.

References

Akhtar SA, Rasul E (2003) Emergence, growth and nutrient composition of sugarcane sprouts under NaCl salinity. *Biology of Plants* 46: 113–117.

Alexander KC, Viswanathan R (2002) Diseases of sugarcane and their rapid diagnosis. In: *Sugarcane crop management*, SB Singh, GP Rao, S Eswaramoorthy (eds.), 10–51. Houston, TX: SCITECH.

Ali A, Naz S, Siddiqui FA, Iqbal J (2008) An efficient protocol for large scale production of sugarcane through micropropagation. *Pakistan Journal of Botany* 40(1): 139–149.

Altpeter F, Oraby H (2010) Sugarcane. In: *Genetic modification of plants: Biotechnology in agriculture and forestry*, vol. 64, F Kempken, C Jung (ed.), 453–472. New York: Springer.

Arencibia AD, Carmona ER, Cornide MT, Castiglione S, O'Relly J, Chinea A, Oramas P, Sala F (1999) Somaclonal variation in insect-resistant transgenic sugarcane (Saccharum hybrid) plants produced by cell electroporation. *Transgenic Research* 8: 349–360.

Arencibia AD, Carmona ER, Tellezs P, Chan MT, Yu SM (1998) An efficient protocol for sugarcane (*Saccharum* spp. L.) transformation mediated by *Agrobacterium tumefaciens*. *Transgenic Research* 7: 213–222.

Arencibia A, Vazquez RI, Prieto D, Tellez P, Carmona ER, Coego A et al. (1997) Transgenic sugarcane plants resistant to stem borer attack. *Molecular Breeding* 3: 247–255.

Arvinth S, Arun S, Selvakesavan RK, Srikanth J, Mukunthan N, Ananda Kumar P, Premachandran MN, Subramonian N (2010) Genetic transformation and pyramiding of aprotinin-expressing sugarcane with cry1Ab for shoot borer (*Chilo infuscatellus*) resistance. *Plant Cell Reports* 29(4): 383–395.

Baker J, Preston C (2003) Predicting the spread of herbicide resistance in Australian canola fields. *Transgenic Research* 12: 731–737.

Bates SL, Zhao ZL, Roush RT, Shelton AM (2005) Insect resistance management in GM crops, past, present and future. *Nature Biotechnology* 23(1): 57–62.

Baulcombe D (1996) Mechanisms of pathogen-derived resistance to viruses in transgenic plants. *Plant Cell* 8: 1833–1844.

Berding N, Moore PH, Smith GR (1997) Advances in breeding technology for sugarcane. In: *Intensive sugarcane production: Meeting the challenges beyond 2000*, BA Keating, JR Wilson (eds.), 189–220. Wallingford, UK: CABI.

Bower R, Birch RG (1992) Transgenic sugarcane plants via microprojectile bombardment. *Plant Journal* 2: 409–416.

Braga DPV, Arrigoni EDB, Burnquist WL, Silva Filho MC, Ulian EC, Hogarth DM (2001) A new approach for control of *Diatraea saccharalis* (Lepidoptera: Crambidae) through the expression of an insecticidal CryIa(b) protein in transgenic sugarcane. *Proceedings of the International Society of Sugarcane Technologists* 24: 331–336.

Braga DPV, Arrigoni EDB, Silva-Filho MC, Ulian EC (2003) Expression of the Cry 1 Ab protein in genetically modified sugarcane for the control of *Diatraea saccharalis* (Lepidoptera: Crambidae). *Journal of New Seeds* 5: 209–221.

Brumbley SM, Purnell MP, Petrasovits LA, Nielsen LK, Twine PH (2007) Developing the sugarcane biofactory for high value biomaterials. *International Sugar Journal* 109: 5–15.

Christy A, Aravith S, Saravanakumar M, Kanchana M, Mukunthan N, Srikanth J, Thomas G, Subramonian N (2009) Engineering sugarcane cultivars with bovine pancreatic trypsin inhibitor (aprotinin) gene for protection against top borer (*Scirpophaga excerptalis* Walker). *Plant Cell Reports* 28: 175–184.

Craveiro KIC, Gomes Júnior JE, Silva MCM, Macedo LLP, Lucena WA, Silva MS et al. (2010) Variant Cry1Ia toxins generated by DNA shuffling are active against sugarcane giant borer. *Journal of Biotechnology* 145(3): 215–221.

Deng ZN, Wei YW, Lu WL, Li YR, Suprasanna P (2008) Fusion insect resistant gene mediated by matrix attachment region (MAR) sequence in transgenic sugarcane. *Sugar Tech* 10(1): 87–90.

Desai NS, Suprasanna P, Bapat VA (2004) Partial desiccation of embryogenic callus improves plant regeneration frequency in sugarcane (*Saccharum* spp.). *Journal of Plant Biotechnology* 6: 229–233.

Elliott AR, Campbell JA, Bretell RIS, Gro CPL (1998) Agrobacterium mediated transformation of sugarcane using GFP as a screenable marker. *Australian Journal of Plant Physiology* 25: 739–743.

Enriquez GA, Trujillo LA, Menndez C, Vazquez RI, Tiel K, Dafhnis F, Arrieta J, Selman G, Hernandez L (2000) Sugarcane (*Saccharum* hybrid) genetic transformation mediated by *Agrobacterium tumefaciens*: Production of transgenic plants expressing proteins with agronomic and industrial value. *Developments in Plant Genetics and Breeding* 5: 76–81.

Enriquez-Obregon GA, Vazquez PRI, Prieto SDL, Riva-Gustavo ADL, Selman HG (1998) Herbicide resistant sugarcane (*Saccharum officinarum* L.) plants by Agrobacterium-mediated transformation. *Planta* 206: 20–27.

Falco MC, Silva-Filho MC (2003) Expression of soybean proteinase inhibitors in transgenic sugarcane plants: Effects on natural defense against *Diatraea saccharalis*. *Plant Physiology Biochemistry* 41: 761–766.

Gallo-Meagher M, Irvine JE (1996) Herbicide resistant transgenic sugarcane plants containing the bar gene. *Crop Science* 36: 1367–1374.

Gilbert RA, Gallo-Meagher M, Comstock JC, Miller JD, Jain M, Abouzid A (2005) Agronomic evaluation of sugarcane lines transformed for resistance to sugarcane mosaic virus strain E. *Crop Science* 45: 2060–2067.

Goebel FR, Way MJ (2007) Crop losses due to two sugarcane stem borers in Reunion and South Africa. *Proceedings of the International Society of Sugarcane Technologists* 26: 805–814.

Gomathi R, Thandapani TV (2005) Salt stress in relation to nutrient accumulation and quality of sugarcane genotypes. *Sugar Tech* 1: 39–47.

Gupta V, Raghuvanshi S, Gupta A, Saini N, Gaur A, Khan MS et al. (2010) The water-deficit stress- and red-rot-related genes in sugarcane. *Functional and Integrative Genomics* 10: 207–214.

Heinz DJ, Mee GWP (1969) Plant differentiation from callus tissue of Saccharum species. *Crop Science* 9: 346–348.

Hendre RR, Iyer RS, Kotwal M, Khuspe SS, Mascarenhas AF (1983) Rapid multiplication of sugarcane by tissue culture. *Sugarcane* 3: 5–8.

Hogarth DM, Cox MC, Bull JK (1997) Sugarcane improvement: Past achievements and future prospects. In: Crop improvement for the 21st century, MS Kang (ed.), 29–56. Baton Rouge, LA: Louisiana State University.

Hotta CT, Lembke CG, Domingues DS, Ochoa EA, Cruz GMQ, Melotto-Passarin DM et al. (2011) The biotechnology roadmap for sugarcane improvement. *Tropical Plant Biology* 3: 75–87.

Ingelbrecht IL, Irvine JE, Mirkov TE (1999) Post transcriptional gene silencing in transgenic sugarcane. Dissection of homology-dependent virus resistance in a monocot that has a complex polyploid genome. *Plant Physiology* 119: 1187–1197.

Jain M, Chengalrayan K, Abouzid A, Gallo M (2007) Prospecting the utility of a PMI/mannose selection system for the recovery of transgenic sugarcane (*Saccharum* spp. hybrid) plants. *Plant Cell Reports* 26: 581–590.

James C (2009) Global status of commercialized biotech/GM crops: 2009, the first fourteen years, 1996 to 2009. ISAAA Brief 41–2009: Executive summary. Ithaca: International Service for the Acquisition of Agri-Biotech Applications.

Jenes B, Moore H, Cao J, Zhang W, Wu R (1993) Techniques for gene transfer in Transgenic Plants. Kung S, Wu R (eds.), San Diego: Academic Press Inc., 1: 125–146.

Joyce PA, McQualter RB, Handley JA, Dale JL, Harding RM, Smith GR (1998) Transgenic sugarcane resistant to sugarcane mosaic virus. *Proceedings of the Australian Society of Sugar Cane Technologists* 20: 204–210.

Kalunke RM, Kolge AM, Babu KH, Prasad DT (2009) *Agrobacterium* mediated transformation of sugarcane for borer resistance using Cry 1Aa3 gene and one-step regeneration of transgenic plants. *Sugar Tech* 11(4): 355–359.

Kenganal, Hanchinal RR, Nadaf HL (2008) Ethyl methanesulfonate (EMS) induced mutation and selection for salt tolerance in sugarcane *in vitro*. *Indian Journal of Plant Physiology* 13: 405–410.

Khaliq A, Ashfaq M, Akram W, Choi JK, Lee JJ (2005) Effect of plant factors, sugar contents, and control methods on the Top Borer (*Scirpophaga nivella* F.) infestation in selected varieties of sugarcane. *Entomological Research* 35: 153–160.

Koch AC, Ramgareeb S, Snyman SJ, Watt MP, Rutherford RS (2010) An *in vitro* induced mutagenesis protocol for the production of sugarcane tolerant to imidazolinone herbicides. *Proceedings of the International Society for Sugar Cane Technology* 27: 1–5.

Lakshmanan P, Geijskes R, Aitken KS, Grof CLP, Bonnett GD, Smith GR (2005) Sugarcane biotechnology: The challenges and opportunities. *In Vitro Cellular and Developmental Biology – Plant* 41: 345–363.

Lakshmanan P, Geijskes RJ, Wang L, Elliott A, Grof CPL et al. (2006) Developmental and hormonal regulation of direct shoot organogenesis and somatic embryogenesis in sugarcane (*Saccharum* spp. interspecific hybrids) leaf culture. *Plant Cell Reports* 25: 1007–1015.

Larkin PJ, Scowcroft WR (1981) Somaclonal variation—A novel source of variability from cell cultures for plant improvement. *Theoretical and Applied Genetics* 60: 197–214.

Legaspi JC, Mirkov TE (2000) Evaluation of transgenic sugarcane against stalk borers. *Proceedings of the International Society of Sugarcane Technologists* 4: 68–71.

Leibbrandt NB, Snyman SJ (2003) Stability of gene expression and agronomic performance of transgenic herbicide-resistant sugarcane line in South Africa. *Crop Science* 43: 671–678.

Limon MC, Pintor-Toro JA, Benitez T (1999) Increased antifungal activity of *Trichoderma harzianum* transformants that overexpress a 33-KDa chitinase. *Phytopathology* 89: 254–261.

Ma HM, Schulze S, Lee S, Yang M, Mirkov E, Irvine J, Moore P, Paterson A (2004) An EST survey of the sugarcane transcriptome. *Theoretical and Applied Genetics* 108: 851–863.

Malathi P, Viswanathan R, Padmanaban P, Mohanraj D, Ramesh Sundar A (2002) Compatibility of biocontrol agents with fungicides against red rot disease of sugarcane. *Sugar Tech* 4: 131–136.

Malathi P, Viswanathan R, Ramesh Sundar A, Padmanaban P, Prakasam N, Mohanraj D, Jothi R (2011) Phylogenetic analysis of *Colletotrichum falcatum* isolates causing red rot in sugarcane. *Journal of Sugarcane Research* 1(1): 69–74.

Manickavasagam M, Ganapathi A, Anbazhagan VR, Sudhakar B, Selvaraj N, Vasudevan A, Kasthurirengan S (2004) *Agrobacterium* mediated genetic transformation and development of herbicide resistant sugarcane (*Saccharum* species hybrids) using axillary buds. *Plant Cell Reports* 23: 134–143.

McQualter RB, Dale JL, Harding RH, McMahon JA, Smith GR (2004a) Production and evaluation of transgenic sugarcane containing a Fiji disease virus (FDV) genome segment S9-derived synthetic resistance gene. *Australian Journal of Agricultural Research* 55: 139–145.

McQualter RB, Fong Chong B, O'Shea M, Meyer K, Van Dyk DE, Walton NJ, Viitanen PV, Brumbley SM (2004b) Initial evaluation of sugarcane as a production platform for p-hydroxybenzoic acid. *Plant Biotechnology Journal* 2: 1–13.

Mirkov TE (2001) Progress and potential for sugarcane improvement through transgenic plant technologies. *Proceedings of the International Society of Sugarcane Technologists* 24: 575–577.

Molinari HBC, Marur CJ, Daros E, de Campos MKP, de Carvalho JFRP (2008) Evaluation of the stress-inducible production of proline in transgenic sugarcane (*Saccharum* spp.): Osmotic adjustment, chlorophyll fluorescence and oxidative stress. *Physiologia Plantarum* 130(2): 218–229.

Moore PH (2005) Integration of sucrose accumulation processes across hierarchical scales: Towards developing an understanding of the gene-to-crop continuum. *Field Crops Research* 92(23): 119–135.

Mukunthan N, Singaravelu B (2005) Evaluation of synthetic sex pheromone of sugarcane top borer (*Scirpophaga excerptalis* Walk). *Sugar Tech* 7(4): 168–170.

Nonato RV, Mantelatto PE, Rossell CEV (2001) Integrated production of biodegradable plastic, sugar and ethanol. *Applied Microbiology and Biotechnology* 57: 1–5.

Nutt KA, Allsopp PG, Geijskes RJ, McKeon MG, Smith GR, Hogarth DM (2001) Canegrub resistant sugarcane. *Proceedings of the International Society of Sugar Cane Technologists* 24: 582–583.

Nutt KA, Allsopp PG, McGhie TK, Shepherd KM, Joyce PA et al. (1999) Transgenic sugarcane with increased resistance to canegrubs. *Proceedings of the Conference of the Australian Society of Sugarcane Technologists*, Townsville, Queensland, Australia, 171–176.

Office of the Gene Technology Regulator (2011) Biology of sugarcane. www.ogtr.gov .au/internet/ogtr/publishing…/sugarcane…/biologysugarcane.

Patade V, Suprasanna P, Bapat VA (2008) Gamma irradiation of embryogenic callus cultures and *in vitro* selection for salt tolerance in sugarcane (*Saccharum offcinarum* L.). *Agricultural Sciences in China* 7(9): 101–105.

Patade VY, Bhargava S, Suprasanna P (2011a) Salt and drought tolerance of sugarcane under iso-osmotic salt and water stress: Growth, osmolytes accumulation and antioxidant defense. *Journal of Plant Interactions*. doi:10.1080/174291 45.2011.557513.

Patade VY, Suprasanna P (2008) An *in vitro* radiation induced mutagenesis-selection system for salinity tolerance in sugarcane. *Sugar Tech* 11(3): 246–251.

Patade VY, Suprasanna P (2010) Short-term salt and PEG stresses regulate expression of MicroRNA, miR159 in sugarcane leaves. *Journal of Crop Science and Biotechnology* 13(3): 177–182.

Patade VY, Rai AN, Suprasanna P (2011b) Expression analysis of sugarcane shaggy-like kinase (SuSK) gene identified through cDNA subtractive hybridization in sugarcane (*Saccharum officinarum* L.). *Protoplasma* 248(3): 613–621.

Prabu G, Kawar PG, Pagariya MC, Theertha Prasad D (2010) Identification of water deficit stress upregulated genes in sugarcane. *Plant Molecular Biology Reporter* 29(2): 291–304.

Rai MK, Kalia RK, Singh R, Gangola MP, Dhawan AK (2011) Developing stress tolerant plants through *in vitro* selection—An overview of the recent progress. *Environmental and Experimental Botany* 71(1): 89–98.

Roach BT (1989) Origin and improvement of the genetic base of sugarcane. *Proceedings of the Australian Society of Sugar Cane Technologists* 11: 34–47.

Romeis J, Meissel M, Bigler F (2006) Transgenic crops expressing *Bacillus thuringiensis* toxin and biological control. *Nature Biotechnology* 4: 63–71.

Rossato JADS Jr, Fernandes OA, Mutton MJR, Higley LG, Madaleno LL (2010) Sugarcane response to two biotic stressors: *Diatraea saccharalis* and *Mahanarva fimbriolata*. *Proceedings of the International Society of Sugar Cane Technologists* 27: 1–5.

Sengar AS, Thind S, Kumar B, Pallavi M, Gosa SS (2009) *In vitro* selection at cellular level for red rot resistance in sugarcane (*Saccharum* sp.). *Plant Growth Regulation* 58(2): 201–209.

Setamou M, Bernal JS, JC Legaspi, Mirkov TE, Legaspi BC (2002) Evaluation of lectin-expressing transgenic sugarcane against stalkborers (Lepidoptera: Pyralidae): Effects on life history parameters. *Journal of Economic Entomology* 95: 469–477.

Singh G, Sandhu S, Meeta M, Singh K, Gill R, Gosal S (2008) *In vitro* induction and characterization of somaclonal variation for red rot and other agronomic traits in sugarcane. *Euphytica* 160(1): 35–47.

Smith GR, Ford R, Frenkel ML, Shukla DD, Dale JL (1992) Transient expression of the coat protein of sugarcane mosaic virus in sugarcane protoplasts and expression in *Escherichia coli*. *Archives of Virology* 125: 15–23.

Smith GR, Gambley RL (1995) Progress in development of a sugarcane meristem transformation system and production of SCMV-resistant transgenics. *Proceedings of the Australian Society of Sugar Cane Technologists* 15: 237–243.

Smith GR, Harding RM (2001) Genetic engineering for virus resistance in sugarcane. In: *Current trends in sugarcane pathology II. Virus and phytoplasma diseases*, GP Rao et al. (eds.), 335–365. Enfield, NH: Science Publishers.

Snyman SJ, Meyer GM, Koch AC, Banasiak M, Watt MP (2011) Applications of *in vitro* culture systems for commercial sugarcane production and improvement. *In Vitro Cellular & Developmental Biology – Plant* 47: 234–249.

Srikanth J, Kurup NK (2011) Damage pattern of sugarcane internode borer Chilo sacchariphagus indicus (Kapur) in Tamil Nadu State, southern India. *Sugar Cane International* 2: 236–241.

Srikanth J, Salin KP, Kurup NK, Subadra Bai K (2009) Assessment of Sturmiopsis inferens (Diptera: Tachinidae) as a natural and applied biological control agent of sugarcane shoot borer *Chilo infuscatellus* (Lepidoptera: Crambidae) in southern India. *Sugar Tech* 11(1): 51–59.

Sunkar R, Zhu JK (2004) Novel and stress-regulated microRNAs and other small RNA from *Arabidopsis*. *The Plant Cell* 16: 2001–2019.

Suprasanna P (2010) Biotechnological interventions in sugarcane improvement: Strategies, methods and progress. *BARC News Letter* 47–53.

Suprasanna P, Bapat VA (2006) Advances in the development of *in vitro* culture systems and transgenics in sugarcane. *Proceedings of International Symposium on Technologies to Improve Sugar Productivity in Developing Countries*, 629–636.

Suprasanna P, Jain SM, Ochatt SJ, Kulkarni VM, Predieri S (2010) Applications of *in vitro* techniques in mutation breeding of vegetatively propagated crops. In *Plant mutation*, Q Shu (ed.), 369–383. Vienna: IAEA.

Suprasanna P, Meenakshi S, Bapat VA (2008a) Integrated approaches of mutagenesis and *in vitro* selection for crop improvement. In *Plant tissue culture molecular markers and their role in crop productivity*, A Kumar, NS Shekhawat (eds.), 73–92. New Delhi: IK International Publishers.

Suprasanna P, Patade YV, Bapat VA (2007) Sugarcane biotechnology—A perspective on recent developments and emerging opportunities. In *Advances in plant biotechnology*, GP Rao, Z Yipeng, VR Volodymyr, S Bhatnagar (eds.), 313–342. Hauppauge, NY: Science Publishers.

Suprasanna P, Rupali C, Desai NS, Bapat VA (2008b) Partial desiccation improves plant regeneration response of gamma irradiated embryogenic callus in sugarcane (*Saccharum* Spp.). *Plant Cell, Tissue and Organ Culture* 92: 101–105.

Tiwari AK, Bharti YP, Tripathi S, Mishra N, Lal M, Rao GP, Sharma PK, Sharma ML (2010) Biotechnological approaches to improve sugarcane crop with special reference to disease resistance. *Acta Phytopathologica et Entomologica Hungarica* 45: 235–249.

Tomov BW, Bernal JS (2003) Effects of GNA transgenic sugarcane on life history parameters of Parallorhogas pyralophagus (Marsh) (Hymenoptera: Braconidae), a parasitoid of Mexican rice borer. *Journal of Economic Entomology* 96(3): 570–576.

Vickers JE, Grof CP, Bonnett GD, Jackson PA, Morgan TE (2005) Effects of tissue culture, biolistic transformation, and introduction of PPO and SPS gene constructs a performance of sugarcane clones in the field. *Australian Journal of Agricultural Research* 56: 57–68.

Viswanathan R, Ramesh Sundar A, Premkumari SM (2003) Mycolytic enzymes of extracellular enzymes of antagonistic microbes to *Colletotrichum falcatum*, red rot pathogen of sugarcane. *World Journal of Microbiology and Biotechnology* 19: 953–959.

Way MJ, Goebel FR (2007) Monitoring Eldana saccharina and other arthropod pests in South African sugarcane. *Proceedings of the International Society of Sugar Cane Technologists* 26: 780–786.

Weng LX, Deng HH, Xu JL, Li Q, Wang LH, Jiang ZD, Zhang HB, Li QW, Zhang LH (2006) Regeneration of sugarcane elite breeding lines and engineering of strong stem borer resistance. *Pest Management Science* 62: 178–187.

Weng LX, Deng HH, Xu JL, Li Q, Zhang YQ, Jiang ZD, Li QW, Chen JW, Zhang LH (2010) Transgenic sugarcane plants expressing high levels of modified cry1Ac provide effective control against stem borers in field trials. *Transgenic Research*. doi:10.1007/s11248-010-9456-8.

Wiedenfeld B (2008) Effects of irrigation water salinity and electrostatic water treatment for sugarcane production. *Agricultural Water Management* 95: 85–88.

Wu LG, Birch RG (2007) Doubled sugar content in sugarcane plants modified to produce a sucrose isomer. *Plant Biotechnology Journal* 51: 109–117.

Zhang LH, Birch RG (1999) Engineered detoxification confers resistance against a pathogenic bacterium. *Nature Biotechnology* 17: 1021–1024.

Zhangsun DT, Luo SL, Chen RK, Tang KX (2007) Improved Agrobacterium-mediated genetic transformation of GNA transgenic sugarcane. *Biologia* 62(4): 386–393.

Zhu C, Gore M, Buckler ES, Yu J (2008) Prospects of association mapping in plants. *Plant Genome* 1: 5–20.

chapter ten

Sugarcane biotechnology towards abiotic stress tolerance

Drought and salinity

Lakshmi Kasirajan, Anuradha Chelliah and Boomiraj Kovilpillai

Contents

Introduction

In India, more than 60% of agricultural land is rainfed and subjected to varying levels of abiotic stresses. Plants are immobile organisms, exposed to many environmental abiotic stresses, thereby they undergo many evolutionary, developmental and physiological strategies to face or avoid these stresses. The abiotic stresses that plants are exposed to include drought, heat, salinity, ultraviolet light, flooding, gaseous pollution, freezing and heavy metals. Drought is often defined in terms of available humidity as compared with a normal value, with the severity correlating in function to the time and magnitude of the exposition to a deficient humidity (Smith and Pethley, 2009). There are different types of drought including meteorological, hydrological, socioeconomic and agricultural (Boken, 2005; Smith and Pethley, 2009). According to Boken (2005), meteorological drought occurs when seasonal or annual precipitation falls below its long-term average. When the meteorological drought is prolonged and causes local shortage of surface and groundwater, it is called a hydrological drought. Socioeconomic drought is a manifestation of continued drought of severe intensity that shatters the economy and sociopolitical situation in a region. In an agricultural drought, due to soil moisture stress, there is a significant decline in crop yields (production per unit area).

When crop plants are exposed to any stress, either abiotic or biotic, they fail to express their full genetic potential for production due to the failure of the resistance system. However, the effect of stresses might vary depending on the developmental stage of the crop, the duration and intensity of the stress, and the genetic nature of the plant species' resistance or susceptibility to that particular stress. Generally, when the stress levels exceed a certain critical level (which varies from crop to crop), the resistance mechanisms imparting tolerance to plants breaks down which results in plant death.

Sugarcane responses to abiotic stress: Drought and salinity

Sugarcane belongs to the Andropogonae tribe of the family Gramineae, order Glumiflorae, class Monocotyledoneae, subdivision Angiospermae, and division Embryophyta siphonogama. The subtribe is Sacharae and the genus is *Saccharum*, derived from the Sanskrit word *sarkara* meaning 'white sugar'. Sugarcane is a C_4 perennial grass which belongs to the genus *Saccharum*. Originating from Southeastern Asia, sugarcane is cultivated in more than 20 million hectares in tropical and subtropical regions of the world, producing up to 1.3 billion metric tons of crushable stems. It is one of the major plant sources for sugar as well as alcohol production. The genus *Saccharum* is composed of hybrids derived from *Saccharum officinarum* (Noble clones), *S. sinense* (Chinese clones), *S. barberi* (North Indian clones), and *S. spontaneum* (Roach, 1995). Commercial sugarcanes are hybrids of polyploidy and aneuploidy, which contain 100 to 120 chromosomes with an estimated somatic cell size of 10,000 Mb (D'Hont and Glaszmann, 2001). Being a C_4 plant, it converts 2% of incident solar energy into biomass. The crop productivity depends on various phases from seed germination to grain development. If there is no stress, the crop would pass through these stages without any loss in productivity. However, one or more of the growth and development phases may be disturbed by drought and its severity, and may lead to low sugarcane productivity because of fewer millable canes. The stages of sugarcane crop growth can be divided into four phases (also see Figure 10.1):

1. *Germination* – The development of buds and roots, taking from 0 to 40 days
2. *Tillering* (formative) – Appearance of secondary and tertiary tillers, beginning approximately on the 40th day after planting and lasting up to 120 days
3. *Grand growth* – Tiller growth and development with height gain and basal sugar accumulation taking up to 9 months after planting
4. *Maturity* – Accumulation of photoassimilates and fast sugar synthesis, lasting until the harvesting period

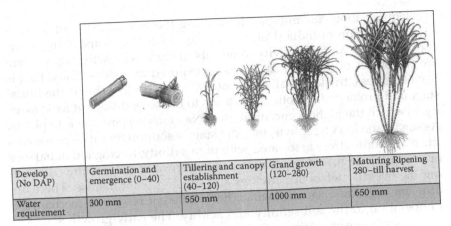

Develop (No DAP)	Germination and emergence (0–40)	Tillering and canopy establishment (40–120)	Grand growth (120–280)	Maturing Ripening 280–till harvest
Water requirement	300 mm	550 mm	1000 mm	650 mm

Figure 10.1 Sugarcane growth stages and their water requirements.

The tillering phase and the grand growth phase have been identified as the critical water demand periods, mainly because during these phases the maximum number of shoots are formed (Gomathi et al., 2011). In most of the sugarcane growing areas, the crop experiences moisture stress during formative and grand growth stages, which results in ultimate reduction of the stalk population, growth, and a yield loss of 30% to 50%. Sugarcane has been rated as moderately tolerant for drought tolerance based on the harvest index proposed by Donald. The relative performance of the genotypes in terms of cane yield and sugar yield was assessed by plotting the cane yield under drought treatment against yield under normal conditions worked out by Gomathi and co-workers in 2010. The identified sugarcane genotypes tolerant to drought are CoC 671, Co 8208, Co 85007, Co 85004, Co 86032, Co 85019 and Co 87263.

Another important abiotic stress the sugarcane crop is exposed to is salinity. An indirect outcome of drought is the higher amount of salts in the soil. Plants that can grow under high salinity (>400 mM NaCl) are called halophytes, and those that are sensitive to this salt concentration are called glycophytes. Based on this study sugarcane (*Saccharum officinarum*) is classified as a highly susceptible crop, with a threshold EC of <2 dSm⁻¹. Other susceptible crops include bean (*Phaseolus vulgaris*), eggplant (*Solanum melongena*), corn (*Zea mays*) and potato (*Solanum tuberosum*), whereas sugar beet (*Beta vulgaris*) and barley (*Hordeum vulgare*) can tolerate an EC up to 7 dSm⁻¹ (Maas, 1990; http://www.ussl.ars.usda.gov/saltoler .htm) and are classified as tolerant crops. By 2050, 50% of all arable lands are expected to be salinized due to irrigation with poor quality water and due to the restricted availability of water resources (Bray et al., 2000). The economic loss due to environmental degradation through these twin problems is about \$37 million, threatening the sustainability of agricultural production in India (Datta and Jong, 2002). Sugarcane being a glycophyte shows high sensitivity to salinity at various growth stages. There are two stages of stresses that plants are exposed to under salinity. First is osmotic stress, the less availability of water molecules. This is the initial stage and prolonged osmotic stress leads to the second stage of ionic toxicity, in which the higher concentration of Na^+ ions becomes toxic to plants. As sugarcane is a glycophyte, the later stage – sodium toxicity – represents the major ionic stress associated with high salinity, forcing ion imbalance or disequilibrium, and hyperionic and hyperosmotic stress, thus disrupting the overall metabolic activities and causing plant demise. There are different sugarcane varieties which have different resistance mechanisms in operation to the soil salinity and acidity. The early growth stages of crop, such as germination, are more sensitive than later stages of crop growth. In the case of sugarcane, the ratoon crop is also important as well as the plant crop for the farmers to harvest the maximum yield out of the crop, but the ratoon crop is more sensitive to salinity than the plant crop. Moreover, in India 3.3 million ha and 2.46 million ha of land are affected due to salinity and waterlogging conditions, respectively (Jain et al., 2008), whereas 1 million ha of irrigation-induced waterlogged saline area is confined to northwest regions of India.

Abiotic stress signaling and stress-responsive mechanisms in plants

Naturally, plants are well equipped with resistance mechanisms for all abiotic and biotic stresses. But due to human selection from the ancient days for specific traits of interest, the genetic base has narrowed and the level of resistance in plants reduced. Plants are equipped with elegant resistant mechanisms by which they escape or avoid or tolerate different stresses. They involve several stress-inducible gene products, such as osmoprotectants, which enable them to maintain homeostasis against these stresses (Hirayama and Shinozaki, 2010; Knight and Knight, 2001; Luis et al. 2009). Stress-inducible genes consists of two major groups: (1) functional protein coding genes that provide direct tolerance to abiotic stresses (such as LEA, CODA, trehalose and glycine betaine proteins, that are involved in osmoprotectant biosynthesis) that do not affect the plant metabolism and are inert in nature; and (2) transcription factor coding genes, the regulatory proteins such as DREBs/CBFs, AREBs (ABA responsive elements) and NACs, which are responsible for downstream signal transduction and induction of functional proteins (Nakashima et al., 2012). When plants are exposed to drought or salinity stress, the extracellular stress signal is first received at the membrane level by the membrane receptors and ion channel receptors, which then activate the large and complex signaling cascade intracellularly, which includes the secondary signal molecules such as Ca^{2+}, inositol phosphates (InsP), reactive oxygen species (ROS) and abscisic acid (ABA). ABA is an important phytohormone which plays a critical role in response to various stress signals. Abiotic stress signals can be induced either through the ABA-dependent pathway or ABA-independent pathway (Figure 10.2). These external stress signals induce several ABA-responsive genes, which ultimately lead to some immediate physiological changes, such as the rapid closure of stomata and start of the synthesis of osmoprotectants (Hubbard et al., 2010). In the last few decades, considerable research has been conducted to identify and characterize various transcription factors (TFs) involved in plant abiotic stress responses either in the ABA-dependent pathway or ABA-independent pathway, such as *AP2/EREBP, MYB, WRKY, NAC* and *bZIP* (Golldack et al., 2011). The *AP2/ERFBP* family includes a large group of plant-specific TFs and is characterized by the presence of the highly conserved *AP2*/ethylene-responsive element-binding factor (ERF) DNA-binding domain that directly interacts with the GCC box and/or dehydration-responsive element (DRE)/C-repeat element (CRT) *cis-*acting elements at the promoter of downstream target genes (Riechmann and Meyerowitz, 1998). A lot of DREB-type TFs have been tested in many plants including *Arabidopsis*, wheat, tomato, soybean, rice, maize and

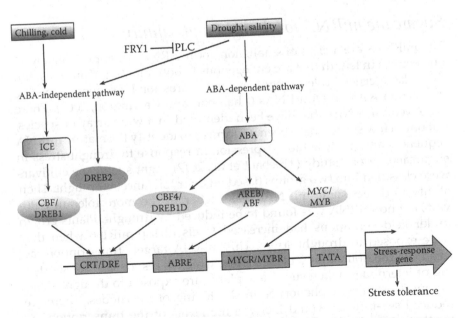

Figure 10.2 Transcriptional regulatory networks for abiotic stress-inducible gene expression. (From Kumar R., *Appl Biochem Biotech*, 174(1), 93–115, 2014.)

barley (Mizoi et al., 2013). A novel sugarcane ethylene responsive factor, SodERF3, which is a member of AP2-type transcription factors, has been identified for genetic manipulation of sugarcane for drought and salinity stress tolerance (Trujillo et al., 2009).

ABA-dependent pathways are known to mediate gene expression in plants during osmotic stress. The distinction is largely based on *cis*-elements that exist in the promoters of ABA-inducible genes. The ABA-dependent pathways are thought to mediate gene expression through an ABRE-element and b-ZIP transcription factors (Busk and Pages, 1998), while the other pathways through MYC and MYB elements and transcription factors. Numerous reports have shown that different pathways are interconnected and coordinately regulate the plant response to biotic and abiotic stresses. ABA-independent abiotic stress signaling mediated by phospholipase C (PLC) activates DREB2 transcription factors that bind to DRE and induce ABA-independent transcription of stress-responsive genes. Inositol polyphosphate 1-phosphatase (FRY1) functions as a negative regulator of drought, salinity and ABA responses which involves in the catabolism of inositol 1, 4, 5-trisphosphate (IP3) (Xiong et al., 2002). ABA-dependent pathway regulates stress-responsive gene expression through CBF4 (DREB1D), AREB (ABF) and MYC/MYB transcription factors, which bind to the CRT/DRE, ABRE, and MYC/MYB recognition sequences (MYCR/MYBR) promoter elements, respectively.

Sugarcane miRNA for drought and salinity

The miRNAs are a class of small, non-coding RNAs of approximately 21 nucleotides in length that are endogenous to both plants and animals. But they play a crucial role to regulate gene expression by sequence-specific interaction with target mRNAs (Chapman and Carrington, 2007). There are several miRNAs that have been identified in a wide array of species, but only a few studies have been performed to identify the mature miRNA sequences and analyse their expression in response to drought stress in sugarcane. In one study (Thiebaut et al., 2012), eight sugarcane cultivars were classified into two groups based on their tolerance to drought. Their ability to detect some miRNAs was higher in the more tolerant cultivars, but no miRNA was found to be induced by drought. Plants grown under field conditions had increased levels of ssp-miR166 when they were exposed to drought stress. This miRNA targets transcription factors from the homeobox-leucine zipper, which is responsible for shortening of internodes, when sugarcane plants are exposed to drought stress, the most prominent phenotype in shortening of internodes. Therefore, reduced ssp-miR166 would increase the levels of the transcription factor that is involved in shortening the internodes. Another miRNA, ssp-miR160, was upregulated in response to drought stress. The target for this miRNA is the VNI2 protein from *Arabidopsis*. This protein represses the activity of a transcription factor, VASCULAR-RELATED NAC-DOMAIN7 (VND7), which is an inducer for xylem formation (Yamaguchi et al., 2010). Although, to our knowledge, there are no works showing xylem differentiation in response to drought in sugarcane, this response has been observed in poplar trees (Arend and Fromm, 2007). Therefore, ssp-miR160 induction could lead to decreased levels of VIN2, releasing the action of VND7, which would work in xylem differentiation.

References

Arend M, Fromm J (2007) Seasonal change in the drought response of wood cell development in poplar. *Tree Physiol* 27: 985–992.

Boken VK (2005) Agricultural drought and its monitoring and prediction: Some concepts. In: *Monitoring and predicting agricultural drought: A global study*, VK Boken, AP Cracknell, RL Heathcote (eds), 3–10.

Bray EA, Bailey-Serres J, Weretilnyk E (2000) Responses to abiotic stresses. In: *Biochemistry and molecular biology of plants*, W Gruissem, B Buchannan, R Jones (eds), 1158–1249. Rockville, MD: American Society of Plant Physiologists.

Busk PK, Pages M (1998) Regulation of abscisic acid-induced transcription. *Plant Mol Biol* 37: 425–435.

Chapman EJ, Carrington JC (2007) Specialization and evolution of endogenous small RNA pathways. *Nat Rev Gen* 8: 884–896.

D'Hont A, Glaszmann JC (2001) Sugarcane genome analysis with molecular markers, a first decade of research. *Proc Int Soc Sugarcane Technol* 24: 556–559.

Datta KK, de Jong C (2002) An adverse effect of waterlogging and soil salinity on crop and land productivity in northwest region of Haryana, India. *Agri Water Manage* 57: 223–238.

Golldack D, Lüking I, Yang O (2011). Plant tolerance to drought and salinity: Stress regulating transcription factors and their functional significance in the cellular transcriptional network. *Plant Cell Rep.* 30: 1383–1391.

Gomathi R, Vasantha S, Hemaprabha G, Alarmelu, S, Shanthi RM (2011) Evaluation of elite sugarcane clones for drought tolerance. *J Sugar Res* 1: 55–62.

Hirayama T, Shinozaki K (2010) Research on plant abiotic stress responses in the post-genome era: Past, present and future. *Plant J* 61: 1041–1052.

Hubbard KE, Nishimura N, Hitomi K, Getzoff ED, Schroeder JI (2010) Early abscisic acid signal transduction mechanisms: Newly discovered components and newly emerging questions. *Genes Dev* 24: 1695–1708.

Jain PK, Ravichandran V, Agrawal RK (2008) Antioxidant and free radical scavenging properties of traditionally used three Indian medicinal plants. *Curr Trends Biotech Pharm* 2: 538–547.

Knight H, Knight MR (2001) Abiotic stress signalling pathways: Specificity and cross-talk. *Trends Plant Sci* 6: 262–267.

Kumar R (2014) Role of MicroRNAs in biotic and abiotic stress responses in crop plants. *Appl Biochem Biotech* 174(1): 93–115.

Maas EV (1990) Crop salt tolerance. In: Agricultural salinity assessment and management, KK Tanji (ed), *ASCE Manual Reports on Engineering Practices* 71: 262–304.

Mizoi J, Ohori T, Moriwaki T, Kidokoro S, Todaka D, Maruyama K et al. (2013). GmDREB2A;2, a canonical DEHYDRATION-RESPONSIVE ELEMENT-BINDING PROTEIN2-type transcription factor in soybean, is posttranslationally regulated and mediates dehydration-responsive element-dependent gene expression. *Plant Physiol* 161: 346–361.

Nakashima K, Takasaki H, Mizoi J, Shinozaki K, Yamaguchi-Shinozaki K (2012) NAC transcription factors in plant abiotic stress responses. *Biochim Biophys Acta* 1819: 97–103.

Riechmann JL, Meyerowitz EM (1998) The AP2/EREBP family of plant transcription factors. *Biol Chem* 379: 633–646.

Roach BT (1995) Case for a core collection of sugarcane germplasm. *Crop Sciences* 339–350.

Smith K, Pethley D (2009) Hydrological hazards: Droughts. In: *Environmental hazards: Assessing risk and reducing disaster*, K Smith, D Pethley (eds), 262–284. New York: Routledge.

Thiebaut F, Grativol C, Carnavale-Bottino M, Rojas CA, Tanurdzic LOS, Farinelli L et al. (2012) Computational identification and analysis of novel sugarcane microRNAs. *BMC Genomics* 13: 290.

Trujillo LE, Menéndez C, Ochogavía ME, Hernández I, Borrás O, Rodríguez R, Coll Y, Arrieta JG, Banguela A, Ramírez R, Hernández L (2009) Engineering drought and salt tolerance in plants using SodERF3, a novel sugarcane ethylene responsive factor. *Biotecnologia Aplicada* 168–171.

Xiong L, Lee H, Ishitani M, Zhu JK (2002) Regulation of osmotic stress-responsive gene expression by the LOS6/ABA1 locus in *Arabidopsis*. *J Biol Chem* 277: 8588–8569.

Yamaguchi M, Ohtani M, Mitsuda N, Kubo M, Ohme-Takagi M, Fukuda H et al. (2010) VND-INTERACTING2, a NAC domain transcription factor, negatively regulates xylem vessel formation in *Arabidopsis*. *Plant Cell* 22: 1249–1263.

chapter eleven

Prospects of biotechnological tools in boosting sugarcane production

Kalpana Sengar, R.S. Sengar and Sanjay Kumar Garg

Contents

Introduction

Several biotechnological achievements have been reported during the last decade for the sugarcane crop. Development of highly efficient *in vitro* regeneration systems coupled with rapid advance in gene delivery techniques led to the establishment of robust genetic transformation technology (Bower and Birch, 1992). This was a good beginning and laid the foundation of sugarcane genetic engineering.

During the past decade, several gene delivery systems for the transformation of sugarcane have also been reported by several researchers (Arencibia et al., 1995; Bower and Birch, 1992; Enriquez-Obregon et al., 1998; Manickavasagam et al., 2004). Of course, the full potential of sugarcane biotechnology is still to be established, because of the technical lacunae in existing transformation systems. In exploiting gene technology for sugarcane improvement, transformation/tissue culture-induced somaclonal variation is a significant limitation (Arencibia et al., 1999), and considerable improvement of current transformation systems are required so as to ensure clonal fidelity of transgenic cultivars. For attaining the practical advantage of transgenic technology, transformation system and gene regulation sequences that produce plants with adequate and stable transgene expression under field conditions are also needed. Similarly, several scientific, legislative and public perception issues would be required to be addressed for the successful release of transgenic sugarcane. Such transformation systems which do not incorporate any non-transgene DNA into plants and utilize non-antibiotic, plant gene-based selection strategies would aid substantially in overcoming regulatory and public perception issues. The ability to control transgenic expression through induction and developmental control would provide an additional platform for the production of a range of new compounds in the sugarcane industry. The development of strategies from the viewpoint of incorporation of polygenic traits, hyperexpression of transgenes and containment of transgenes within the transgenic plants (Daniell, 1999; Daniell et al., 2001; Maliga, 2004) together with the possibility of engineering native genes without significant genetic rearrangements (Beetham et al., 1999) are some of the valuable research which may be of immense importance in boosting sugarcane production in years to come.

Similar to other crops, transgenic sugarcane plants were engineered with input traits such as resistance to herbicides, diseases and pests in the first generation. By now, significant gains have been achieved in this field and much more progress can be made by exploiting various resistant genes, which are already available from other plants (McDowell and Woffenden, 2003). Sincere efforts are also required to understand the molecular basis of pathogenicity and the natural disease-resistance quality found in *Saccharum* germplasm so as to develop targeted biotechnological approaches for better control of sugarcane pathogens and pests.

Tactful manipulation of sugar metabolism with the objective of enhancement of yield remains a major challenge both to the conventional and molecular breeder to achieve a success. The plasticity of plant metabolism, with the existence of alternative pathways of sugar metabolism, makes the manipulation of metabolic pathways unpredictable to some extent. Even than, the molecular modification of sugar metabolism is potent and can be attempted due to ever-increasing knowledge of biochemistry and the molecular biology of carbohydrate metabolism. Modern information on sugarcane biochemistry, transgenic expression and transgene product storage may help a lot in successfully development of the sugarcane as a viable biofactory for the production of several commercially valuable products.

Remarkable increase in sugarcane crop productivity has been recorded in the last century worldwide (Keating and Wilson, 1997) due to widespread use of improved disease- and pest-resistant cultivars, better management of input sources, and the availability of cheaper chemical fertilizers. Sustaining this rate of improvement in sugarcane productivity with the application of innovative and improved practices of agriculture, and at the same time ensured minimum environmental influence, are likely to be one of the major challenges in maintaining the profitably level of the sugar industry in years to come.

By the end of the 19th century, most of the cultivated sugarcanes were the clones of high sucrose (i.e. *Saccharum officinarum*) containing $2n = 80$ chromosomes. The development of the first hybrids between *S. officinarum* and wild vigorous species (e.g. *Saccharum spontaneum*) resulted in a major breakthrough in sugarcane breeding first time in the history. Sugarcane cultivars with the desirable traits of higher yields, improved ratooning ability and disease-resistant quality were obtained as a series of break crosses to *S. officinarum*. Modern cultivars developed from these initial hybrids have between $2n = 100$ and $2n = 130$ chromosomes. In the development of these early hybrids only a few clones (i.e. *S. officinarum* and *S. spontaneum*) are thought to have been involved (Roach, 1989). Most of the sugarcane breeding programmes being followed nowadays rely on extensive intercrossing of elite cultivars which have derived from the aforementioned early hybrids.

It has been recently reported that *in situ* hybridization analysis of two ribosomal RNA gene families determined that *S. officinarum* has a basic chromosome number of $X = 10$, meaning that these are octoploid plants (D'Hont et al., 1998). Following the same method it was revealed that *S. spontaneum* has a basic chromosome number of $X = 8$, and that the ploidy level of this species varies between 5 and 10 (D'Hont et al., 1998; Ha et al., 1999). The co-existence of two distinct chromosome organizations in modern cultivars was established on the basis of these studies. Using genomic *in situ* hybridization (GISH), D'Hont et al. (1996) and Cuadrado et al. (2004) have demonstrated that modern cultivars contain around 15% to 20% *S. spontaneum* chromosomes and also that less than 5% are recombinant or translocated chromosomes. The high ploidy and complex genome structure of these modern cultivars creates challenges for both transgenesis and the development of molecular markers.

According to Berding et al. (1997) and Hogarth et al. (1997), much of the progress in increasing sugarcane productivity has occurred as a consequence of genetic improvement in sugarcane by conventional breeding methods. Substantial economic gains achieved in the past by controlling major sugarcane diseases with clones selected for disease resistance are some of the important examples of this aspect. The estimates of Cox and Hansen (1995) also supported the significant contribution of breeding in enhancing crop production.

As a matter of fact, sugarcane biotechnology started in the 1960s with *in vitro* plant regeneration research, and serious efforts to improve this crop by molecular technique have recently commenced (Heinz and Mee, 1969; Nickell, 1964). Invention of an efficient transformation system made by Bower and Birch (1992) was the first major achievement towards developing an integrated molecular/conventional breeding system for sugarcane. Following this development, much effort has been made to engineer important agronomical traits into various cultivated genotypes (Allsopp and Manners, 1997; Arencibia et al., 1997; Enriquez-Obregon et al., 1998; Gallo-Meagher and Irvine, 1996; Manickavasagam et al., 2004; Setamou et al., 2002; Smith et al., 1996; Vickers et al., 2005a, 2005b; Zhang et al., 1999). Considerable attention has also been given to the development of molecular marker technologies for sugarcane breeding and variety identification (Butterfield et al., 2003; D'Hont et al., 1995; D'Hont and Glaszmann, 2001; McIntyre et al., 2001; Ming et al., 2002a, 2002c; Oropeza and de Garcia, 1997) and also to the structural and functional genomics (Da Silva et al., 2003; Grivet and Arruda, 2002).

Worldwide, sugarcane crop productivity has progressively increased to remarkable levels in the last century (Keating and Wilson, 1997). This increase in productivity has been ascribed to the development and widespread use of improved cultivars with increased resistance to diseases and pests; better management of water, nutrients and other resources;

and the availability of relatively cheap chemical fertilizers and pesticides. Sustaining this pace of improvement in crop productivity by innovative and intensive agriculture, whilst ensuring minimal environmental impact, will be one of the major challenges to maintain a profitable sugar industry in the future.

Much of the progress in increasing sugarcane productivity has come from the genetic improvement of sugarcane by conventional breeding (Berding et al., 1997; Hogarth et al., 1997). For instance, substantial economic gains were made in the past by controlling major sugarcane diseases with clones selected for disease resistance (Hogarth et al., 1997). The significant contribution of breeding to crop yield is further supported by Cox and Hansen (1995), who reported that sugarcane productivity could be greatly increased by releasing elite cultivars. These past achievements are impressive, but it is becoming increasingly apparent that only an integrated approach combining both conventional and molecular breeding strategies will enable the sugar industry worldwide to face the challenges ahead. Important developments in sugarcane biotechnology; identification of major challenges in this area; and evaluation of biotechnology so as to overcome the various constraints in order to develop more useful, profitable and productive cultivars of sugarcane will be presented and discussed in this chapter.

Current status on sugarcane biotechnology

Biotechnology offers excellent opportunities for sugarcane crop improvement. Commercial sugarcane, mainly the interspecific hybrids of S. *officinarum* and S. *spontaneum* (Bakker, 1999), would greatly benefit from biotechnological improvements due to its complex polyploidy–aneuploid genome, narrow genetic base, poor fertility, susceptibility to various diseases and pests and long duration (12–15 years) required to breed elite cultivars. More important, there is an ongoing need to provide durable disease and pest resistance in combination with superior agronomic performance in the commercially exploited clones. This led to considerable research in different areas of biotechnology pertinent to sugarcane breeding and disease control. Thus far, biotechnology research for sugarcane improvement has been in the areas of (1) cell and tissue culture techniques for molecular breeding and propagation; (2) engineering novel genes into commercial cultivars; (3) molecular diagnostics for sugarcane pathogens to improve exchange of *Saccharum germplasm* and the germplasm of related species such as *Miscanthus* and *Erianthus*; (4) developing genetic maps using molecular marker technology; and (5) understanding the molecular basis of sucrose accumulation in the stem. Remarkable achievements have been made in the first three areas and significant advancements have recently occurred in the other areas.

Cell and tissue culture techniques and their applications

History of in vitro cultures

The first attempt made by G. Haberlandt, a German botanist, in 1902 to culture isolated mesophyll and palisade cells of *Lamium purpureum* on a nutrient medium containing inorganic salts and sucrose was not completely successful due to poor knowledge on hormonal and nutritional requirements at the time. However, between 1907 and 1909, Harrison, Burrows and Carrel succeeded in culturing animal and human cells *in vitro*. Although earlier workers had achieved *in vitro* culture of orchid seedlings, embryos and even plant organs, in 1939 Nobécourt, Gautheret and White (cf. Street, 1973) succeeded in raising the first plant tissue culture. The development in this field was rapid after the Second World War and numerous results important for agriculture, forestry and horticulture have been published (Bhojwani et al., 1986; Pierik, 1979).

Due to the late discovery of plant hormones (growth regulators), the progress in plant tissue culture lagged behind animal and human tissue culture. With the discovery of the auxin (IAa), the first plant growth regulator, great opportunities for the *in vitro* culture of plant tissues were created. The discovery of the regulator Kinetin (a cytokinin) in 1955 was a further stimulus to plant tissue culture. Since that time tremendous developments have taken place initially in France and the United States and later in other countries.

The work carried out by Steward (1958) provided further insight of the totipotency of plant cells. It has been shown that certain specialized cells of the plant body can revert back to the meristematic state and subsequently regenerate into whole plants under defined nutritional, hormonal and physical conditions. The discovery of Skoog and Miller (1957) that shoot and root initiation was basically regulated by interaction between two hormonal substances (auxins and cytokinins) was a breakthrough in the *in vitro* propagation of diverse plants. They also demonstrated that both substances are necessary for tissue growth, and the pattern of morphogenesis of determined by their relative concentration and sequence of application in the nutrient medium. Later, Murashige (1974) suggested that a relatively high concentration of auxin favours root initiation while suppressing shoot formation and vice versa.

Another dimension to the *in vitro* plant culture technique was added by Murashige and Skoog (1962) who standardized and developed the basic nutrient medium using tobacco tissues as the experimental material. This medium, popularly known as 'MS', is a widely used media for *in vitro* experiments.

The substantial increase in the number of research workers during the last 20 years can be accounted for the benefits which the *in vitro* culture of higher plants offer agricultural scientists, plant breeders, molecular biologists, biotechnologists and so forth. Consequently, the number of research workers in the field is continuously on the increase. In the following, an effort has been made to summarize the progress made in the field of sugarcane tissue culture during the past few decades.

Since the pioneering work on induction of callus at the Hawaiian Sugar Planters Association Experiment Station and the production of roots on callus by Nickell (1964), cell and tissue culture of sugarcane has emerged as a valuable tool for various research activities. A few years later, the first report of sugarcane plant regeneration from callus cultures was published independently by Heinz and Mee (1969) and Barba and Nickell (1969). Heinz and Mee (1971) then demonstrated the occurrence of large variations in both chromosome number and morphological characters in the plants regenerated from callus. These initial but important findings led to the exploitation of *in vitro* cell and tissue culture for various applications such as micropropagation (Hendre et al., 1983), breeding (Krishnamurthi and Tlaskal, 1974), germplasm conservation (Taylor and Dukic, 1993), the elimination of systemic pathogens (Parmessur et al., 2002; Wagih et al., 1995) and genetic engineering (Arencibia et al., 1995; Bower and Birch, 1992) of sugarcane.

Sugarcane was first cultured *in vitro* by Nickell (1964) in 1961 using mature parenchyma of internodal tissues for some physiological studies. Later, after demonstration of totipotency in callus cultures of sugarcane (Barba and Nickell, 1969; Heinz and Mee, 1969), rapid progress was made in cell and tissue cultures of this crop, and it was found that the cultures could be raised from any part of the plant and that the highest regenerable callus could be produced from immature leaves and inflorescences (Heinz 1977). It is now well established that the morphogenetic responses of different tissues are under the influence of hormones.

In the past few years many laboratories around the world where sugarcane research programmes are in progress have utilized tissue culture techniques in sugarcane improvement. The major potential areas identified so far include (1) somaclonal variation for crop improvement, (2) micropropagation for seed cane multiplication of new varieties, (3) rejuvenation of older elite varieties, (4) development of interspecific and intergeneric hybrids, (5) disease resistance, (6) selection for biotic and abiotic stresses, (7) protoplast culture, (8) molecular biology, (9) genetic transformation, (10) *in vitro* germplasm conservation and (11) artificial seed. Considerable progress has been made in almost all these areas in India (Sreenivasan and Jalaja, 1985).

Micropropagation

Sugarcane is a perennial grass that normally propogates vegetatively through nodal buds and rhizomes for commercial purpose culturing through seed propagation also occurs (Bakker, 1999). Micropropagation is an *in vitro* method for clonal multiplication of plants using meristematic or non-meristematic cells/tissues as the explant. Plants can be regenerated directly (adventitiously) from the explant (Geijskes et al., 2003) or indirectly (*de novo*) through the callus derived from the explant (Heinz and Mee, 1969). In sugarcane, plants have been produced by direct regeneration from both apical and axillary meristems (Taylor and Dukic, 1993) and from immature leaf tissues (Geijskes et al., 2003; Lakshmanan et al., 2002). As with other plant species, sugarcane plants propagated *in vitro* from meristems are considered to be more genetically and phenotypically stable than those produced from callus (Hendre et al., 1983; Lee, 1987). Thus, considerable effort has been expended to investigate the adaptability of meristem culture to commercially grown elite sugarcane cultivars (Burner and Grisham, 1995; Taylor and Dukic, 1993). Some recent reports indicate that direct regeneration can be obtained from a number of sugarcane genotypes using a thin cell layer culture of immature leaf of inflorescence tissue (Geijskes et al., 2003; Lakshmanan et al., 2003), which would also reduce the time taken for *in vitro* propagation.

Although considerable effort was initially directed to establish procedures for direct plant regeneration, the development of callus-based propagation methods was not neglected. From the published results, it is evident that essentially every part of the sugarcane plant is capable of callus production (Liu, 1981); however, only immature leaves (Irvine and Brenda, 1987) and the inflorescence (Liu, 1993) are capable of producing morphogenic callus to any appreciable level. Interest in callus-based regeneration increased significantly when the experimental results substantiated earlier predictions that *in vitro*-induced mutation could be used for sugarcane improvement (Price and Warner, 1959). However, only a few such examples have appeared since then (Krishnamurthi and Tlaskal, 1974; Oropeza et al., 1995), and little useful *in vitro*-induced variability for important phenotypic characteristic was observed in callus-derived plants.

A serious problem in conventional sugarcane breeding is that it takes several years (8–10 years) before any newly identified variety can be commercially released. It further takes about 10 years to reach the growers in remote villages for general cultivation if multiplied through conventional methods. By that time the varieties start declining. Lack of adequate procedures for rapid multiplication of newly released varieties of sugarcane has been experienced by scientists for a long time (Singh et al., 1995) and for this reason, micropropagation offers a practical and fast method for mass production of clonal material. Various protocols for *in vitro* micro-propagation of sugarcane have been described by several investigators (Hendre et al., 1975, 1983; Jadhav et al., 2001; Kumari and Verma, 2001; Lal and Singh, 1994; Lee, 1987; Pawar et al., 2002a, 2002b; Sauvaire and Glozy, 1978; Shukla et al., 1994; Sreenivasan and Sreenivasan, 1992).

Morel (1963) applied *in vitro* culture for micropropagation of orchids. The technique was standardized as an important source of rapid multipli-cation of sugarcane and also for obtaining disease-free plants. Traditional methods of vegetative propagation exhibit a slow rate, taking 10 or more years to build sufficient stock of seed cane of a new variety for general cul-tivation. The success of a commercial method depends on sequential steps and establishment of optimum conditions for each step. Similarly, micro-propagation methods go through a series of stages with specific require-ments. Murashige and Skoug (1962) and Hammerschlag (1982) reported three stages: establishment of aseptic culture, multiplication and rooting of *in vitro* developed shoots. Another stage, acclimatization, was later added. Thus the different stages involved in micropropagation methods include establishment of *in vitro* culture, propagule multiplication, rooting of shoots, acclimatization, field transfer and development/establishment of micropropagation protocols.

Establishment of *in vitro* culture is the first stage in any micropropaga-tion programme. Efforts have been made to identify the suitable explants and various media combinations for establishment of aseptic cultures. The choice of explants for micropropagation depends on the system adopted. For producing virus-free plants, tips (meristems) are selected (Quark, 1977; Waithaka, 1982; Walkey, 1980), whereas nodal segments and large vegetative buds are the best explants for general micropropagation. The size of the explants, besides determining the rate of survival, also deter-mines its morphogenetic competence (Lim and Sio, 1985). The physiologi-cal status of the donor plant also influences the response of the explant (Anderson, 1980). About 85% to 90% percent of the shoot tip cultures suc-cessfully reached to the shoot proliferation stage (Grisham and Bourg, 1989). Explants taken from field grown plants created the problem of microbial contamination, as total sterilization of these explants was generally diffi-cult. Callus induction and shoot regeneration in several tropical plant spe-cies have been studied and it was found that sterilization efficiency was satisfactory (82%–93%) for maize, *Citrus* sp., *Brassica* sp. and winged bean (*Psophocarpus tetragonolobus*), except sugarcane (53%), which was thought to be due to the presence of surface hairs on leaves and stalks. Victoria and Guzman (1993) suggested a method to eliminate systemic pathogens of sug-arcane propagule by treatment with hot water (51°C, 1 hr) and subsequent transfer to a growth chamber at 41°C continuous temperature.

A serious problem associated with this stage is the oxidation of phe-nolic substances that leach from the explant resulting in browning of the explants and the medium, and thus killing the whole culture. A pre-treatment of the explant with a solution of an antioxidant such as ascorbic acid, citric acid, or polyvinylpyrrolidone (PVP), culturing them in liquid medium for few days or incorporation of these antioxidants and/or acti-vated charcoal in the medium, have helped in overcoming the problem (Kumar et al., 1998; Skirvin and Chu, 1979; Shukla et al., 1994; Upadhyay, 1994). Preece and Compton (1991) observed browning as a major prob-lem associated with the *in vitro* culture of sugarcane. When tissues are injured, some phenolics are released and oxidized, which are highly reac-tive, form bands with proteins and inhibit enzymes activity. Presence of phenolic compounds in higher plants and their role in plant growth, dif-ferentiation and defence response against pathogens is well documented (Kefeli and Kutlácek, 1977). *In vitro* biosynthesis of phenolic compounds is influenced by light, temperature, injury and other stress factors including nutrients (Monacco et al., 1977). Production and extracellular secretions of phenolic compounds known to affect the establishment and maintenance of cultures in many species, particularly coffee and sugarcane (Heinz et al., 1977; Monacco et al., 1977), sometimes results in the loss of entire stock. The increase in phenolics at 4% sucrose gave maximum tissue growth and shoot multiplication, indicating that phenols stimulated organ pro-liferation, as they are reported to regulate IAa metabolism and lignifica-tion in newly formed cells (Mäder and Füssl, 1982). Tissue fresh weight and endogenous phenolics peaked at 5% sucrose, whereas subsequent increase in sucrose levels showed a decline in fresh weight and phenolic content. Phenols are generally not as harmful to sugarcane tissue. The browning and death of plant tissues in culture are due to quinines, which are oxida-tion products of phenols (Monacco et al., 1977), and cause loss of enzy-matic activities. Oxidation of phenols to quinines can be controlled by the use of suitable chemical antioxidants. Maintenance of stock cultures in micropropagation can be achieved by using a combination of antioxi-dants along with sucrose concentration supporting optimal growth and multiplication in liquid media (Lal, 1990). Various chemical treatments, media conditions (i.e. liquid or gel) and rapid subculturing are reported to minimize the problems of phenols, but better a solution is to optimize a medium allowing minimum production and extracellular secretion of phenolic compounds (Lal, 1993).

The rate and mode of propagule multiplication determines the suc-cess of the micropropagation system. The morphogenesis and prolifera-tion rate of cultures depends on the various factors influencing the relative incidence of organogenesis or embryogenesis. Photoautotrophic micro-propagation of sugarcane was successfully done by Erturk et al. (1977). Sugarcane shoots in the multiplication stage were able to grow photoauto-trophically when transferred from multiplication media to sugarless MS medium 10 days after subculturing. Photoautotrophic micropropagation in an aseptic system was considered as an alternative for the multiplication stage of micropropagation in sugarcane. Conventional (aseptic) micropropagation of sugarcane is expensive and culture loss occurs due to biological contamination. Hussey (1983) suggested that repeated subculturing of the buds and shoots enhance the number of plants and all plant products have the same genetic characteristics of the original plants.

Micropropagation in sugarcane has been reported via induction of axillary shoot formation, adventitious shoot formation (Chowdhury and Vasil, 1993; Hendre et al., 1983) or through both shoot apex culture and somatic embryogenesis from callus (Lee, 1986, 1987). Grisham and Bourg (1989) compared the efficiency of *in vitro* propagation of sugarcane plants by direct regeneration from leaf tissues and by shoot tip culture. The rate of propagation by leaf roll sections was low (i.e. 4–18 plants/leaf roll) but by shoot tips, the number of shoots after six transfers in shoot proliferation medium was >27,500 plants for CP. 65-357 and >7000 plants for CP. 70-321. The rate of propagule multiplication depends upon the auxin–cytokinin balance in the culture medium, genotype and age of *in vitro* cultures. Dhumale et al. (1994) used sterilized, 2 to 3 mm long shoot tips of sugarcane cultivar CoC 671 and cultured them on modified MS medium containing 3 or 4 mgl^{-1} BAP and 0 to 2 mgl^{-1} IAA (indole-3-acetic acid). Percent induction of multiple shoots and mean multiple shoot numbers were highest at 90% and 16.5, respectively, with 3 mgl^{-1} BAP + 1 mgl^{-1} IAA. Sucrose had a significant role in proliferation and greening of *in vitro* shoot cultures. Biomass production, multiplication rate and shoot vigour were highest at 4% sucrose, while shoot chlorophyll content was highest at 2% sucrose (Lal, 1993). Evaluation of gelling and supporting methods for *in vitro* shoot multiplication in sugarcane was carried out by Lal and Singh (1993). They found the greatest fresh weight (4.93 g/explant), greatest weight increase (160.2%) and optimal multiplication rate with shaken liquid culture. Static liquid culture with or without absorbent cotton or filter paper bridge had faster multiplication rates but produced shoots with less vigour.

Victoria and Guzman (1993) developed a rapid multiplication method for sugarcane varieties, which helped in elimination of systemic pathogens of sugarcane. They also reported that 1500 plantlets were produced from each original meristem. Alam et al. (1995) found high frequency *in vitro* plant regeneration in sugarcane by using MS medium containing 0.5 to 1.0 mgl^{-1} BA and 4 mgl^{-1} NAA (naphtalene acetic acid). Casein hydrolysate (400 mgl^{-1}) favoured shoot growth. Once the propagules have multiplied in sufficient numbers, they were either left on the multiplication medium or transferred to another medium without or with lower concentrations of phytohormones for shoot elongation. Sometimes addition of GA$_3$ in the medium was required for shoot elongation in Litchi (Kumar et al., 1995, 1998). Patel et al. (1999) reported the effect of medium composition on establishment and growth of *in vitro* sugarcane meristems using variety CoLk 8001. They used different levels of BAP, NAA and IBA (indole-3-butyric acid). MS media supplemented with the combination of 0.5 mg1⁻¹ BAP + 1 mg1⁻¹ IBA and 5 mg1⁻¹ BAP + 0.75 mg1⁻¹ NAA were found to be best for establishment and shoot proliferation from meristems, taking minimum days for shoot initiation.

Sucrose is an important media component which serves as an energy source and osmoticum, and regulates *in vitro* shoot proliferation and rooting in sugarcane tissue cultures (Maretzki and Hiraki, 1980). Rooting could be accomplished under non-sterile *in vitro* conditions. The *in vitro*-raised shoots were taken out of culture vessels, and their base was treated with a standard rooting hormone solution and planted directly in potting mixture of soil. Leaf trimming resulted in more vigorous root growth in cold treatment. Islam et al. (1982) reported plantlet differentiation but found it difficult to obtain well developed roots. Generally 60% to 70% rooting was achieved while using a nutrient-rich medium containing a high amount of auxin and sucrose for a longer period (Hendre et al., 1983; Lee, 1986, 1987; Murashige and Skoog, 1962; Nadar and Heinz, 1977). Root induction in the remaining 30% shoots required water culture for 3 to 4 weeks, plantlets kept for 4 to 6 days on moist cotton placed in water-filled trays exhibited quick elongation of roots. In an experiment, Gill et al. (1989) found that the best root induction and high survival rate of transplanted plants were obtained on MS media supplemented with IAA (7.4 μm) and 4% sucrose. NAA is reported to be the strongest auxin followed by IBA and indole-3-acetic acid (IAA) for *in vitro* rooting and plantlet growth in sugarcane. Naphthaleneacetic acid was thus adjudged most suitable auxin for rooting in sugarcane tissue culture (Lal, 1992, 1999) since natural auxins are subjected to degradation by sugarcane tissues. Lal and Singh (1994) studied the role of different gelling agents and found that root induction was slowest with agar and most rapid with 0.2% gelrite. Gelrite was adjudged the cheapest of the gelling agents which produced tall multiple shoots within 4 weeks. Mannan and Amin (1999) used half-strength MS medium supplemented with different combinations of NAA, IBA and IAA, and found that a combination of 0.5 mg1⁻¹ IBA and 0.5 mg1⁻¹ NAA was the best treatment for induction of rooting.

Krishnamurthi (1981) reported potential benefits of the tissue culture technique for sugarcane improvement. Gradual acclimatization with decreasing humidity is necessary for tissue culture plants to survive the transition from *in vitro* condition to greenhouse and field conditions (Sommer et al., 1986). Sreenivasan and Sreenivasan (1992) suggested that both leaves and excess roots should be trimmed before planting the *in vitro*-raised plantlets in potting mixture and kept under shade and high humidity conditions to get up to 95% survival. Savangikar et al. (1991) noted that the elongation of nodes of tissue culture-derived planting material occurred in the third month after planting, and transplant survival was over 80%. Shukla et al. (1994) obtained 100% survival of plantlets when transplanted in a hydroponic system and 90% survival when transplanted to pots containing FYM + potting mixture. Since uniformity in tissue culture plants, both genetically and phenotypically, is an important aspect of micropropagation, Burner and Grisham (1995) found that sugarcane micropropagated from shoot tips were generally less phenotypically variable than those derived from callus cultures. Thus, during evaluation of micropropagated plants, selection of phenotypically uniform plants was essential.

Jimenez et al. (1991) conducted on experiment to compare the performance of micropropagated and conventionally raised plants and found that the former had improved vigour expressed by an increase in the number of stalks by 29.4% which was maintained in ratoon (22.8%). The impact of micropropagation of sugarcane varieties for increasing the cane yield was studied by Sreenivasan and Sreenivasan (1992) using White's and MS media for shoot multiplication. Frequent meristem subculture onto fresh medium was essential for growth and elongation. New shoots arose at frequencies of 10% to 17% after 45 days on multiplication medium. Generally four to five shootlets were obtained within 20 days which later multiplied fourfold every 2 weeks by subculturing in the same medium. Further shoot growth enhanced on solid MS medium containing 1 mgl^{-1} 2.4-D and 15% (V/V) coconut water while 1 mgl^{-1} IBA was sufficient for root stimulation in sugarcane. Shukla et al. (1994) carried out *in vitro* clonal propagation of sugarcane on modified MS media and indicated that both liquid and solid media were good for establishment of cultures. The optimum concentration of various phytohormones in different media was IAA, BAP and Kinetin (0.5 mgl^{-1} each) for establishment; IAA 0.1 mgl^{-1}, BAP 2 mgl^{-1} and Kinetin 1 mgl^{-1} for proliferation; and NAA 5.0 mgl^{-1} for rooting. They added 0.05% pyrrolione in the media for checking the release of polyphenolic compounds. Gosal et al. (1998) established an efficient protocol for mass multiplication of sugarcane using three commercial varieties, namely CoJ 64, CoJ 83 and CoPant 84211, in which shoot cultures were established from spindle (shoot tip) explants (0.5–1.0 cm) on semi-solid MS medium supplemented with IAA (0.5 mgl^{-1}), BAP (0.5 mgl^{-1}) and Kinetine 0.5 mgl^{-1}, and rapid shoot multiplication was achieved through several subculturing of shoot clumps in liquid MS medium supplemented with Kinetin and BAP (0.5 mgl^{-1} each).

Somaclonal variation for crop improvement

Sugarcane callus cultures show a considerable variation from cell to cell and among differentiated plantlets. This phenomenon of somaclonal variation was recognized very early by Heinz and Mee (1969) and it was speculated that this variation arose through (1) inherent chromosome numerical variation existing in the donor plants, (2) changes caused by *in vitro* cultural conditions which include physical (temperature, photo-period, light intensify, humidity etc.) and chemical (chemical and hor-mones used in the medium) factors, and (3) disruption of intact tissue and organelles causing mutations. Larkin and Scowcroft, who coined the term *somaclonal variation* (1981), have discussed in detail various factors respon-sible for somaclonal variation which include karyotype changes, cryptic changes associated with chromosome rearrangement, transposable ele-ments, somatic gene rearrangements, gene amplification and depletion, somatic crossing over and sister-chromatoid exchanges. Several experi-mental evidences are now available to show that some of these factors operate simultaneously in sugarcane.

The chromosomal mosaicism existing in commercial sugarcane hybrids is viewed as an important source of variation (Sreenivasan and Jalaja, 1995). Research conducted in Hawaii, Fiji, Taiwan, Australia and India clearly demonstrated that somaclonal variation is enormous and pro-vides an opportunity for improvement of highly adapted elite sugarcane varieties with one or two defect(s) by affecting many important agronomic characters. Besides variations observed in morphological characters such as stalk height, girth, stalk colour, leaf, foliar characters, auricle length, bud groove, bud missing, bud shape and size, and flowering, variations were also observed in tillering, high silicate deposits on leaf surface and differences in growth habits (Heinz and Mee, 1969).

Liu et al. (1972) in Taiwan observed morphological variation in stool-ing and erectness among sugarcane somaclones. Significant variations were recorded in characters like cane yield, sugar yield, stalk number, length, diameter, volume, density and weight, percent tibre, auricle length, dewlap shape and hair group among subclones from eight sugarcane cul-tivars (Chen, 1986; Kresovich, 1983; Liu and Chen, 1976, 1978; Liu et al., 1983). Somaclone 70-6132 was superior to its donor for stalk number, cane yield and sugar yield by 6%, 32% and 34%, respectively. This somaclone showed an improvement for aforesaid three characteristics over Taiwan's best variety (F-160) by 8%, 20% and 16%, respectively. However, a contrary report was given by Gonzalez et al. (1987) who said that somaclones did not show significant variations for stalk number and diameter from their respective donors.

Isolation of somaclones with Fiji and downy mildew disease resis-tance (Krishnamurthi, 1975; Krishnamurthi and Tlaskal, 1974; Tlaskal, 1975), eyespot disease resistance (Heinz et al., 1977; Larkin and Scowcroft, 1981), smut resistance (Liu and Chen, 1978; Sreenivasan et al., 1987), rust resistance (Sreenivasan et al. 1987), improvement in yield (Liu et al., 1983) and sucrose content (Liu and Chen, 1976, 1978; Liu et al. 1972) are important examples. Leaf drying characteristic of variety Co 7704, another complex trait, could be corrected through somaclonal variation (Sreenivasan and Jalaja, 1982). These observations have clearly shown that variability for a number of characters can be obtained by passing the genotypes through the tissue culture cycle.

To emphasize the utility of somaclonal variation in crop improvement, sugarcane is often cited as an example, along with the potato. Sugarcane is considered to be an ideal plant material for such studies because of the existence of high polyploidy, the capacity of the plants to tolerate chromosomal aberrations and the capacity of the deviant cells to differentiate into plants. However, when a somaclone with a desirable trait is identified and selected, for example disease resistance, it has to pass through all the stages of testing and adaptive trials, and thus the time taken for their ultimate release as a cultivar will remain almost the same as that of sexual populations, since there are no precise *in vitro* screening methodologies available so far for most of the characteristics. Jalaja and Sreenivasan (1995) were of the view that selection among the somaclonal population would be effective mainly for rectifying one or two specific defect(s) and not for an overall improvement of a cultivar.

Cytological analyses of somaclones derived from a number of genera belonging to family gramineae such as *Triticum durum*, Hordeum and *Triticum crassum* × *Hordeum vulgare* indicated the occurrence of chromosome translocations in somaclonal populations. In sugarcane where meiotic chromosome pairing is mostly autosyndetic, the methodology of restructuring the chromosome by passing the genotypes through a tissue culture cycle and isolating the plants with repatterned chromosome complements and their further utilization in sexual breeding will be useful for obtaining variability. The somaclonal variants will thus provide a novel gene pool for further exploitation through sexual breeding (Jalaja and Sreenivasan, 1995).

Somaclonal variation in sugarcane has been well documented (Larkin and Scowcroft, 1981), and reports of somaclones with improvement in desired traits are on the increase. Although the use of tissue culture technique in sugarcane improvement, as an aid to the conventional system, was emphasized very early (Heinz et al., 1977; Krishnamurthi, 1977), the utility of this technique will depend on the judicious selection for breeding objectives (Jambhale et al., 1995).

Somatic embryogenesis

Somatic embryogenesis is probably the most intensively investigated method of *in vitro* regeneration in sugarcane (Guiderdoni et al., 1995). Although developed originally as an alternative regeneration system to meristem culture, somatic embryogenesis has achieved prominence as an integral part of the genetic transformation system (Bower and Birch, 1992). Somatic embryogenesis has been reported from a large number of commercial sugarcane clones (Guiderdoni et al., 1995; Manickavasagam and Ganapathi, 1998), and can be obtained directly (Manickavasagam and Ganapathi, 1998) or indirectly (Guiderdoni and Demarly, 1988) from the leaf tissue. Embryogenic callus can be maintained for several months without losing its embryogenic potential to any significant level (Fitch and Moore, 1993).

Genetic stability of in vitro *plants*

The application of cell and tissue culture to clonal propagation and *in vitro* germplasm storage is significantly influenced by the genetic stability of the regenerated plant. *In vitro* culture-induced variability, although infrequently beneficial, is undesirable for both commercial propagation and germplasm storage. Genetic variability has been frequently reported in tissue-cultured sugarcane (Burner and Grisham, 1995; Heinz and Mee, 1971; Hoy et al., 2003; Lourens and Martin, 1987; Taylor et al., 1995). Studies were conducted to assess the extent of variability arising from *in vitro* regeneration and its transmission into successive generations via vegetative propagation (Burner and Grisham, 1995; Lourens and Martin, 1987). These investigations demonstrated that substantial somaclonal variability occurred in *in vitro*-derived propagules, irrespective of the method of regeneration. However, with extensive field experiments, Lourens and Martin (1987), Burner and Grisham (1995), and Irvine et al. (1991) showed that the phenotypic variations in tissue-cultured sugarcane were frequently temporary as the majority of variants reverted to the original parental phenotype in the first ratoon crop.

Interestingly, results from several other studies showed little evidence of somaclonal variation in tissue-cultured plants. For example, Chowdhury and Vasil (1993) could not identify any significant DNA variation in sugarcane plants regenerated from protoplasts, cell suspension and callus cultures. In another study, Taylor et al. (1995) only found a few polymorphisms following random amplified polymorphic DNA (RAPD) analysis of sugarcane plants produced from embryogenic cultures. Taken together, the widely disparate results of genetic variation reported by various researchers may be a reflection of different genotypes and experimental conditions employed, as the effect of the tissue culture varies considerably with the genotype and method used (Lourens and Martin, 1987).

Other applications of cell and tissue cultures

Tissue culture techniques have been employed with variable success to recover disease-free sugarcane plants from infected lines. Meristem culture was successfully used to eliminate sugarcane mosaic virus (SCMV) (Kristini, 2004), chlorotic streak disease, ratoon stunting disease and white leaf disease (Leu, 1978). In combination with heat treatment, meristem and callus cultures were effective in producing pathogen-free stocks from plants infected with Fiji disease virus (FDV) (Wagih et al., 1995), downy mildew (Leu, 1978) and SCMV (Kristini, 2004). Recent research in our laboratory showed that direct plant regeneration using thin cell layer culture could be used for rapid production of disease-free plants from sugarcane infected with FDV, SCMV and *Leifsonia xyli* subsp. *Xyli* (Kirstini, 2004). In the United States and Brazil, *in vitro* culture techniques are extensively used for the production of disease-free planting material for commercial planting.

In vitro *germplasm conservation*

Another area that has attracted researchers' interest is the application of *in vitro* techniques to germplasm conservation (Bajaj and Jain, 1995). *In vitro* storage of sugarcane germplasm has been developed at the Sugarcane Breeding Institute in India, the Centre de Cooperation International en Recherche Agronomique pour le Development (CIRD) in France and BSES Limited (formerly the Bureau of Sugar Experiment Station) in Australia. Essentially, for *in vitro* storage, apical or axillary meristem is maintained on minimal growth medium at 18°C or 25°C with transfer to fresh media every 6 to 24 months, depending on the genotype. Evidence to date indicates that *in vitro* storage of sugarcane on a low-maintenance medium for extended periods causes little genetic change, suggesting its potential use for long-term conservation and international exchange of germplasm.

Feasibility of the tissue culture technique

The tissue culture technique seems to be feasible for rapid seed multiplication of newly released varieties of sugarcane if the three-tier seed multiplication programme is followed. Tissue culture plantlets should not be distributed directly to the farmers. Initially, the plantlets should be transplanted at the laboratory's farms for raising breeder seed nurseries. The breeder seed cane obtained from tissue culture raised nurseries should be subsequently used for raising foundation and certified seed nurseries. Two budded sets obtained from about 10-month-old tissue culture raised breeder seed cane should be used for planting foundation seed nurseries. Only newly released varieties should be multiplied on the basis of demand. Micropropagation, sugarcane tissue culture protocol (STP) and such other techniques should be a part of seed production programmes. Earning profit by selling the plants should not be an objective of tissue culture labs. More labs need to be established at sugar factories to fulfil the local demand for seed cane of new varieties.

Genetic engineering

Systems of DNA delivery in sugarcane

Besides being an important food and energy crop, there are several other reasons that make sugarcane a candidate for crop improvement via genetic engineering. First, genetic improvement of elite sugarcane clones by conventional breeding is difficult due to the complex polyploidy–aneuploid genome, poor fertility and the long time period (12–15 years) required for creating new cultivars by conventional breeding. Traditional backcrossing to recover elite genotypes with desired agronomic traits is very time-consuming in sugarcane. In this context, genetic engineering is a very valuable tool to introduce commercially important traits into elite germplasm. Second, practically useful sugarcane transformation systems are available (Arencibia et al., 1998; Birch, 1997; Enriquez-Obregon et al., 1998; Manickavasagam et al., 2004) and useful transgenic lines can be maintained indefinitely by vegetative propagation.

Initial transformation technologies

In the past decade, substantial research effort has been expended to develop efficient genetic transformation systems for sugarcane (Arencibia et al., 1998; Birch, 1997; Birch and Maretzki, 1993; Bower and Birch, 1992; Chen et al., 1987; Enriquez-Obregon et al., 1998; Gambley et al., 1993, 1994; Rathus and Birch, 1992; Smith et al., 1992). Different transformation techniques using electroporation (Rathus and Birch, 1992), polyethylene glycol (PEG) treatment (Chen et al., 1987), particle bombardment (Franks and Birch, 1991) and *Agrobacterium* (Arencibia et al., 1998; Elliott et al., 1998) were used to introduce marker genes in sugarcane cells and callus. The first transgenic sugarcane cells/callus were obtained following PEG-mediated DNA transfer into protoplasts (Chen et al., 1987). Despite being a simple procedure requiring little specialized equipment, PEG-mediated transformation did not receive much attention due to its low efficiency and poor reproducibility. A transformation efficiency of one per 10^6 treated protoplasts was reported for the cultivar F164 (Chen et al., 1987).

Sugarcane transformation by electroporation was found to be more efficient and reproducible than the PEG method. Stably transformed callus has been obtained from electroporated protoplasts of different sugarcane cultivars (Chowdhury and Vasil, 1992; Rathus and Birch, 1992). Rathus and Birch (1992) documented one transformation event per 10^2 to 10^4 electroporated protoplasts for the cultivars Q63 and Q96, a frequency sufficient for the generation of useful transgenic plants. However, a lack of regeneration from protoplasts prevented production of transgenic plants with this approach. Nonetheless, transgenic sugarcane plants have been obtained following electroporation of intact embryogenic cells (Arencibia et al., 1995) and of meristematic tissues of *in vitro* grown plants (Arencibia et al., 1992).

Agrobacterium-mediated transformation

Early attempts to transform sugarcane using *Agrobacterium*, with or without virulence gene inducers and other treatments that enhance infection, were met with little success (Birch and Maretzki, 1993). More recently, Arencibia et al. (1998) were able to regenerate morphologically normal transgenic sugarcane plants following co-cultivation of calluses with *Agrobacterium tumefasiens* strains LBA 4404 and EHA 101. Almost simultaneously, Enriquez-Obregon et al. (1998) reported the production of herbicide-resistant sugarcane plants by *Agrobacterium*-mediated transformation using a different bacterial strain, C58C1 Rif^R (At 2260). By manipulating various steps of co-cultivation and plantlet regeneration, Enriquez-Obregon et al. (1998) attained a relatively high transformation frequency in the commercial cultivar Ja60-5, with nearly 35% of the co-cultivated meristematic tissues being transformed. These results indicated that both prevention of cell death due to *Agrobacterium*-induced hypersensitive reactions and the use of young regenerable callus as target tissue are critical to achieve high-frequency transformation. On the other hand, Arencibia et al. (1998) found that inducing *A. tumefaciens* virulence in the presence of sugarcane cell cultures and priming the target cells for organogenesis or embryogenesis prior to co-cultivation were crucial for enhancing the T-DNA transfer process. Transformation of another commercial sugarcane cultivar Q 117 with the *A. tumefaciens* strain AGLO using green fluorescent protein as a visual marker, allowing simple visual selection of transgenic lines, has also been reported (Elliott et al., 1998, 1999). Manickavasagam et al. (2004) used axillary buds as target tissue to transform two *Saccharum officinarum* accessions at high efficiency with *A. tumefaciens* strain EHA 105. Using the axillary bud system they produced herbicide BASTA-resistant sugarcane in about 5 months, with transformation efficiency close to 50% (50 out of 100 buds co-cultivated were transformed), the highest rate ever reported for sugarcane.

Microprojectile-mediated transformation

To date microprojectile-mediated transformation, a technique for introducing DNA by bombarding the target tissue with DNA-coated microprojectiles, is the most widely exploited method for sugarcane transformation

(Birch, 1997; Moore, 1999). Developed in the 1960s for the inoculation of intact plants with infectious viral particles (Mackenzie et al., 1966), particle bombardment was used by Hawaiian researchers to produce transgenic plants of a *Saccharum* species (Maretzki et al., 1990). Investigation on microprojectile-mediated transformation by Franks and Birch (1991) in Australia led to the development of the first transgenic sugarcane plants from a commercial cultivar in 1992 (Bower and Birch, 1992). Subsequently, microprojectile-mediated transformation of several commercially cultivated sugarcane genotypes was reported from a number of laboratories worldwide (Birch, 1997; Birch and Maretzki, 1993; Bower et al., 1996; Irvine and Mirkow, 1997; Joyce et al., 1998a, 1998b; Moore, 1999; Nutt et al., 1999).

There were two principal technologies driving the rapid development of microprojectile-mediated methods for sugarcane transformation: (1) the availability of the equipment required for microprojectile-mediated transformation and (2) the ability to produce somatic embryogenic systems from a range of sugarcane genotypes. Among the different tissues tested, embryogenic callus appears to be the preferred target due to its high transformability and regenerability. Nonetheless, regenerable cell suspension (Chowdhury and Vasil, 1992; Franks and Birch, 1991), apical meristems (Gambley et al., 1993), and immature leaf whorls and inflorescences (Elliott et al., 2002; Lakshmanan et al., 2003) have also been successfully used for microprojectile-mediated sugarcane transformation. The applicability to a wide range of target tissues and genotypes, and the simplicity of operation, make the microprojectile approach the preferred method for sugarcane transformation.

Development of commercially useful transgenic sugarcane

Following the development of a microprojectile transformation system (Bower and Birch, 1992), considerable efforts were directed to engineer economically important traits into commercially grown sugarcane cultivars (Table 11.1). In the past few years, transgenic sugarcane plants with improved resistance to a number of microbial pathogens, such as SCMV (Joyce et al., 1998a, 1998b), sorghum mosaic potyvirus (SrMV) (Ingelbrecht et al., 1999), FDV (McQualter et al., 2001, 2003b, 2004a) and *Xanthamanas albalinians* (Zhang et al., 1999) were reported. Considerable success in developing resistance to pests such as canegrubs (Allsopp et al., 2000; Nutt et al., 1999), Mexican rice borer and sugarcane borer (Arencibia et al., 1999; Legaspi and Mirkow, 2000; Setamou et al., 2002), and plants containing other agronomically useful traits such as herbicide resistance (Enriquez-Obregon et al., 1998; Gallo-Meagher and Irvine, 1996; Leibbrandt and Snyman, 2003) has also been reported. Efforts are also underway to engineer sugarcane for increased sugar accumulation (Ma et al., 2000), low colour raw sugar (McQualter et al., 2004b; Roberts et al., 1996), although

Table 11.1 Examples of markers and traits engineered into sugarcane

Trait	Gene	Transformation method	Sugarcane variety	Field trial	Reference
		Metabolic engineering and alternative products			
Sucrose accumulation	Antisense soluble acid invertase	Microprojectile	H62-4671 (suspension cells)	No	Ma et al., 2000
Soluble acid invertase	Microprojectile	Q117	No	Botha et al., 2001	
Fructooligosaccharide	IsdA	Agrobacterium	B4362	No	Enriquez et al., 2000
Polyphenol oxidase	Ppo	Microprojectile	Q117	Yes	Vickers et al., 2005a
Polyhydroxbutyrate	phaA, phaB, and phaC	Microprojectile	Q117	No	Brumbley et al., 2003
p-Hydroxybenzoic acid	hchl and cpl	Microprojectile	Q117	No	McQualter et al., 2004b
Phosphinothricin acety transferase	Bar	Agrobacterium	Q117	No	Elliott et al., 1998
		Herbicide resistance			
Bialaphos	Bar	Microprojectile	NCo310	Yes	Gallo-Meagher and Irvine, 1996
Phosphinothricine	Bar	Agrobacterium	Ja60-5	Yes	Enriquez-Obregon et al., 1998
Glufosinate ammonium	Pat	Microprojectile	NCo310	Yes	Leibbrandt and Snyman, 2003

(Continued)

Table 11.1 (Continued) Examples of markers and traits engineered into sugarcane

Trait	Gene	Transformation method	Sugarcane variety	Field trial	Reference
		Disease resistance			
SCMV	SCMV-CP	Microprojectile	Q155	No	Joyce et al., 1998a, 1998b
SrMV	SrMV-CP	Microprojectile	CP72-121, CP65-357	No	Ingelbrecht et al., 1999
SCYLV	SCYLV-CP	Microprojectile	CC 84-75	No	Rangel et al., 2003
Fiji leaf gall	FDVS9 ORF1	Microprojectile	Q124	No	McQualter et al., 2004a
Sugarcane leaf scald	albD	Microprojectile	Q63, Q87	Yes	Zhang et al., 1999
Puccinia melanocephala	Glucanase, chitanase and ap24	Agrobacterium	B4362	No	Enriquez et al., 2000
		Pests resistance			
Sugarcane stem bore	crylA	Electroporation	Ja60-5	Yes	Arencibia et al., 1999
Sugarcane canegrub resistance	gna or pinall	Microprojectile	Q117	No	Nutt et al., 1999
Mexican rice borer	gna	Microprojectile	CP65-357	Yes	Legaspi and Mirkov, 2000
Sugarcane stem borer	gna	Microprojectile	CP65-357	Yes	Setamou et al., 2002

current field data are only available for sugarcane plants engineered to resist sugarcane borer (Arencibia et al., 1999) and Mexican rice borer (Legaspi and Mirkov, 2000) infestation, raw sugar colour improvement by manipulating polyphenol oxidase activity (Vickers et al., 2005a), and herbicide resistance (Gallo-Meagher and Irvine, 1996; Leibbrandt and Snyman, 2003). Results obtained from glasshouse trials with transgenic sugarcane producing bioplastics (McQualter et al., 2004b) and those resistant to leaf scald (Zhang et al., 1999), SCMV (Joyce et al., 1998a, 1998b), SrMV (Ingelbrecht et al., 1999) and canegrubs (Allsopp et al., 2000) are encouraging. For instance, the Q117 transgenic lines engineered with the potato proteinase inhibitor II (Murray and Christeller, 1994) or snowdrop lectin genes (van Damme et al., 1987) exhibited levels of antibiosis to the canegrub species, *Antitrogus consanguineus* (Allsopp et al., 2000). Similarly, a significant increase in resistance to SCMV was shown by plants engineered with the SCMV coat protein gene in glasshouse trials (Joyce et al., 1998a, 1998b).

Polymerase chain reaction

Singh et al. (2004) demonstrated that by using polymerase chain reaction (PCR) to amplify the bE mating-type gene of *Ustilago scitaminea*, it is possible to detect this pathogen in sugarcane. Different parts of smut-inoculated tissue-cultured plantlets were assayed to determine whether plant resistance to smut infection could be effectively evaluated by PCR. The PCR assay was significantly better for smut detection than microscopy. Genetic transformation of sugarcane (*Sacchrum* spp.) holds promise for increasing yields and disease resistance. However, the tissue culture and transformation process may produce undesirable field characteristics in transgenic sugarcane. Agronomic evaluation of sugarcane lines transformed for resistance to SCMV strain (Gilbert et al., 2005) was done to evaluate variability in agronomic characteristic and field disease resistance of sugarcane transformed for resistance to SCMV strain E. One hundred plants derived from cultivars CP 84-1198 ($n = 82$) and cp 80-1827 ($n = 18$), consisting of independent virus resistant lines VR 1 ($n = 14$), VR 4 ($n = 24$), VR 14 ($n = 14$) and VR 18/$n = 58$ were evaluated. Transgenics derived from CP 84-1198 had significantly greater tones of sucrose per hectare (TSH) and significantly lower SCMV disease incidence than those from CP 80-1827 in the plant cane (PC), first ratoon and second ratoon crops. Plants from the VR 18 line had significantly greater economic indices and lower SCMV disease incidence than the UR + lines in all three crops. Phenotypic variation was high with tonnes of cane per hectare (TCH) ranging from 26 to 211 and TSH from 3.2 to 28.9 in the PC crop. Agronomic trait variation decreased with increased selection pressure, evaluating 30 UR 18 lines, with TCH ranging from

70 to 149 and TSH from 8.5 to 19.0 in PC. The large variability in yield characteristics and disease resistance encountered demonstrate the necessity of thorough field evaluation of transgenic sugarcane while selecting genetically stable and agronomically acceptable material for commercial use.

In a series of field experiments (Vickers et al., 2005c), it was demonstrated that PPO activity among clones correlated significantly with juice colour. In the laboratory, crystallization of raw sugar using juice derived from clones with high and low PPO activity, the juice with the higher PPO activity produced darker coloured crystals. PPO activity was elevated and juice colour was darker in all types of transgenic plants. However, clones derived from a sense construct had higher PPO activity than the other transgenic clones, tissue culture control clones or cultivars. PCR analysis demonstrated that for set 1 and set 2, all sense and antisense plants had the respective transgene present. The sense approach sometimes leads to downregulation of the target gene through co-suppression (Napoli et al., 1990). It is not clear why PPO activity is elevated in the juice of transgenic plants. Either more PPO activity is released into the juice from the transgenic plants or the transgenic plants have a higher level of activity then non-transformed controls. Roles have been proposed for PPO activity such as defence against pathogens or regulating oxygen evolution in photosynthesis. It is not clear how the production of transgenic plants could cause the regulation of PPO activity to be altered. There is evidence, however, for raised PPO activity in the process of dedifferentiation and redifferentiation of shoots from *Feronia limonia* (L.) Swingle and altered PPO activity in callus of *Ranunculas asiticus* L. The similarity between the PPO activities in the TC and AC plant suggest that elevation could have been caused by the tissue culture process. Viswanathan (2005) standardize PCR for the detection of phytoplasmas causing grass disease (GSD) in sugarcane using two sets of forward and reverse primers of 16 Sr RNA/235 spacer region and 165 of rRNA were specifically amplified by PCR. PCR-based diagnosis can be used for the detection of GSD phytoplasm in sugarcane.

There is increasing pressure worldwide to enhance the productivity of sugarcane cultivation in order to sustain profitable sugar industries (Handlon et al., 2000). The significant challenges to increase crop productivity could be addressed by innovative strategies focused on developing new and improved cultivars, improving farming systems, reducing input, and enhancing the efficiency of product recovery systems. Recent advances in biotechnology offer several opportunities to address issues related to the development of novel, high-yielding cultivars (Briggs and Koziel, 1998; Ellis et al., 2000).

Transgenic technology

Despite the advancements in DNA delivery systems, current sugarcane transformation technology is far from ideal. The availability of suitable gene expression elements, the phenomenon of transgene silencing, and the limited knowledge about the performance and heritability of transgenes in sugarcane are of immediate concern.

Promoters for transgene expression

The availability of strong constitutive and tissue-specific promoters is critical for the development of commercially useful transgenic plants, especially for biofactory programs. The CaMV 35S promoter, though very active in dicots, shows low levels of transgene expression in most of the sugarcane studies reported to date. In sugarcane, the maize ubiquitin 1 promoter (Ubi-1) (Christensen and Quail, 1996) produced significantly higher expression than other tested promoters (Gallo-Meagher and Irvine, 1996; Rathus et al., 1993), including the CaMV 35S promoter, the synthetic Emu promoter (Last et al., 1991) and the rice actin Act 1 promoter (McElroy et al., 1991). From the current research trend it appears that Ubi-1 is emerging as the promoter of choice for constitutive expression of transgenes in sugarcane.

In a previous study using transient expression assays, two sugarcane Ubi promoters, Ubi-4 and Ubi-9, were found to produce high levels of transgene expression in callus from a number of species including sugarcane and rice (Wei, 2001; Wei et al., 1999). Further characterization of these promoters showed that Ubi-9 could drive high transgene expression in stably transformed rice but not in sugarcane due to post-transcriptional gene silencing (Wei et al., 2003). Similar silencing of sugarcane promoters following reintroduction into sugarcane has been noted earlier (Birch et al., 1995; Hansom et al., 1999), but these studies also reported endogenous promoters that remained active when introduced back into sugarcane. Recently, in an effort to identify promoters that confer high levels of gene expression in sugarcane, Yang et al. (2003) used an approach of direct screening of genomic libraries for highly expressed genes. A total of 11 promoters were isolated with this strategy and 2 of them, obtained from sugarcane elongation factor 1α and sugarcane proline-rich protein-encoding genes, drove β-glucuronidase (GUS) expression in sugarcane callus and wheat embryos at levels equivalent to that obtained with maize Ubi-1 promoter. In another study, the GUS gene driven by a promoter of a sugarcane UDP-glucose dehydrogenase gene showed developmentally regulated transgene expression in young leaves and developing internodes (van der Merwe et al., 2003). Notably, this promoter on its own was ineffective in driving transgene expression in sugarcane but showed a very strong GUS activity when fused with an intron (980 bp) derived from the 5'UTR of the same gene. Together, these observations indicate that an endogenous promoter may or may not provide stable, high-level trans-gene expression in sugarcane, but a detailed evaluation of the activity of these potentially useful promoters in driving transgene expression at dif-ferent phases of sugarcane development, from seedling through mature plant, is still missing.

In the recent past, research has been focused on increasing the range of promoters available for sugarcane transformation. Liu et al. (2003) have recently shown that the rice ubiquitin promoter RUBQ2 has increased transgene expression by about 1.6-fold over the maize Ubi-1 promoter in sugarcane. Promoter elements derived from plant viruses have also been recently explored. Braithwaite et al. (2004) compared four genomic regions of sugarcane bacilliform virus that have promoter activity in sugarcane. One of these promoters drove GUS expression in meristems, leaves and roots of glasshouse-grown sugarcane plantlets at levels equal to or higher than that obtained with the maize Ubi-1 promoter. In another study, Schenk et al. (2001) isolated a promoter from banana streak virus capable of driving green fluorescent protein (GEP) expression up to three-fold higher than that reported with the maize Ubi-1 promoter. Very little is known about the tissue specificity of promoters in sugarcane. Strong tissue-specific promoters, particularly for stem and root tissues, would be very valuable for biofactory and crop improvement research. For instance, manipulation of product accumulation in the stem or successful control of stem borer, canegrub, or soldier fly infestation by a transgenic approach will largely depend on the availability of promoters that are specific to or highly active in stem or root tissues. A sugarcane stem-specific pro-moter, UQ67P, has been isolated and shown to be able to drive reporter gene expression in stem tissues (Hansom et al., 1999). Recently, Mirkov and co-workers at Texas A&M University were successful in cloning two stem-specific promoters from sugarcane. These promoters showed strong reporter gene activity in stem tissues of transgenic sugarcane (T.E. Mirkov et al., unpublished results). Unfortunately, these is no published informa-tion on sugarcane root–specific promoters.

Transgene integration and expression

Commercial exploitation of transgenic crops is determined by the level of transgene expression and the maintenance of all the traits that char-acterize an elite germplasm. Microprojectile-mediated transforma-tion is the principal method of introducing transgenes into sugarcane. Inherently, transgenic lines produced by this method show large variation in transgene expression. Recent investigations suggest the mechanisms of transgene integration and rearrangements appear to be the same in transformation systems based on direct DNA delivery and *Agrobacterium* (Somers and Makarevitch, 2004). However, microprojectile-mediated transformation is different from *Agrobacterium* in that it often results in plants with fragmented and multiple copies of transgenes, which is attributed to gene silencing. Although the exact mechanism is not known in sugarcane, transgene silencing has been shown to occur at both transcriptional and post-transcriptional levels by promoter methylation and RNA degradation, respectively (Ingelbrecht et al., 1999). The same study also showed that RNA-mediated, homology-dependent post-transcriptional gene silencing (PTGS) is the basis of transgene-mediated virus resistance observed in sugarcane.

Interestingly, molecular characterization of transgenic sugarcane lines engineered with FDV S9 did not show my correlation between FDV resistance and transgene copy number (McQualter et al., 2004a). The FDV-resistant line, the only one out of 64 transgenic lines tested, contained eight copies of FDV S9, whereas the copy number in all other susceptible lines ranged from 3 to 21, mostly more than 10. It is important to note that no FDV S9 transcripts or the corresponding expressed proteins were detected in any of the transgenic lines tested, including the resistant clone, indicating that a transgenic approach based on PTGS may not be effective against double-stranded (ds) RNA viruses such as FDV. This is probably due to the inherent feature of reoviruses retaining its dsRNA within the viral particle, preventing its exposure to RNA degradation mechanisms of the host cytoplasm. A number of transgenic sugarcane plants with different engineered traits were tested for their expression in the field (Arencibia et al., 1999; Gallo-Meagher and Irvine, 1996; Leibbrandt and Snyman, 2003; Vickers et al., 2005b), but in the absence of detailed molecular characterization of these plants, it is impossible to draw any meaningful conclusion on aspects for transgene integration/expression and phenotype under field conditions.

Whereas unraveling the pathways of transgene silencing would offer long-term strategies for successful engineering of complex genomes, linear DNA (L-DNA) and *Agrobacterium*-mediated transformation are two practical approaches that are likely to address transgene silencing in sugarcane. *Agrobacterium*-mediated transformation results in 'perfect' transgene loci often with complete and single copy transgene insertions in actively transcribed regions and with no damage to the recipient genome (Somers and Makarevich, 2004). Methods for *Agrobacterium*-mediated sugarcane transformation have already been developed in Australia, Cuba, India, and Taiwan. Considerable effort is now under way in Australia to develop this method as a routine transformation system for the local commercial varieties.

A major bottleneck that limits the application of the *Agrobacterium* method is its genotype dependence. L-DNA transformation, however, combines the advantages of both microprojectile and *Agrobacterium*-mediated transformation. Being a direct DNA delivery system, L-DNA transformation is less genotype dependent and has been shown to produce a large proportion of rice plants with single or low copy-number insertions without compromising transformation efficiency (Fu et al., 2000). Experiments conducted with two Australian commercial sugarcane varieties (Q117 and Q205) showed that L-DNA transformation slightly increased the efficiency of transformation, as compared to that obtained with supercoiled plasmids (Geijskes et al., 2003).

Evaluation of transgenic plants and transgene inheritance

Transgenic sugarcane plants with a variety of introduced genes have been field tested to evaluate the expression of the new phenotypes under field conditions. Collectively, these studies are providing some useful insights into the nature of transgene expression and somaclonal variation in relation to the transformation strategies employed. In one of the first field trials, bialaphos-resistant NCo310 clones produced by microprojectile-bombardment were screened for herbicide resistance (Gallo-Meagher and Irvine, 1996). As expected, different transgenic lines showed stable integration of 3 to 10 copies of the *bar* gene in their genome and displayed varying levels of resistance to the commercial herbicide Ignite. The most striking finding of this study was the retention of similar levels of herbicide resistance in one of the best characterized clones even after three rounds of vegetative propagation in the field. While this study did not report information on morphological and agronomic characters of transgenic lines grown in the field, Leibbrandt and Snyman (2003) recorded agronomic performance of transgenic clones of the same variety, NCo310, containing the *pat* gene, which confers resistance to another herbicide Buster (glufosinate ammonium). As observed with bialaphos-resistant plants, different transgenic lines showed wide variation in herbicide resistance, and one of the best performing clones exhibited stable expression of herbicide resistance during three rounds of vegetative propagation. Notably, important agronomic traits such as height, diameter and the number of stalks, fibre content, disease resistance, and yield of transgenic clones were not significantly different from that of untransformed controls.

Contrary to the aforementioned report, results of a field trial involving insect-resistant transgenic sugarcane plants produced by cell electroporation presented evidence for the occurrence of limited but consistent morphological, physiological and phytopathological variation caused by transformation (Arencibia et al., 1999). As many as 51 polymorphic DNA bands were detected in five selected elite transgenic clones by amplified fragment length polymorphism (AFLP) analysis, and detailed molecular characterization of this population pointed towards minor genomic changes arising from events associated with plant regeneration. This finding, however, is not completely consistent with the recent report that tissue culture and transformation events have significant, independent negative effects on yield parameters of transgenic sugarcane with altered polyphenol oxidase activity (Vickers et al., 2005b). Based on the data presented in this study, the tissue culture process is not solely responsible for the poor performance of transgenic lines and the yield depression may not be significantly improved with ratooning or by further vegetative propagation.

In order to investigate the inheritance and segregation pattern of transgenes in sugarcane, Butterfield et al. (2002) crossed transgenic sugarcane harbouring multiple copies of herbicide resistance (*bar*) and SrMV resistance genes with non-transgenic sugarcane varieties. Depending on the linkage relationships between transgenes in the parent plants, distinct segregation patterns were observed in the progeny. For example, in two of the transgenic parents where all copies of the transgenes were linked to one position, progenies with the same copy number and organization as the parents were recovered at a relatively high proportion. However, progenies with rearranged, mutated transgene DNA fragments were observed with parents carrying independent, unlinked transgenic loci. Evidently, most transgenic progenies containing the *bar* gene showed herbicide resistance, implying that plants with stable inheritance of transgene could be used for breeding.

Transgenics for resistance to disease and pests

Diseases and pests cause considerable economic losses to sugar industries throughout the world. A recent estimate in Australia found that two-thirds of the loss is caused by soil-borne pathogens (McLeod et al., 1999). Sugarcane is susceptible to a host of viral, bacterial, fungal and phytoplasma diseases, and there are also at least seven recognized sugarcane diseases of unknown aetiology (Rott et al., 2000). In most sugar industries, diseases are controlled by an integrated approach involving the use of disease-resistant cultivars, disease-free planting material, appropriate farm management practices and strict quarantine measures. Some of the commercially grown sugarcane cultivars are susceptible to more than one pathogen. Further, a number of elite, high-yielding lines developed in breeding/selection programs have not been released commercially due to susceptibility to pathogens. Thus, there is an ongoing requirement to retain, maintain or introduce resistance to various pathogens in the currently cultivated genotypes as well as other valuable sugarcane germplasm with commercial or breeding potential.

The major viral pathogens of sugarcane include SCMV, FDV SrMV, sugarcane streak virus (SSV) and sugarcane yellow leaf virus (SCYLV). There is considerable resistance to different viral diseases in the '*Saccharum* complex' gene pool (Hogarth et al., 1997), and current sugarcane breeding programs have, in many instances, successfully exploited these sources of natural resistance. However, as mentioned earlier, selective inclusion of a particular resistance trait into agronomically elite clones by conventional breeding has proven to be very difficult due to the complexity of the sugarcane genome.

One of the most attractive and practically demonstrated molecular approaches to develop resistance against viral diseases is the use of pathogen-derived resistance (PDR) genes, the genetic elements originating from and conferring resistance against the pathogen itself. The PDR genes that have been engineered into plants to develop virus resistance include full length, truncated, or mutated virus coat-proteins; antisense sequences to sections of the viral genome; viral replicases; satellite RNA; and viral movement proteins. With the genomes of a number of sugarcane viruses fully or partly sequenced (Bouhida et al., 1993; Geijskes et al., 2002; Hughes et al., 1993; McQualter et al., 2003a, 2003c; Moonan et al., 2000; Soo et al., 1998), it is highly likely that molecular breeding will emerge as an important strategy to expand viral disease resistance in sugarcane germplasm. The use of PDR genes to develop resistance against SCMV (Joyce et al., 1998a, 1998b), SrMV (Ingelbrecht et al., 1999), SCYLV (Rangel et al., 2003), and FDV (McQualter et al., 2001, 2003b, 2004a) has already been demonstrated in sugarcane.

There are about 160 fungal and 8 bacterial pathogens of sugarcane reported to date (Rott et al., 2000). Similar to the viral pathogens, there is substantial natural resistance present within the *Saccharum* complex against many of these fungal and bacterial pathogens (Comstock et al., 1996; Hogarth et al., 1997; Vijaya, 1997). However, many elite sugarcane clones are susceptible to different fungal and/or bacterial diseases, limiting their commercial exploitation. Unfortunately, with the exception of leaf scald disease (Birch and Patil, 1987a, 1987b; Zhang et al., 1999), little is known about the molecular basis of pathogenesis of the numerous fungal and bacterial diseases of sugarcane. This currently limits opportunities to specifically engineer resistance against these pathogens. Some valuable information about *Leifsonia xyli* subsp. *Xyli*, the causal agent of ratoon stunting disease (RSD), has been obtained from transposon tagging (Brumbley et al., 2002) and functional genomics approaches (Brumbley et al., 2004a), and the identification of pathogenesis-related genes is only in its infancy. Recently, a number of molecular strategies to confer resistance against various fungal and bacterial pathogens have been conceived and tested in many crop species (Bent and Yu, 1999; Cao et al., 1999; Hancock and Lehrer, 1998; Mourgues et al., 1998; Osusky et al., 2000). These include the use of antimicrobial proteins (Harrison et al., 1996), genes that inactivate pathogenicity (Zhang et al., 1999), natural disease-resistance genes (Staskawiez et al., 1995), phytoalexins (Glaszebrook et al., 1997) and antimicrobial peptides (Hancock and Lehrer, 1998; Osusky et al., 2000). Many of these genes have been successfully used to control plant fungal and bacterial diseases (Chakrabarti et al., 2003; Krishnamurthy et al., 2001; Terras et al., 1995). A large number of natural antimicrobial peptides, particularly cationic antimicrobial peptides (CAPs), are evolving as potent antimicrobial compounds with substantial practical application. CAPs vary greatly in their biological activity spectrum, killing bacteria, fungi, protozoa, and even viruses at very low concentrations (Hancock and Lehrer, 1998). In a recent study, transgenic potatoes expressing CAPs exhibited broad-spectrum resistance to bacterial and fungal pathogens (Osusky et al., 2000, 2004). There are also numerous examples of R genes effective against pathogens in other plant species. For instance, the *Bs2* gene from pepper, Vel and Ve2 genes from tomato, and the *Rpg1*gene from barley have provided strong, durable resistance to *Xanthomonas campestris*, Verticillium, and stem rust, respectively, in a number of species (McDowell and Woffenden, 2003). These examples suggest that the genes encoding CAPs and R genes appear to be suitable candidates to augment the existing fungal and bacterial disease resistance in sugarcane.

Sugarcane pests are another major cause of economic loss in all cane-growing countries, including Australia. Presently, sugarcane pests such as canegrubs, rodents, borers, earthpearls and other insects are controlled by integrated pest management (IPM) approaches comprising cultural, biological and insecticidal controls (Allsopp and Manners, 1997; Allsopp and Suasa-ard, 2000). Although IPM strategies are complementing the already existing tolerance in sugarcane, increasing pest resistance by the introduction of novel insecticidal genes by a transgenic approach would be another technology to assist in maximizing and sustaining crop productivity (Allsopp and Manners, 1997; Falco and Silva-Filho, 2003; Legaspi and Mirkov, 2000).

Considerable progress has already been made in this direction in Australia (Allsopp et al., 2000; Nutt et al., 1999), Cuba (Arencibia et al., 1997, 1999; Setamou et al., 2002) and the United States (Legaspi and Mirkov, 2000). In a recent study, transgenic sugarcane engineered with the *Nicotiana alata* proteinase inhibitor gene (Atkinson et al., 1993) or snowdrop lectin gene (van Damme et al., 1987) exhibited marked antibiosis to canegrubs (Allsopp et al., 2000; Nutt et al., 2001). In another investigation, Legaspi and Mirkow (2000) observed considerable growth inhibition of sugarcane stalk borers when they were fed on transgenic sugarcane engineered with lectin genes. Remarkable tolerance to the borer *Diatraea saccharalis* Fab was also reported in transgenic sugarcane expressing a Bt *cry* lA (b) gene (Arencibia et al., 1999), and some resistance reported with soybean proteinase inhibitors (Falco and Silva-Filho, 2003). These examples clearly indicate that transgenes can be a valuable alternative source of resistance, which could be exploited to enhance IPM strategies as well as providing opportunities to pyramid natural and transgenic pest-resistance genes into breeding and commercial germplasm.

Transgenics for sucrose accumulation and plant development

Sucrose is currently the major product from sugarcane worldwide. Increasing the yield of sucrose can come from either increasing biomass at the same concentration of sucrose or increasing sucrose concentration. The latter approach would be more profitable, as there are lower increased costs of harvesting, transport and milling compared with increasing biomass (Jackson et al., 2000). However, yields of sucrose continue to increase through increased biomass, whereas increasing sucrose content, at least where the levels are already relatively high such as in Australia, has not occurred recently through conventional breeding (Jackson, 2005).

Grof and Campbell (2001) reviewed approaches to identify and overcome the rate-limiting steps, which, broadly, may be the leaf reaction, rate of phloem loading and transport, sucrose transport to the site of storage, and loss of sucrose to support vegetative growth. Increasing biomass in more modern cultivars indicates that leaf reactions may not be limiting, and consequently there have been several attempts to manipulate enzyme activity either to increase sucrose synthesis or reduce hydrolysis in the stem.

A primary target for metabolic manipulation to increase sucrose synthesis is the key enzyme sucrose phosphate synthase (SPS), which catalyses the penultimate step in the pathway, the synthesis of sucrose-6 phosphate from UDP-glucose and fructose-6 phosphate. Successful overexpression of maize SPS in tomato resulting in greater dry weight, number of fruit and higher sucrose concentration was first reported more than 10 years ago (Laporte et al., 1997; Worrell et al., 1991). Overexpression of SPS has been attempted in a number of crop species since, including sugarcane (Grof et al., 1996), with the primary aim of increasing yield, with limited success. However, more recent application of this approach in monocots by overexpression of maize SPS in rice resulted in a threefold increase in SPS activity in the transgenic rice plants leading to significantly taller plants, a measure of biomass yield, identifiable from an early growth stage (Ishimaru et al., 2004). The role of SPS in increased growth in this case was determined through a combination of quantitative trait loci (QTL) and candidate gene approaches. With increasing knowledge of sugarcane genome structure, expressed sequence tag (EST) collections and functional analysis, this approach may also be worth pursuing for several genes in sugarcane.

Sucrose unloaded from the phloem may pass through three cellular compartments: (1) the apoplastic compartment or cell wall and intercellular spaces, (2) the metabolic compartment or cytoplasm, and (3) the storage compartment or vacuole. Different invertase isoforms are associated with each of these metabolic compartments. Invertases have long been considered as principal regulators of sugarcane growth and more specifically sucrose accumulation (Gayler and Glasziou, 1972). Downregulation of soluble acid invertase in the vacuole of sugarcane cells in liquid culture increased the concentration of sucrose twofold (Ma et al., 2000). However, up to a 70% reduction of soluble acid invertase activity in the immature internodes of transgenic sugarcane plants had no significant impact on sucrose concentration (Botha et al., 2001).

In a more systematic attempt to determine which targets when manipulated may lead to higher sucrose concentration, a kinetic model of sucrose accumulation was developed (Botha and Vorster, 1999). Rohwer and Botha (2001) pinpointed cytosolic neutral invertase, the putative vacuolar sucrose import protein and hexose transporters as promising targets for genetic manipulation to increase sucrose concentration. The expression of neutral invertase has been studied (Bosch et al., 2004) but as yet there are no reports of attempts to manipulate it. The vacuolar transporter has yet to be isolated but progress has been made in understanding the other sugar transporters in sugarcane.

The most highly represented ESTs involved in sucrose metabolism in maturing internodes of the sugarcane stem have been identified as putative sugar transporters (Casu et al., 2003). Localization and functional characterization of a range of putative hexose and sucrose transporters are currently in progress (Rae et al., 2005a, 2005b). The key role that each of these transporters may play in the overall process will only be definitively determined by gene knockout using an RNA interference (RNAi) approach. Investigation of genetic variation for such transporters as well as other genes involved in sucrose metabolism may also provide a significant correlative pointer to potential targets for manipulation to achieve higher sucrose concentrations.

A thorough understanding of the balance of sucrose metabolism leading either to growth, storage or respiration would provide the basis for informed targeted manipulation leading to improved cultivars. Internode development can be considered in terms of elongation, dry matter accumulation and plateau phases, and in broad terms sucrose may be directed either towards utilization or storage (Lingle, 1999). If one of the first two stages varies between cultivars, the number of internodes which reach maturity prior to harvest may also vary. An understanding of the processes that underpin the partitioning of sucrose in the uppermost immature internodes may provide the means to direct increased sucrose towards storage leading to a higher sucrose concentration overall.

Sucrose content delivered to the mill, where mechanical harvesting is employed, is not only a function of sucrose content in the stalk but also of how much extraneous matter is harvested. This is influenced by the physiological processes of suckering and lodging (Hurney and Berding, 2000; Singh et al., 2000). Many current commercial cultivars have a high propensity to sucker and lodge, and the physiological and genetic bases of these developmental processes are now being investigated (Hurney and Berding, 2000; Salter and Bonnett, 2000; Singh et al., 2000). Plant development is regulated by hormones, and manipulation of hormone biosynthesis or modulation of hormone activity should result in the control of plant growth and development. Hormone biosynthetic pathways have been largely educated, and genes encoding many of the hormone biosynthetic enzymes have been cloned and characterized (Hidden and Phillips, 2000). Manipulation of gene activity, especially those involved in gibberellin metabolism, has resulted in dramatic alteration in plant growth and development (Hidden and Phillips, 2000). Appropriated modulation of endogenous levels or activity of gibberellins either by transgenic or molecular marker technologies is a logical approach to control shoot growth and architecture in sugarcane. This will be an exciting challenge for biotechnologists and breeder teams to address in the near future.

An understanding of bud outgrowth in sugarcane is also relevant to achieving more reliable bud germination and shoot architecture, and therefore crop establishment. Remarkable progress has been made in our knowledge about the physiology and molecular biology of bud outgrowth in model plants such as *Arabidopsis*, rich, maize and pea (Shimizu-Sato and Mori, 2001). To date, more than half a dozen genes directly involved in bud outgrowth and shoot architecture have been cloned and characterized (Shimizu-Sato and Mori, 2001). We have recently cloned and partially characterized sugarcane homologs of *TB1* and *RMS1*, two genes involved in lateral branching in other species. It is expected that a better understanding of the physiological and molecular bases of bud outgrowth may aid in the selection of cultivars with reliable bud germination, shoot architecture and rationing traits.

While the absence of flowering and the lack of synchronized flowering of desired breeding lines are two important issues related to sugarcane breeding, elimination of flowering in commercial cultivars has considerable economic significance (Bakker, 1999). It is interesting to note that flowering is one of the most intensively investigated developmental phenomena in plants, and a wealth of information is now available for practical application. In the recent past, several genes and genetic loci controlling flowering in *Arobidopsis* and other species have been cloned (Hempel et al., 2000). This molecular information on flowering genes and molecular markers in sugarcane may be used advantageously to manipulate flowering in sugarcane.

Molecular markers and their applications

Molecular markers have great potential for improving the efficiency of the breeding process not only by targeting traits to be selected in one generation but also by the precision and efficiency with which the genotypes can be selected. Information obtained with these markers has contributed to a better understanding of origin, evaluation, genetics and QTLS. Molecular markers are having an impact in several areas of sugarcane improvement and disease management. For each area their impact to date and likely contribution into the future are discussed in turn next.

Understanding the origin of commercial sugarcane

Sugarcane belongs to the informal taxonomical group the *Saccharum* complex, which contains species from the genera *Saccharum, Erianthus, Sclerostachya, Miscanthus* and *Narenga* (Daniels et al., 1975). This gave rise to the suggestion that sugarcane may have emerged from hybridization involving different genera including *Saccharum, Erianthus* and *Miscanthus* (Daniels and Roach, 1987). Various molecular marker systems including isozymes (Glaszmann et al., 1989), restriction fragment length polymorphism (RFLP) (Lu et al., 1994), ribosomal RNA (Glaszmann et al., 1990), mitochondria and chloroplast genes (D'Hont et al., 1993), RAPDs (Nair et al., 1999), and simple sequence repeats (SSRs) (Cordeiro et al., 2003; Selvi et al., 2003) have shown that the three genera have highly contrasting patterns and can be easily differentiated. Further evidence of the distinction of the genera is provided by the occurrence of repeated elements specific to *Erianthus* and *Miscanthus*, which are not detected in the *Saccharum* genus (Alix et al., 1999). Relatively recently, *in situ* hybridization analysis of two ribosomal RNA gene families determined that *S. officinarum* has a basic chromosome number of $x = 10$, meaning that these plants are octoploid (D'Hont et al., 1998). Using the same method it was shown that *S. spontaneum* has a basic chromosome number of $x = 8$ and that the ploidy level of this species varies between 5 and 16 (D'Hont et al., 1998; He et al., 1999). These studies established the coexistence of two distinct chromosome organizations in modern sugarcane cultivars. Using genomic *in situ* hybridization (GISH), D'Hont et al. (1996) and Cuadrado et al. (2004) have demonstrated that modern cultivars contain around 15% and 20% *S. spontaneum* chromosomes and less than 5% are recombinant or translocated chromosomes. Their high ploidy and complex genome structure creates challenges for both transgene expression and the development of molecular markers.

Sugarcane diversity and variety identification

Investigation of diversity within sugarcane cultivars is highly heterozygous with many distinct alleles at a locus (Jannoo et al., 1999b). This has also been demonstrated with SSRs (Selvi et al., 2003). The majority of the diversity is due to the more polymorphic and smaller contribution of the *S. spontaneum* portion of the genome (Jannoo et al., 1999b). Recently, AFLP markers have shown that they can be used to determine genetic similarity among sugarcane cultivars (Lima et al., 2002). In the development of modern cultivars only a small number of meiotic divisions have levels of linkage disequilibrium expected among modern cultivars. A study by Jannoo et al. (1999a) on Mauritian cultivars confirmed this where chromosome haplotypes were conserved over regions as long as 10 cM. This could have major implications on detection and location of genes involved in traits of interest using association-mapping techniques (Butterfield et al., 2003).

One of the major ways sugarcane industries have already benefited from molecular markers is the use of SSRs for cultivar identification. As sugarcane is highly heterozygous, it is vegetatively propagated and goes through a number of stages of selection, propagation and replanting to produce a new cultivar. Consequently, there are many opportunities for mistakes to occur, and varieties to be mislabeled or propagated unintentionally. SSR markers have been used to fingerprint 180 sugarcane varieties and the data stored in a database (Piperidis et al., 2004). This provides a source of information to identify varieties of unknown or disputed origin. SSR profiles of sugarcane varieties can also be used as additional information in a Plant Breeding Rights application and for quality assurance for delivery of new cultivars to the industry (Piperidis et al., 2004).

Introgression and identification of quantitative trait loci (QTL)

As sugarcane cultivars have a narrow genetic base, sugarcane breeders have been interested in incorporating new germplasm into the breeding pool. *Erianthus arundinaceus* is one species that has been of interest to sugarcane breeders because of its good ratooning performance and tolerance to environmental stresses. Hybrid progeny have been produced from crossing *E. arundinaceus* and *S. officinarum*, and molecular markers and GISH have been used to verify these crosses (D'Hont et al., 1995; Piperidis et al., 2000). These hybrids were not themselves fertile. However, Deng et al. (2002) have generated fertile hybrids between *E. arundinaceus* and *S. officinarum*, which have been confirmed by analysis of species-specific SSRs (Gai et al., 2005). This significant development provides a new genetic stock for exploitation in commercial sugarcanes. Introgression of genes from these types of hybrids would be greatly facilitated by identification of molecular markers linked to genes of interest. Molecular genetic mapping of *Saccharum* species has been limited due to its polyploid nature and the resulting mix of single dose and multidose alleles. But using single-dose markers (Wu et al., 1992) linkage maps have been constructed with RAPD (Da Silva et al., 1995; Guimaraes et al., 1997a; Ming et al., 1998, 2002b) markers for *S. spontaneum, S. officinarum* and *S. robustum*. Single-dose linkage maps have also been constructed in cultivars using RFLP (Grivet et al., 1996), AFLP (Hoarau et al., 2001), and AFLP and SSR markers (Rossi et al., 2003; Aitken et al., 2005). Using SSRs, double-dose markers, and repulsion phase linkage, Aitken et al. (2005) succeeded in forming 127 of the 136 LGs into eight homology groups (HGs). Two HGs were each represented by two sets of LGs. These sets of LGs potentially correspond to *S. officinarum* chromosomes, with each set aligning to either end of one or two larger LGs. The larger chromosomes in the two HGs potentially correspond to *S. spontaneum* chromosomes. For two of these groups. two sets of linkage groups align to one larger linkage group giving 10 homology groups corresponding to the basic chromosome number of *S. officinarum*. There was large variation in map coverage for the different homology groups, indicating that genome coverage was not complete. Chromosome pairing in sugarcane is still not fully understood. It does not appear to have either complete disomy or complete polysomy, but some disomic-like pairing behaviour has been observed (Aitken et al., 2005; Hoarau et al., 2001). Due to the size of the sugarcane genome (2C = 7440 Mbp), a large number of markers is required. Even the most complete of these linkage maps, which have over 1000 markers, still only cover about two-thirds of the genome. Development of new high-throughput marker systems like single nucleotide polymorphisms (SNPs) (Grivet et al., 2003) and diversity array technology (DArT) markers (Wenzl et al., 2004) for sugarcane are expected to have a major impact on this area in the future. New theoretical models are also being developed to improve accuracy and extract more information from mapping experiments in complex polyploids (Que and Hancock, 2001).

Mapping of single gene traits are relatively easy and thus far two single major gene traits – eyespot susceptibility (Mudge et al., 1996) and rust resistance (Daugrois et al., 1996) – have been mapped in sugarcane. Apparently, due to the polyploidy nature of the sugarcane genome, the majority of traits of interest are quantitatively inherited. The initial QTL studies were done on small populations with sparse map coverage (Guimaraes et al., 1997a; Sills et al., 1995). More recent studies carried out on two interspecific *S. officinarum* X *S. spontaneum* crosses were used to identify QTL for sucrose content and related traits (Ming et al., 2001, 2002c). In total, 102 significant associations were detected; of these, 61 could be assigned map locations and 50 clustered in 12 genomic regions on 7 sugarcane homologous groups. The corresponding location of the QTLs suggested the presence of several alleles at a locus that contributed to variation of a trait. Using map comparisons to a number of grasses, some of the alleles were given a candidate identity. Another similar study on the self-crossed progeny of a cultivar again identified numerous QTL of small effect for yield traits (Hoarau et al., 2002). The size of the effects was not consistent across crop cycles and the putative genetic factors revealed here explain from 30 to 55% of the total phenotypic variance depending on the trait. Comparative analysis of QTLs affecting plant height and flowering between sorghum and sugarcane identified QTL clusters for both these traits in sugarcane, which corresponded closely to QTL previously mapped in sorghum (Ming et al., 2002a). The large number of alleles with small effect detected in all these studies is largely due to high ploidy and high levels of heterozygosity in sugarcane, which results in numerous alleles at a locus. Although theory is being developed for QTL detection in polyploids (Doerge and Craig, 2000), there is still a need for improved biometrical methods and tools to extract the maximum amount of information from QTL studies.

The conservation of chromosome organization across the grasses also has the potential to impact location and identification of genes of interest in sugarcane. The colinearity between sugarcane and sorghum chromosomes (Guimaraes et al., 1997b; Ming et al., 1998) means that genes identified in this diploid relative could be of use in locating the same genes in sugarcane. Although markers linked to QTL for a number of agronomically important traits have been identified in sugarcane, the implementation of these markers has yet to commence. Work is under way to use marker-assisted selection introgress QTL for high sucrose identified in *S. officinarum* into commercial sugarcane (Aitken et al., 2002). da Silva and Bressiani (2005) described the development of an EST-derived RFLP marker for sugarcane elite genotypes which can be used for QTL tagging for sugar content. It is likely that cultivars produced with molecular marker input will initially arise through introgression studies. Use of markers in routine breeding activities will occur only when high-throughput marker detection and more robust statistical methods are developed.

Molecular markers for diagnostics

Development of new and improved molecular assays to detect various sugarcane pathogens is progressing in different laboratories worldwide (Braithwaite and Smith, 2001). Molecular diagnostic tests, based mostly on nucleic acids, are highly sensitive and relatively easy to use compared with the traditional detection method such as histology, electron microscopy, sap transmission onto indicator plants, or the isolation and culture of the causal agents. Due to their high sensitivity, molecular diagnostic methods are capable of detecting pathogens even in asymptomatic plants with an extremely low pathogen titer. Molecular tests for diseases such as ratoon stunting (Fegan et al., 1998; Pan et al., 1998), Fiji disease (Smith and van de Velde, 1994; Smith et al., 1994), mosaic (Smith and van de Velde, 1994), striate mosaic (Thompson et al., 1998), yellow leaf syndrome (Chatenet et al., 2001; Irey et al., 1997), smut (Albert and Schenck, 1996), sorghum mosaic (Yang and Mirkow, 1997) and SCBV (Braithwaite et al 1995) have been developed and are being applied to screen quarantine germplasm. With advances in genome sequencing and the continued refinement of various techniques in recombinant DNA technology, development of more reliable, faster and cost-effective molecular diagnostic tests for all the important sugarcane pathogens can be expected in the near future. Molecular markers have been used to assess genetic diversity in sugarcane germplasm (Selvi et al., 2003). Among these, microsatellites have been found to be the markers of choice.

Structural and functional genomics

The rapid evolution of structural and functional genomics will have a major influence on future sugarcane crop improvement programs. Technological and conceptual developments in these areas are occurring at an astonishing pace. These advances include the development of highly efficient DNA sequencing techniques, innovative approaches to the identification of SNPs and genome mapping, a variety of powerful DNA microarray technologies for the analysis of gene expression, RNAi technology, and the rapid improvements in data mining tools to organize and analyse the complex data obtained from these technologies. These technological advancements are now offering an unprecedented analytical power in studying various agriculturally important biological processes and traits.

Deciphering gene function will be the most challenging, exciting and valuable phase of sugarcane genomics research. At present, both micro- and macroarrays are being used for the identification of genes expressed specifically in stems, disease-resistance genes, and those involved in carbohydrate metabolism (Casu et al., 2005; Grivet and Arruda, 2002; Ulian, 2000). A large collection of sugarcane ESTs, nearly 260,000, generated mainly by Brazilian researchers (Vettore et al., 2003), are now in the public domain and will remain as a highly valuable resource for genomics research for a long time. More than 33,000 unique genes have been identified from the EST collection (Vettore et al., 2003). Recently, using a new approach called 'genetical genomics,' which combines the power of genomics and the genetics of segregating populations, 62 differentially regulated transcripts that are linked to sugar content were determined (Casu et al., 2005). The study demonstrated that cDNA microarray technology represents a powerful method to identify regulatory genes associated with a particular process or trait in sugarcane. In another example, using a cDNA subtractive hybridization and cDNA macroarray strategy, Carson et al., (2002) identified sugarcane ESTs differentially expressed in immature and maturing internodal tissue, proving that a combination of cDNA subtraction with macroarray screening is an effective strategy to identify developmentally regulated genes in sugarcane. With the rapid advances now occurring in microbial, plant and animal systems, there is little doubt that functional genomics will soon emerge as one of the most powerful approaches to address the complex issues related to genetic improvement of sugarcane.

Sugarcane as a biofactory: Production of high-value alternative products

Development of sugarcane as a biological system for large-scale production of value-added products is now entering into an exciting phase. Alternative products to sucrose can provide not only higher-value products but make sugar industries more sustainable and competitive. Sugarcane has all the features needed for a natural biofactory: it grows rapidly, has a very efficient carbon fixation pathway, produces a large biomass, possesses a well-developed storage system (stem) with a large pool of hexose sugar and is cultivated in different parts of the world. Industrial and agricultural wealth from sugarcane is also developed in India (Figure 11.1).

Significant background technology for biofactory research has already been developed in other plant species, which could possibly be adapted to sugarcane. Development of plant-based expression systems has resulted in over 100 recombination proteins successfully produced in plant species. These expressed proteins include a wide range of products from viral proteins, vaccines, antimicrobial peptides, antibodies, pharmaceuticals and industrial compounds (Twyman et al., 2003). Recently, a human pharmaceutical protein, human granulocyte macrophage colony stimulating factor (GM-GSF), used in clinical applications for the treatment of neutropenia and aplastic anaemia, was successfully produced in sugarcane (Wong and Fletcher, 2004). Although successful experimentally, accumulation of GM-GSF was limited to 0.02% of the total soluble protein in transgenic plants under field conditions.

Sugarcane has also proved to be a model system for the production of industrial products such as poly-3-hydroxybutyrate (PHB). PHBs are present in many bacteria as a carbon storage compound; they are biodegradable and possess thermoplastic properties. Several attempts were made in the past to produce PHB in tobacco and *Arabidopsis* (Bohmert et al., 2000; Nawrath et al., 1994; Poirier et al., 1992). PHB accumulation at a commercially viable level without penalizing plant growth and development was not achieved in any of the systems reported to date. Brumbley et al. (2003) engineered the genes encoding the enzymes of PHB production in a commercial sugarcane cultivar, and the results obtained are encouraging. PHB accumulated up to 1.2% of the dry weight in leaves without affecting plant growth. However, the amount of PHB in stem pith was only 0.004% on a dry weight basis. Another example of sugarcane as a biofactory for industrial compounds comes from the successful production of *P-hydroxybenzoic* acid (pHBA) from transgenic sugarcane expressing the bacterial enzymes chorismate pyruvate-lyase (GPL) and 4-hydroxycinnamoy-CoA hydratase/yase (HCGL) (McQualter et al., 2004b). As with PHB, pHBA was also produced in other plant species, again with growth penalties (Mayer et al., 2001). Interestingly, no such growth or developmental anomalies

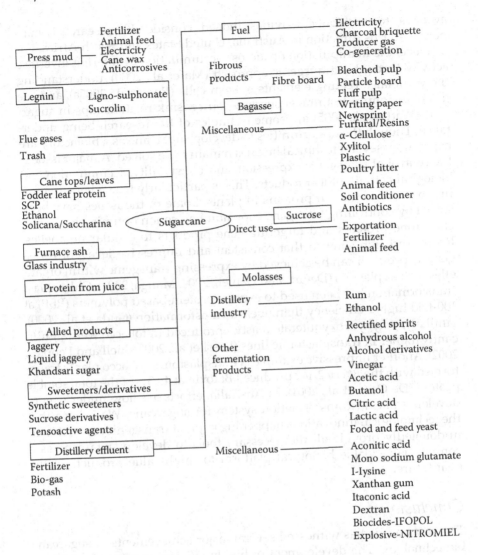

Figure 11.1 Industrial and agricultural wealth from sugarcane.

were noted in sugarcane despite pHBA accumulation in leaves reaching about 7% of dry weight. Similar to the trend observed with PHB, accumulation of pHBA in stem was much lower than that in leaf, attaining a maximum of 1.5% dry weight.

Clearly, the two aforementioned examples indicate that a better understanding of physiology and molecular biology is needed to channel PHB, pHBA or other similar biopolymer production to the stem, the storage organ of sugarcane where a large pool of carbon reserve is readily available for metabolic activities. Indeed, considerable research is happening in this direction in Australia, United States and South Africa. For instance, the manipulation of sucrose accumulation by altering invertase activity through a transgenic approach (Ma et al., 2000), understanding the vacuolar targeting elements in stem cells (Rae et al., 2005a), and the use of genomics to unravel sucrose–source–sink relationships in sugarcane (Watt et al., 2005) are some examples of the research being undertaken. Though the opportunities to develop sugarcane as a biofactory are expanding, several technical hurdles remain to be solved. A major issue is the control of transgenic expression, and the stability, accumulation and biological activity of its products. This is particularly true when attempting to produce foreign proteins in plants. Some of the issues have been solved by controlling transgene copy number, codon optimization, using chaperone proteins and targeting gene products to cellular organelles. Recent research showed that consistent and impressive levels of transgenic expression can be achieved by expressing transgenic within organelles such as plastids (Devine and Daniell, 2004; Maliga, 2004). Chloroplast transformation has been used to express protein-based polymers (PBP) at 100-fold higher efficiency than nuclear transformation (Guda et al., 2000). Similarly, polyhydroxybuterate plastics produced in tobacco were significantly higher in transplastomic lines (Lossl et al., 2003; Snell and Peoples, 2002). Another impressive example is transplastomic tobacco plants transformed with the *Bt cry 2Aa2* produced Bt toxin at 45.3% of the total soluble protein (De Cosa et al., 2001). In Australia, efforts are now under way to develop a plastid transformation system for sugarcane. We believe that the technological innovations happening in plant transgenic research will undoubtedly provide all the necessary tools to develop sugarcane as a commercially viable biological platform for high-value products in the near future.

Conclusions

The past decade has witnessed several major achievements in sugarcane biotechnology. The development of highly efficient *in vitro* regeneration systems coupled with rapid advances in gene delivery techniques led to the establishment of a robust genetic transformation technology for sugarcane (Bower and Birch, 1992). This was an impressive beginning, and it laid the foundation for genetic engineering of sugarcane worldwide.

Over the past 10 years, different gene delivery systems for sugarcane transformation have been developed (Arencibia et al., 1995; Bower and Birch, 1992; Enriquez-Obregon et al., 1998; Manickavasagam et al., 2004). Nonetheless, the full potential of sugarcane biotechnology has yet to be realized as a result of technical limitations in the existing transformation systems. Transformation/tissue culture-induced somaclonal variation remains a significant bottleneck in exploiting gene technology for sugarcane improvement (Arencibia et al., 1999), and considerable refinements of current transformation systems are required to ensure clonal fidelity of transgenic cultivars. Similarly, transformation systems and gene regulation sequences that produce plants with adequate and stable transgene expression under field conditions are needed if the practical benefits of transgenic technology are to be delivered. For the successful release of transgenic sugarcane, various scientific, legislative, and public perception issues must be addressed. Transformation systems that do not incorporate any non-transgene DNA into the plant, and utilize non-antibiotic, plant gene-based selection strategies would greatly aid in overcoming regulatory and public perception issues. In addition, the ability to control or tissue specificity will provide a platform for the production of a range of new compounds in sugarcane at commercially useful levels. The development of strategies for incorporating polygenic traits, hyper expression of transgenes and containment of transgenes within the transgenic plants (Daniell, 1999; Daniell et al., 2001; Maliga, 2004) together with the possibility of engineering native genes without significant genetic rearrangements (Beetham et al., 1999) are valuable innovations that could be utilized for improvement of sugarcane in the near future.

As with most other crops, the first generation of transgenic sugarcane plants were engineered with input traits such as resistance to herbicides, diseases and pests. Significant gains have been made in this area, and much more progress will be achieved by exploiting various resistance genes already available from other plants (McDowell and Woffenden, 2003). Some of these genes could be readily adapted to sugarcane, delivering faster benefits to the worldwide industries. At the same time, efforts should also be directed to understand the molecular basis of pathogenicity and the natural disease resistances existing in the *Saccharum* germplasm in order to develop more targeted biotechnological approaches for the control of pathogens and pests.

Successful manipulation of sugar metabolism to enhance yield remains a major challenge for both conventional and molecular breeding. The plasticity of plant metabolism, with the existence of alternative pathways, makes the manipulation of metabolic pathways somewhat unpredictable. Nonetheless, the potential exists for the molecular modification of sugar metabolism as our knowledge about the biochemistry and molecular biology of carbohydrate metabolism continues to increase. Building on the increasing information on sugarcane biochemistry, transgenic expression and transgene product storage, sugarcane could be developed as a viable biofactory for the production of a range of high-volume and high-value products.

Considerable practical benefits are beginning to occur from the applications of the various marker technologies and molecular diagnostics to sugarcane genetic improvement and germplasm exchange. The potential of the current genomics programmes, aimed at elucidating the structure, function and interactions of the sugarcane genes, are significant, and they will revolutionize the application of biotechnology to crop improvement. All the conceptual and technological advancements that are occurring in the different domains of biotechnology make the prospects and opportunities for improving sugarcane by molecular breeding virtually unlimited.

References

Abel P, Nelson RS, De B, Holffmann N, Rogers SG, Fraley RT, Beachy RN (1986). Delay of disease development in transgenic plants that express the tobacco mosaic virus. *Science* 232:738.

Aitken KS, Jackson PA, Mclntyre CL (2005). A combination of AFLP and SSR markers provides extensive map coverage and identification of homo(eo) logous linkage groups in a sugarcane cultivar. *Theor. Appl. Genet.* 110:789.

Aitken KS, Jackson PA, Mclntyre CL, Piperidis G (2002). Marker assisted introgressing of high sucrose genes in sugarcane. *Australasian Plant Breed. Conf.* 12:120.

Alam MZ, Haider SA, Islam R, Joarder DI (1995). High frequency in vitro plant regeneration in sugarcane. *Sugarcane*, 6:20–21.

Albert HH, Schenck S (1996). PCR amplification from a homolog of the bE mating-type gene as a sensitive assay for the presence of *Ustilago scitaminea* DNA. *Plant Dis.* 80:452–457.

Alix K, Paulet F, Glaszmann JC, D'Hont A (1999). Inter-Alu-like species specific sequences in the *Saccharum* complex. *Theor. Appl. Genet.* 99:962–968.

Allsopp PG, Manners JM (1997). Novel Approaches for managing pests and diseases in sugarcane. In: BA Keating, JR Wilson, eds, *Intensive sugarcane production: Meeting the challenge beyond 2000*, 173–188. Wallingford, UK: GAB International.

Allsopp PG, Nutt KA, Geijskes RJ, Smith GR (2000). Transgenic sugarcane with increased resistance to canegrubs. In: PG Allsopp, W Suasa-ard, eds, *Proceedings of 4th International Society of Sugar Cane Technologists Entomology Workshop*, Khon Kaen, Thailand, pp. 63–67.

Allsopp PG, Suasa-ard W (2000). Sugarcane pest management strategies in the new millennium. In: *Proceedings of International Society of Sugar Cane Technologists, Entomology, Sugarcane Entomology Workshop*, Khon Kaen, Thailand, p. 4.

Anderson WC (1980). Mass propagation by tissue culture. Principle and practices. In: RH Zimmerman, ed, *Proc. of the Conf. on Nursery Production of Fruit Plants through Tissue Culture-Applications and Feasibility*, Agric Res. Sci. Edu. Admn. USDA, Beltsville.

Arencibia A, Garmona E, Cornide MT, Gastiglione SO, Relly J, Cinea A, Oramas P, Sala F (1999). Somaclonal variation in insect resistant transgenic sugarcane (*Saccharum hybrid*) plants produced by cell electroporation. *Transgenic Res.* 8:349–360.

Arencibia A, Garmona, E, Tellez P, Ghan MT, Yu SM, Trujillo L, Oramas P (1998). An efficient protocol for sugarcane (*Saccharum* spp.) transformation mediated by *Agrobacterium tunefaciens*. *Transgenic Res.* 7:213–222.

Arencibia A, Molina P, De la Riva G, Selman-Houssein G (1995). Production of transgenic sugarcane (*Saccharum officinarum* L.) plants by intact cell electroporation. *Plant Cell Rep.* 14:305–309.

Arencibia A, Molina P, Cutierrez C, Fuentes A, Greenidge V, Menendez E, De la Riva G, Selman G (1992). Regeneration of transgenic sugarcane (*Saccharum officinarum* L.) plants from intact meristematic tissues transformed by electroporation. *Biotechnol. Aplicada* 9:156–165.

Arencibia A, Vazquez RI, Prieto D, Tellez P, Garmona ER, Goego A, Hernandez L, De la Riva GA, Selman HG (1997). Transgenic sugarcane plants resistant to stem borer attack. *Mol. Breed.* 3:247–255.

Atkinson AH, Heath RL, Simpson RJ, Clarke AE, Anderson MA (1993). Proteinase inihitors in *Nicotiana alata* stigma are derived from a precursor protein which is processed into five homologous inhibitors. *Plant Cell* 5:203–211.

Bajaj YPS, Jain LG (1995). Cryopreservation of germplasm of sugarcane (*Saccharum* species). In: YPS Bajaj, ed, *Biotechnology in agriculture and forestry*, vol. 32. *Cryopreservation of plant germplasm I*, 256–266. Berlin: Springer-Verlag.

Bakker H (1999). *Sugar cane cultivation and management*. New York: Kluwer Academic/Plenum.

Barba R, Nickell LG (1969). Nutrition and organ differentiation in tissue culture of sugarcane: A monocotyledon. *Planta* 89:299–302.

Beetham PR, Kipp PB, Sawycky XL, Arnzen CJ, May GD (1999). A tool for functional plant genomics: Chimeric RNA/DNA oligonicleotides cause *in vitro* gene-specific mutations proc. *Natl. Acad. Sci. USA* 96:8774–8778.

Bent AF, Yu IC (1999). Applications of molecular biology to plant disease and insect resistance. *Adv. Agron.* 66:251.

Berding N, Moore PH, Smith GR (1997). Advances in breeding technology for sugarcane. In: BA Keating, JR Wilson, eds, *Intensive sugarcane production: Meeting the challenge beyond 2000*, 189–220. Wallingford, UK: CAB International.

Bhojwani SS, Dhawan V, Cocking EC (1986). Plant tissue culture – A classified bibliography, 399–432. Amsterdam: Elsevier.

Birch RG (1997). Transgenic sugarcane: Opportunities and limitations. In: BA Keating, JR Wilson, eds, *Intensive sugarcane production: Meeting the challenge beyond 2000*, 125–140. Wallingford, UK: GAB International.

Birch RG, Bower R, Elliot A, Potier B, Franks T, Cordeiro G (1995). Expression of foreign genes in sugarcane. *Proc. Int. Soc. Sugar Cane Technol.* 22:368–373.

Birch RG, Maretzki A (1993). Transformation of sugarcane. In: YPS Bajaj, ed, *Plant protoplasts and genetic engineering IV. Biotechnology in agriculture and forestry*, vol. 23, 348–360. Heidelberg: Springer-Verlag.

Birch RG, Patil SS (1987a). Correlation between albicidin production and chlorosis induction by *Xanthomonas albilineans*, the sugarcane leaf scald pathogen. *Physiol. Mol. Plant Pathol.* 30:199–206.

Birch RG, Patil SS (1987b). Evidence that an albicidin-like phytotoxin induces chlorosis in sugarcane leaf scald disease by blocking plastid DNA replication. *Physiol. Mol. Plant Pathol.* 30:207–214.

Bohmert K, Balbo I, Kopka J, Mittendorf V, Nawrath G, Poirier Y, Tishcendorf G, Trethewey RN, Willmitzer L (2000). Transgenic *Arabidopsis* plants can accumulate polyhydroxybutyrate to up to 4% of their fresh weight. *Planta* 211:841–845.

Bosch S, Grof CPL, Botha FC (2004). Expression of neutral invertase in sugarcane. *Plant Sci.* 166:1125–1133.

Botha FC, Sawyer BJB, Birch RG (2001). Sucrose metabolism in the culm of transgenic sugarcane with reduced soluble acid invertase activity. *Proc. Int. Soc. Sugar Cane Technol.* 24:588–591.

Botha FC, Vorster DJ (1999). Analysis of carbon partitioning to identify meta-bolic steps limiting sucrose accumulation. In: V Singh, V Kumar, eds, *Proceedings of International Society of Sugar Cane Technologists*, New Delhi 23:259–272.

Bouhida M, Lockhart BEL, Olszewski NN (1993). An analysis of the complete sequence of a sugarcane bacilliform virus genome infectious to banana and rice. *J. Gene Virol.* 74:1–8.

Bower R, Birch RG (1992). Transgenic sugarcane plants via microprojectile bombardment. *Plant J.* 2:409–416.

Bower R, Elliott AR, Potier AM, Birch RG (1996). High-efficiency, microprojectile-mediated cotransformation of sugarcane, using visible or selectable markers. *Mol. Breed.* 2:239–249.

Braithwaite KS, Egeskov NM, Smith GR (1995). Detection of sugarcane bacilliform virus using the polymerase chain reaction. *Plant Dis.* 79:792–796.

Braithwaite KS, Geijskes RJ, Smith GR (2004). A variable region of the SCBV genome can be used to generate a range of promoters for transgene expression in sugarcane. *Plant Cell Rep.* 23:319–326.

Braithwaite KS, Smith GR (2001). Molecular-based diagnosis of sugarcane virus diseases. In: GP Rao, RE Ford, M Tosic, DS Teakle, eds, *Sugarcane pathology*, vol. II. *Viruses and phytoplasma diseases*, 175–192. New Delhi: Oxford and IBH Publishing Co. Pvt. Ltd.

Briggs SP, Koziel M (1998). Engineering new plant traits for commercial markets. *Curr. Opin. Biotechnol.* 9:233–235.

Brumbley SM, Petrasovits LA, Birch RG, Taylor PWJ (2002). Transformation and transposon mutagenesis of *Leifsonia xyli* subsp. Xyli, causal organism of ratoon stunting disease of sugarcane. *Mol. Plant-Microbe Interact.* 15:262–268.

Brumbley SM, Petrasovits LA, Bonaventura PA, O'Shea MJ, Purnell MP, Nielsen LK (2003). Production of polyhydroxyalkanoates in sugarcaneproc. *Proc. Int. Soc. Sugar Cane Technol.* 4:31.

Brumbley SM, Petrasovits LA, Murphy RM, Nagel RJ, Candy JM, Hermann SR (2004a). Establishment of a functional genomics platform for *Leifsonia xyli* subsp. *Xyli. Mol. Plant-Microbe Interact.* 17:175–183.

Brumbley SM, Purnell MP, Petrasovits LA, O'Shea MJ, Nielsen LK (2004b). Development of sugarcane as a biofactory for biopolymers. *Plant and Animal Genomes XII Conference*, San Diego, January 10–14 (Abstract W 117).

Burner DM, Grisham MP (1995). Induction and stability of phenotypic variation in sugarcane as affected by propagation procedure. *Crop Sci.* 35:875–880.

Butterfield MK, Barnes JM, Heinze BS, Rutherford RS, Huckett BI (2003). RFLP markers for resistance to eldana and smut from an unstructured sugarcane population. *Proc. Int. Soc. Sugar Cane Technol., Mol. Biol. Workshop*, 4:14.

Butterfield, MK, Irvine JE, Valdez G M, Mirkov TE (2002). Inheritance and segregation of virus and herbicide resistance transgenes in sugarcane. *Theor. Appl. Genet.* 104:797–803.

Cao H, Li X, Dong X (1999). Generation of broad spectrum disease resistance by over expression of an essential regulatory gene in systemic acquired resistance. *Proc. Natl. Acad. Sci. USA* 95:6531–6536.

Carson DL, Huckett BI, Botha FC (2002). Sugarcane ESTs differentially expressed in immature and maturing internodal tissue. *Plant Sci.* 162:289–300.

Casu RE, Grof CPL, Rae AL, Meintyre CL, Dimmock CM, Manners JM (2003). Identification of a novel sugar transporter homologue strongly expressed in maturing stem vascular tissues of sugarcane by expressed sequence tag and microrray analysis. *Plant Mol. Biol.* 52:371–386.

Casu, RE, Manners JM, Bonnet GD, Jackson PA, Meintyre CL, Dunne R, Chapman SC, Rae AL, Grof CP (2005). Genomic approaches for the identification of genes determining important traits in sugarcane. *Field Crop Res.* 92:137–147.

Chakrabarti A, Ganpathi TR, Mukherjee PK, Bapat VV (2003). MSI-99, a maganin analogue, imparts enhanced disease resistance in transgenic tobacco and banana. *Planta* 216:587–596.

Chatenet M, Delage C, Ripolles M, Irey M, Lockhart BEL, Rott P (2001). Detection of sugarcane yellow leaf virus in quarantine and production of virus-free sugarcane by apical meristem culture. *Plant Dis.* 85:1177–1180.

Chen WH, Garland KMA, Davey MR, Sotak R, Gatland JS, Mulligan BJ, Power JB, Cocking EC (1987). Transformation of sugarcane protoplasts by direct uptake of a selectable chimaeric gene. *Plant Cell Rep.* 6:297–301.

Chowdhury MKU, Vasil I (1992). Stably transformed herbicide resistant callus of sugarcane via microprojectile bombardment of cell suspension cultures and electroporation of protoplasts. *Plant Cell Rep.* 11:494–498.

Chowdhury MKU, Vasil IK (1993). Molecular analysis of plants regenerated from embryogenic cultures of hybrid sugarcane cultivars (*Saccharum* spp.). *Theor. Appl. Genet.* 86:181–188.

Christensen AH, Quail PH (1996). Ubiquitin promoter-based vectors for high-level expression of selectable and/or screenable marker genes in monocotyledonous plants. *Transgenic Res.* 5:213–318.

Comstock JC, Shine JM, Davis MJ, Dean JL (1996). The relationship between resistance to *Clavibacter xyli* subsp. *Xyli* colonization in sugarcane and spread of ratoon stunting disease in the field. *Plant Dis.* 80:704–708.

Cordeiro GM, Pan YB, Henry RJ (2003). Sugarcane microsatellites for the assessment of genetic diversity in sugarcane germplasm. *Plant Sci.* 165:181–189.

Cox MC, Hansen PB (1995). Productivity traits in southern and central regions and the impact of new varieties. *Proc. Aust. Soc. Sugarcane Technol.* 17:1–7.

Cuadrado A, Acevedo R, Moreno Diaz de la Espina S, Jouve N, de la Torre C (2004). Genome remodeling in three modern *S. officinarum X s. spontaneun* sugarcane cultivars. *J. Exp. Bot.* 55:847–854.

Daniell H (1999). Environmentally friendly approaches to genetic engineering. *In Vitro Cell. Dev. Biol. Plant* 35:361–368.

Daniell H, Streatfield SJ, Wycoff K (2001). Medical molecular farming production of antibodies, biopharmaceuticals and edible vaccines in plants. *Trend Plant Sci.* 6:219–225.

Daniels J, Roach BT (1987). Taxonomy and evolution. In: DJ Heinz, ed, *Sugarcane improvement through breeding,* 7–84. Amsterdam: Elsevier.

Daniels J, Smith P, Paton N, Williams CA (1975). The origin of the genus *Saccharum. Sugarcane Breed. Newsl.* 35:19–20.

da Silva JA, Bressiani JA (2005). Sucrose synthase molecular marker associated with sugar content in elite sugarcane progeny. *Genet. Mol. Biol.* 28(2):294–298.

Da Silva, Honeycutt J, Burnquist RJ, Al-Janabi SM, Sorrells ME, Tanksley SD, Sobral BWS (1995). *Sccharum spontaneum* L. 'SES 208' genetic linkage map combining RFLP- and PCR-based markers. *Mol. Breed.* 1:165–179.

Da Silva, Telles FR, Arruda GP (2003). Clustering 300,000 sugarcane ESTs, the challenge of dealing with orthologs, peralogs, homologs and some rubbish. *Proc. VII Int. Conger. Plant Mol. Biol.* 58:23–28.

Daugrois JH, Grivet L, Roques D, Hoarau JY, Lombard H, Glaszmann JC, D'Hont A (1996). A putative major gene for rust resistance linked with a RFLP marker in sugarcane cultivar 'R570'. *Theor. Appl. Genet.* 92:1059–1064.

De Cosa, Moar B, Lee W, Miller SB, Daniell M (2001). Overexpression of the BtCry2Aa2 operon in chloroplasts leads to formation of insecticidal crystals. *Nature Biotechnol.* 19:71–74.

Deng HH, Liao ZZ, Li QW, Loa FY, Fu C, Chen XW, Zhang CM, Liu SM, Yang YH (2002). Breeding and isozyme marker assisted selection of F2 hybrids from *Saccharum spp.* X *Erianthus arundinaceous. Sugarcane Canesugar* 1:1–5.

Devine AL, Daniell H (2004). Chloroplast genetic engineering for enhanced agronomic traits and expression of proteins for medical/industrial applications. In: S Moller, ed, *Chloroplast genetic engineering in plastids*, 283–320. Oxford: Blackwell.

D'Hont A, Glaszmann JC (2001). Sugarcane genome analysis with molecular markers: A first decade of research. *Proc. Int. Soc. Sugar Cane Technol.* 24:556–559.

D'Hont A, Grivet L, Feldmann P, Rao S, Berding N, Glaszmann JC (1996). Characterisation of the double genome structure of modern sugarcane cultivars (*Saccharum spp.*) by molecular eytogenetics. *Mol. Gen. Genet.* 250:405–413.

D'Hont A, Ison D, Alix K, Roux C, Glaszmann JC (1998). Determination of basic chromosome numbers in the genus *Saccharum* by physical mapping of ribosomal RNA genes. *Genome Res.* 41:221–225.

D'Hont A, Lu YH, Feldmann P, Glaszmann JC (1993). Cytoplasmic diversity in sugar cane revealed by heterologous probes. *Sugar Cane* 1:12–15.

D'Hont A, Rao PS, Feldmann P, Grivet L, Islam FN, Taylor P, Glaszmann JC (1995). Identification and characterization of sugarcane intergeneric hybrids, *Saccharum officinarum* X *Erianthus arundinaceus*, with molecular markers and DNA *in situ* hybridization. *Theor. Appl. Genet.* 91:320–326.

Dhumale DB, Ingole GC, Durge DV (1994). In vitro regeneration of sugarcane by tissue culture. *Ann. Plant Physiol.* 8(2):192–194.

Doerge RW, Craig BA (2000). Model selection for quantitative trait locus analysis in polyploids. *Proc. Natl Acad. Sci. USA* 97:7951–7956.

Elliott AR, Campbell JA, Bretell RIS, Crof CPL (1998). *Agrobacterium mediated transformation of sugarcane using* GFP as a screenable marker. *Aust. J. Plant Physiol.* 25:739–743.

Elliott AR, Campbell JA, Dugdale B, Brettell RIS, Grof CPL (1999). Green-fluorescent protein facilitates rapid in vivo detection of genetically plant. *Cell Rep.* 18:707–714.

Elliott AR, Geijskes RJ, Lakshmann P, McKeon MG, Wang LF, Berding N, Crof CPL, Smith GR (2002). Direct regeneration of transgenic sugarcane following microprojectile transformation of regenerable cells in thin transverse section explants. In: IK Vasil, ed, *Proceedings 10th International Association for Plant Tissue Culture and Biotechnology*, Orlando, Florida, June 23–28. (Abstract P-1376).

Ellis J, Dodds P, Pryor T (2000). The generation of plant disease resistance gene specificities. *Trends Plant Sci.* 5:373–379.

Enriquez CA, Trujillo LE, Menendez C, Vazquez RI, Tiel K, Arieta J, Selman C, Hernandez I (2000). Sugarcane (*Saccharum hybrid*) genetic transformation mediated by *Agrobacterium tumefaciens*: Production of transgenic plants expressing proteins with agronomic and industrial value. In: AD Arencibia, ed, Plant genetic engineering: Towards the third millennium, 76–81. Amsterdam: Elsevier Science.

Enriquez-Obregon CA, Vazquez PRI, Prieto SDL, Riva-Custavo ADL, Selman HG (1998). Herbicide resistant sugarcane (*Saccharum officinarum* L.) plants by Agrobacterium-mediated transformation. *Planta* 206:20–27.

Erturk H, Walker PN, Escribens J (1997). In: *Am. Soc. Agric. Engg. Annual Int. Meeting*. St. Joseph. Minnesota, USA.

Falco MC, Silva-Filho MC (2003). Expression of soybean proteinase inhibitors in transgenic sugarcane plants: Effects on natural defense against *Diatraea saccharalis*. *Plant Physiol. Biochem.* 41:761–766.

Fegan M, Croft BJ, Teakle DS, Hayward AC, Smith CR (1998). Sensitive and specific detection of *Clavibacter xyli subsp. Xyli,* causal agent of ratoon stunting disease of sugarcane, with a polymerase chain reaction-based assay. *Plant Pathol.* 47:495–504.

Fitch MM, Moore PH (1993). Long term culture of embryogenic sugarcane callus. *Plant Cell Tiss. Organ Cult.* 32:335–343.

Franks T, Birch RG (1991). Gene transfer into intact sugarcane cells using microprojectile bombardment. *Aust. J. Plant Physiol.* 18:471–480.

Fu X, Due LT, Fontana S, Bong BB, Tinjuangjun P, Sudhakar D, Twyman RM, Christou P, Kohli A (2000). Linear transgene constructs lacking vector backbone sequences generate low-copy-number transgenic plants with simple integration patterns. *Transgenic Res.* 9:11–19.

Gai I, Aitken K, Deng HH, Chen XW, Cheng F, Jackson PA, Fan YH, McIntyre CL (2005). Verification of intergeneric hybrids (FI) from *Saccharum officinarum* X *Erianthus arundinaceus,* and BCI from FI X sugarcane (*Saccharum* spp.) clones using molecular markers. *J. Plant Breed.* 11:231.

Gallo-Meagher M, Irvine JE (1996). Herbicide resistant transgenic sugarcane plants containing the *bar* gene. *Crop Sci.* 36:1367–1374.

Gambley RL, Bryant JD, Masel NP, Smith GR (1994). Cytokinin enhanced regeneration of plants from microprojectile bombarded sugarcane meristematic tissue. *Aust. J. Plant Physiol.* 21:603–612.

Gambley RL, Ford R, Smith GR (1993). Microprojectile transformation of sugarcane meristems and regeneration of shoots expressing β-glucuronidase. *Plant Cell Rep.* 12:343–346.

Gautheret R (1939). Sur la possibilité de réaliser la culture indéfinie des tissues de tubercules de carotte. *C. R. Soc. Biol. Paris* 208:118–120.

Gayler KR, Glasziou KT (1972). Physiological functions of acid and neutral invertases in growth and sugar storage in sugarcane. *Physiol. Plant.* 27:25–31.

Geijskes RJ, Braithwaite KS, Dale JL, Harding RM, Smith GR (2002). Sequence analysis of an Australian isolate of sugarcane bacilliform badnavirus. *Arch. Virol.* 147:2393–2404.

Geijskes RJ, Wang L, Lakshmanan P, McKeon MG, Berding N, Swain RS, Elliott AR, Grof CPL, Jackson JA, Smith GR (2003). Smartsett™ seedlings: Tissue culture seed plants for the Australian sugar industry. *Sugarcane Int.* May/June, 13–17.

Gilbert RA, Gallo-Meagher M, Comstock JC, Miller JD, Jain M, Abouzid A (2005). Agronomic evaluation of sugarcane lines transformed for resistance to sugarcane mosaic virus strain E. *Crop Sci* 45:2060–2067.

Gill D, Gonzalez V, Morejon A, Peroz L, Herrera O, Farrilla I (1989). Improvement of rooting induction in sugarcane hybrid (Saccharum spp) obtained by micropropagation. *Centro Agric.* 16(4):17–23.

Glaszmann JC, Fautret A, Noyer JL, Feldmann P, Lanaud C (1989). Biochemical genetic markers in sugarcane. *Theor. Appl. Genet.* 78:537–543.

Glaszmann JC, Lu Y, Lanaud C (1990). Variation of nuclear ribosomal DNA in sugarcane. *J. Genet. Breed.* 44:191–198.

Glaszebrook J, Zook M, Mert F, Kagan I, Rogers EE, Crute IR, Holub EB, Hammerschmidt R, Ausubel FM (1997). Phytoalexin deficient mutants of *Arabidopsis* reveal that *PAD* genes contribute to downy mildew resistance. *Genetics* 146:381–392.

Gonzalez RV, Manzano CA, Ordosgoitti FA, Salazar PY (1987). Genetics of the reaction of sugar cane (*Saccharum* spp.) To Puccinia melanocephala, causing rust. *Tropical Agronomy* 37:99–116.

Gosal SS, Thind KS, Dhaliwal HS (1998). Micropropagation of sugarcane – An efficient protocol for commercial plant production. *Crop Improv.* 25(2):167–170.

Grisham MP, Bourg D (1989). Efficiency of in vitro propagation of sugarcane plants by direct regeneration from leaf tissue and by shoot tip culture. *J. Am. Soc. Sugarcane Technol.* 9:97–102.

Grivet L, Arruda P (2002). Sugarcane genomics: Depicting the complex genome of an important tropical crop. *Curr. Opin. Plant Biol.* 2:122–127.

Grivet L, D'Hont A, Roques D, Feldmann P, Lanaud C, Glaszmann JC (1996). RFLP mapping in cultivated sugarcane (*Saccharum* spp.): Genome organization in a highly polyploidy and aneuploid interspecific hybrid. *Genetics* 142:987–1000.

Grivet L, Glaszmann JC, Vineentz M, Da Silva F, Arruda P (2003). ESTs as a source for sequence polymorphism discovery in sugarcane example of the *Adh* genes. *Theor. Appl. Genet.* 106:190–197.

Grof CPL, Campbell JA (2001) Sugarcane sucrose metabolism: Scope for molecular manipulation. *Aust. J. Plant Physiol.* 28:1–12.

Grof CPL, Glassop D, Quick WP, Sonnewald U, Campbell JA (1996). Molecular manipulation of sucrose phosphate synthase in sugarcane. In: JR Wilson, DM Hogarth, JA Campbell, AL Garside, eds, *Sugarcane: Research towards efficient and sustainable production*, 124–126. Brisbane: CSIRO Division of Tropical Crops and Pastures.

Guda C, Lee SB, Daniell H (2000). Stable expression of a biodegradable protein-based polymer in tobacco chloroplasts. *Plant Cell Rep.* 19:257–262.

Guiderdoni E, Demarly Y (1988). Histology of somatic embryogenesis in cultured leaf segments of sugarcane plantlets. *Plant Cell Tiss. Organ Cult.* 14:71–88.

Guiderdoni E, Merot B, Eksomtramage T, Paulet F, Feldmann P, Glaszmann JC (1995). Somatic embryogenesis in sugarcane (*Saccharum species*). In: YPS Bajaj, ed, *Biotechnology in agriculture and forestry*, vol. 31. *Somatic embryogenesis and synthetic seed II*, 92–113. Berlin: Springer.

Guimaraes CT, Honeycutt RJ, Sills CR, Sorbal BWS (1997a). Genetic maps of *Saccharum officinarum* L. and *Saccharum robustum* Brandes & Jesw. Ex Grassl. *Genet. Mol. Biol.* 22:125–132.

Guimaraes CT, Sills GR, Sobral BWS (1997b). Comparative mapping of Andropogoneae: *Saccharum* L. (sugarcane) and its relation to sorghum and maize. *Proc. Natl. Acad. Sci. USA* 94:14261–14266.

Ha S, Moore PH, Heinz D, Kato S, Ohmido N, Fukui K (1999). Quantitative chromosome map of the polyploidy *Saccharum spontaneum* by multicolor fluorescence *in situ* hybridization and imaging methods. *Plant Mol. Biol.* 39:1165–1173.

Hammerschlag FA (1982). Factors influencing in vitro multiplication and rooting of the plum rootstock Myrobalan (*Prunus cerasifera* Ehrh.). *J. Am. Soc. Hort. Sci.* 107:44–47.

Hancock REW, Lehrer R (1998). Cationic peptides: A new source of antibiotics. *Trends Biotechnol.* 16:82–88.

Handlon D, MacMahon GG, McGuire P, Beattie RN, Stringer JK (2000). Managing low sugar prices on farms – Short term and long term strategies. *Proc. Aust. Soc. Sugarcane Technol.* 22:1–8.

Hansom S, Bower R, Zhang L, Potier B, Elliot A, Basnayake S, Cordeiro C, Hogarth DM, Cox M, Berding N, Birch RC (1999). Regulation of transgene expression in sugarcane. *Proc. Int. Soc. Sugar Cane Technol.* 23:278–289.

Harrison SJ, Marcus JP, Goulter KC, Brumbley S, Green JL, Maclean DJ, Manners JM (1996). Antimicrobial proteins new options for disease control in sugarcane. In: JR Wilson, DM Hogarth, JA Campbell, AL Garside, eds, *Sugarcane: Research towards efficient and sustainable production*, 135–137. Brisbane: CSIRO Division of Tropical Crops and Pastures.

Heinz, DJ (1977). Cell, tissue and organ culture in sugarcane improvement. In: J Reinert, YPS Bajaj, eds, *Applied and Fundamental Aspects of Plant Cell, Tissue and Organ Culture*, 3–17. Berlin Heidelberg-New York: Springer.

Heinz DJ, Krishnamurthi M, Nickell LG, Maretzki A (1977). Cell, tissue and organ culture in sugarcane improvement. In: J Reinert, YPS Bajaj, eds, *Applied and Fundamental Aspects of Plant Cell, Tissue and Organ Culture*, 3–17. Berlin: Springer.

Heinz DJ, Mee GWP (1969). Plant differentiation from callus tissue of *Saccharum* species. *Crop Sci.* 9:346–348.

Heinz DJ, Mee GWP (1971). Morphologic, cytogenetic, and enzymatic variation in *Saccharum* species hybrid clones derived from callus tissue. *Am. J. Bot.* 58:257–262.

Hempel FD, Welch DR, Feldman LJ (2000). Floral induction and determination: Where is flowering controlled? *Trends Plant Sci.* 5:17–21.

Hendre RR, Iyer RS, Kotwal M, Khuspe SS, Mascarenhas AF (1983). Rapid multiplication of sugarcane by tissue culture. *Sugarcane*, May/June, 5–8.

Hendre RR, Mascarenhas AF, Nadgir AL, Pathak M, Jagannathan V (1975). Growth of mosaic virus-free sugarcane plants from apical meristem. *Indian Phytopathol.* 28:175–178.

Hidden P, Phillips A (2000). Manipulation of hormone biosynthetic genes in transgenic plants. *Curr. Opin. Biotechnol.* 11:130–137.

Hoarau JY, Grivet L, Offmann B, Raboin LM, Diorflar JP, Payet J, Hellmann M, D'Hont A, Glaszmann JC (2002). Genetic dissection of a modern sugarcane cultivar (*Saccharum* spp.) II. Detection of QTLs for yield components. *Theor. Appl. Genet.* 105:1027–1037.

Hoarau JY, Offmann B, D'Hont A, Risterucci AM, Roques D, Glaszmann JC, Grivett L (2001). Genetic dissection of a modern sugarcane cultivar (*Saccharum spp.*) I. Genome mapping with AFLP markers. *Theor. Appl. Genet.* 103:84–97.

Hogarth DM, Gox MC, Bull JK (1997). Sugarcane improvement: Past achievements and future prospects. In: MS Kang, ed, *Crop improvement for the 21st century*. Trivandrum, India: Research Signpost.

Hoy JW, Bischoff KP, Milligan SB, Gravois KA (2003). Effect of tissue culture explant source on sugarcane yield components. *Euphytica* 129:237–240.

Hughes FL, Rybicki EP, Kirby R (1993). Complete nucleotide sequence of sugarcane streak monogeminivirus. *Arch. Virol.* 132:171–182.

Hurney AP, Berding N (2000). Impact of suckering and lodging on productivity of cultivars in the wet tropics. *Proc. Aust. Soc. Sugarcane Technol.* 22:328–333.

Hussey G (1983). In: SH Mantell, H Smith, eds, *Plant Biotechnology*, 111–138. Cambridge Univ. Press.

Ingelbrecht IL, Irvine JE, Mirkov TE (1999). Post transcriptional gene silencing in transgenic sugarcane. Dissection of homology-dependent virus resistance in a monocot that has a complex polyploidy genome. *Plant Physiol.* 119:1187–1197.

Irey MS, Baucum LE, Derrick KS, Manjunath KL, Lockhart BE (1997). Detection of the luteovirus associated with yellow leaf syndrome of sugarcane (YLS) by a reverse transcriptase polymerase chain reaction and incidence of YLS in commercial varieties in Florida. Proceedings of 5th International Society of Sugar Cane Technologists Pathology and 2nd International Society of Sugar Cane Technologists Molecular Biology Workshop, Umhlanga Rocks, South Africa, May.

Irvine JE, Benda GTA (1987). Transmission of sugarcane diseases in plants derived by rapid regeneration from diseased leaf tissue. *Sugar Cane* 6:14–16.

Irvine JE, Benda GTA, Legendre BL (1991). The frequency of marker changes in sugarcane plants regenerated from callus culture. *Plant Cell Tissue Organ Cult.* 26(2):115.

Irvine JE, Mirkov TE (1997). The development of genetic transformation of sugarcane in Texas. *Sugar J.* 60:25–29.

Ishimaru K, Ono K, Kashiwagi T (2004). Identification of a new gene controlling plant height in rice using the candidate gene approach. *Planta* 218:388–395.

Islam AS, Begam HA, Haque MM (1982). Studies on regeneration of *Saccharum officinarum* for disease th resistant varieties. Proc of 5 Intl. Congr. on Plant Tissue and Callus Culture. pp. 709–710.

Jackson P, Bonnett G, Chudleigh P, Hogarth M, Wood A (2000). The relative importance of cane yield and traits affecting CCS in sugarcane varieties. In: DM Hogarth, ed, *Proc. Aust. Soc. Sugarcane Technol.* 22: 23-29.

Jadhav AB, Vaidya ER, Aher VB, Pawar AM (2001). In vitro multiplication of co–86032 sugarcane (S. officinarum) hybrid. *Indian J. Agric. Sci.* 71:113–115.

Jalaja NC, Sreenivasan TV (1995). Production of somaclonal variants. Screening for salt tolerance. Sugarcane Breeding Institute Annual Report, 1994–95, p. 33.

Jambhale ND, Patil SC, Pradeep T (1995). Tissue culture in sugarcane improvement-A review. *Ann Conv. D.S.T.A., Pune.* 44:10–20.

Jannoo N, Grivet L, Dookun A, D'Hont A, Glaszmann JC (1999a). Linkage disequilibrium among modern sugarcane cultivars. *Theor. Appl. Genet.* 99:1053–1060.

Jannoo N, Grivet L, Seguin M, Paulet F, Domaingue R, Rao PS, Dookun A, D'Hont A, Glaszmann JC (1999b). Molecular investigation of the genetic base of sugarcane cultivars. *Theor. Appl. Genet.* 99:171–184.

Jimenez E, Perez-Ponce J, Martin D, Garcia I (1991). Study of sugarcane (*Saccharum* spp.) populations obtained by in vitro micropropagation. *Centro Agricol.* 18(2):74–78.

Joyce PA, McQualter RB, Bernad MJ, Smith GR (1998a). Engineering for resistance to SGMV in sugarcane. *Acta Hort.* 461:385–391.

Joyce PA, McQualter RB, Handley JA, Dale JL, Harding RM, Smith GR (1998b). Transgenic sugarcane resistant to sugarcane mosaic virus. *Proc. Aust. Soc. Sugarcane Technol.* 20:204–210.

Keating BA, Wilson JR (1997). *Intensive sugarcane production: Meeting the challenge beyond 2000.* Wallingford, UK: CAB International.

Kefeli VI, Kutlácek M (1977). Phenolic substances and their possible role in plant growth regulation. In: PE Pilet, ed, *Plant Growth Regulation*, 181–188. Springer, Berlin, Heidelberg.

Krishnamurthy K, Balconi C, Sherwood JE, Giroux MJ (2001). Wheat puroindolines enhance fungal disease resistance in transgenic rice. *Mol. Plant-Microbe Interact.* 14:1255–1260.

Krishnamurthi M (1975). Notes on disease resistance of tissue culture sub-clones and fusion of sugarcane protoplasts. *Sugarcane Breed Newsl.* 35:24–26.

Krishnamurthi M (1977). Sugarcane improvement through tissue culture. *Proc. ISSCT.* 16(1):23–28.

Krishnamurthi M (1981). Sugarcane tissue culture, an example for crop improvement. Proc. Int. Workshop on Improvement of Tropical Crop through Tissue Culture, March 7–14, 1981, Dhaka Bangladesh. 12–22.

Krishnamurthi M, Tlaskal J (1974). Fiji disease resistance *Saccharum officinarum* var. pindar sub-clones from tissue culture. *Proc. Int. Soc. Sugar Cane Technol.* 15:130–137.

Kristini A (2004). The use of tissue culture to eliminate some important disease in sugarcane. MSc thesis, University of Queensland, Australia.

Kumar H, Upadhyaya Kumar M, Nasar SKT (1995). Micropropagation in litchi: The technology in development. In: SKT Nasar, ed, *Entrepreneurial Opportunities in Food and Agro Based Industries*, 46–49. Dept. Sci. Tech. Govt. of India, Bhagalpur.

Kumar H, Upadhyaya Kumar M, Shahi VK, Nasar SKT (1998). Effect of antioxidant in controlling browing of culture during micropropagation of litchi. In: Proc. of National Seminar on Plant Biotechnology for Sustainable Hill Agriculture, DARL, DRDO. Pithoragar (accessed December 14 2017).

Kumari R, Verma DK (2001). Development of micropropagation protocol sugarcane (*Saccharum officinarum*) A review. *Agric. Rev.* 22:87–94.

Lakshmanan P, Geijskes RJ, Elliott AR, Wang, LF, McKeon MG, Swain RS, Borg Z, Berding N, Grof CPL, Smith GRA (2002). Thin cell layer culture system for the rapid and high frequency direct regeneration of sugarcane and other monocot species. In: IK Vasil, ed, *Proceedings of 10th International Association for Plant Tissue Culture and Biotechnology*, Orlando, June 23–28. (Abstract p-1441).

Lakshmanan P, Geijskes RJ, Elliott AR, Wang LF, McKeon MG, Swain RS, Borg Z, Berding N, Grof CPL, Smith GR (2003). Direct regeneration tissue culture and transformation systems for sugarcane and other monocot species. *Proc. Int. Soc. Sugar Cane Technol. Mol. Biol. Workshop* 4:25.

Lal N (1990). *Indian Sugar* XLVIII:961–965.

Lal N (1992). Assessment of indole-3-acetic acid, indole-3-butyric acid and alpha naphthalene acetic acid for in vitro rooting and plantlet growth in sugarcane. *Indian Sugar* 42(4):205–208.

Lal N (1993). Proliferation and greening effects of sucrose on sugarcane shoot cultures in vitro. *Sugarcane* 6:13–15.

Lal N (1999). Sucrose induced changes in growth and production of phenolics in sugarcane shoot cultures. *Indian Sugar* March:961–965.

Lal N, Singh HN (1993). Evaluation of gelling and support materials for in vitro shoot multiplication in sugarcane. *Sugarcane* 2:2–3.

Lal N, Singh HN (1994). Rapid clonal multiplication of sugarcane through tissue culture. *Plant Tissue Cult.* 4:1–7.

Laporte MM, Galagan JA, Shapiro JA, Boersing MR, Shewmaker CK, Sharkey TD (1997). Sucrose-phosphate synthase activity and yield analysis of tomato plants transformed with maize sucrose-phosphate synthase. *Planta* 203:253–259.

Larkin PJ, Scowcroft WR (1981). Somaclonal variation a novel source of variability from cell cultures for plant improvement. *Theor. Appl. Genet.* 60:197–214.

Last KI, Brettell RIS, Chamberlain DA, Chaudhury AM, Larkin PJ, Marsch EL, Peacock WJ, Dennis ES (1991). pEmu: An improved promoter for gene expression in cereal cells. *Theor. Appl. Genet.* 81:581–588.

Lee TSG (1986). Multiplication of sugarcane by apex culture. *Turrialba.* 36:231–236.

Lee TSG (1987). Micropropagation of sugarcane (*Saccharum spp.*). *Plant Cell Tiss. Organ Cult.* 10:47–55.

Legaspi JC, Mirkov TE (2000). Evaluation of transgenic sugarcane against stalkborers. *Proc. Int. Soc. Sugar Cane Technol. Sugarcane Entomol. Workshop* 4:68–71.

Leibbrandt NB, Snyman SJ (2003). Stability of gene expression and agronomic performance of a transgenic herbicide-resistant sugarcane line in South Africa. *Crop Sci.* 43:671–678.

Leu LS (1978). Apical meristem culture and redifferentiation of callus masses to free some sugarcane systemic diseases. *Plant Protect. Bull. (Taiwan)* 20:77–82.

Lim HO, Sio CL (1985). Micropropagation of Lagerstroemia speciosa (L) Pers Lythranae. *Gardens Bull.* 38:175–178.

Lima MLA, Garcia AAF, Oliveira KM, Matsuoka S, Arizono H, De Souza CL, De Souza AP (2002). Analysis of genetic similarity detected by AFLP and coefficient of parentage among genotypes of sugarcane (*Saccharum spp.*). *Theor. Appl. Genet.* 104:30–38.

Lingle SA (1999). Sugar metabolism during growth and development in sugarcane internodes. *Crop Sci.* 39:480–486.

Liu DW, Oard SV, Oard JH (2003). High transgene expression levels in sugarcane (*Saccharum officinarum* L.) driven by the rice ubiquitin promoter RUBQ2. *Plant Sci.* 165:743–750.

Liu MC (1981). *In vitro* methods applied to sugarcane improvement. In: TA Thorpe, ed, *Plant tissue culture: Methods and applications in agriculture*, 299–323. New York: Academic Press.

Liu MC (1993). Factors affecting induction, somatic embryogenesis and plant regeneration of callus from cultured immature inflorescences of sugarcane. *J. Plant Physiol.* 141:714–720.

Liu LJ, Marquez ER, Biascoechia MC (1983). Variation in degree of rust resistance among plantlets derived from callus cultures of sugarcane in Puerto Rico. *Phytopathology* 73(5):797 (Abstract).

Liu MC, Chen W (1976). Tissue and cell cultures as aids to sugarcane breeding I. creation of genetic variability through callus cultures. *Euphytica*. 25:393–403.

Liu MC, Chen W (1978). Tissue and cell culture as aids to sugarcane breeding – II. Performance and field potential of Callus derived lines. Euphytica. 27:273–282.

Liu MC, Huang VJ, Shih SC (1972). The in vitro production of plants from several tissues of *Saccharum* species; *Agric. Abst. China.* 77:52–58.

Lossl A, Eibl C, Harloff HJ, Jung C, Koop HU (2003). Polyester synthesis in transplastomic tobacco (*Nicotiana tabacum* L.): Significant contents of polyhydroxybutyrate are associated with growth reduction. *Plant Cell Rep.* 21:891–899.

Lourens AG, Martin FA (1987). Evaluation of *in vitro* propagated sugarcane hybrids for somaclonal variation. *Crop Sci.* 27:793–796.

Lu YH, D'Hont A, Paulet F, Grivet L, Arnaud M, Glaszmann JC (1994). Molecular diversity and genome structure in modern sugarcane varieties. *Euphytica* 78:217–226.

Ma H, Albert HH, Paull R, Moore PH (2000). Metabolic engineering of invertase activities in different subcellular compartments affects sucrose accumulation in sugarcane cells. *Aust. J. Plant Physiol.* 27:1021–1030.

MacKenzie DR, Anderson PM, Wernham CC (1966). A mobile air blast inoculator for pot experiments with maize dwarf mosaic virus. *Plant Dis. Rep.* 50:363–367.

Mäder M, Füssl R (1982). Role of peroxidase in lignification of tobacco cells: II. Regulation by phenolic compounds. *Plant Physiol.* 70:1132–1134.

Maliga P (2004). Plastid transformation in higher plants. *Annu. Rev. Plant Biol.* 55:289–313.

Manickavasagam M, Ganapathi A (1998). Direct somatic embryogenesis and plant regeneration from leaf explants of sugarcane. *Indian J. Exp. Biol.* 36:832–835.

Manickavasagam M, Ganapathi A, Anbazhagan VR, Sudhakar B, Selvaraj N, Vasudevan A, Kasthurirengan S (2004). *Agrobacterium* mediated genetic transformation and development of herbicide resistant sugarcane (*Saccharum* species hybrids) using axillary buds. *Plant Cell Rep.* 23:134–143.

Mannan SKA, Amin MN (1999). Callus and shoot formation from leaf sheath of sugarcane (*Saccharum officinarum* L.) In vitro. *Indian Sugar.* 49(3):187–192.

Maretzki A, Hiraki P (1980). Sucrose promotion of root formation in plantlets regenerated from callus of *Saccharum* species. *Phyton.* 38:85–88.

Maretzki A, Sun SS, Nagai C, Bidney D, Houtchens KA, Dela Cruz A (1990). Development of a transformation system for sugarcane. *Proc. VII Int. Congr. Plant Tiss. Cell Cult.*, 68.

Mayer MJ, Narbad A, Parr AJ, Parker ML, Walton NJ, Mellon FA, Michael AJ (2001). Rerouting the plant phenylpropanoid pathway by expression of a novel bacterial enoyl-CoA hydratase/lyase enzyme function. *Plant Cell* 13:1669–1682.

McDowell JM, Woffenden BJ (2003). Plant disease resistance genes: Recent insights and potential application. *Trends Biotechnol.* 21:178–183.

McElroy D, Blowers AD, Jenes B, Wu R (1991). Construction of expression vectors based on the rice actin I (ActI) 5 region for use in monocot transformation. *Mol. Gen. Genet.* 231:150–160.

McIntyre L, Aitken K, Berding K, Casu R, Drenth J, Jackson P et al. (2001). Identification of DNA markers linked to agronomic traits in sugarcane in Australia. *Proc. Int. Soc. Sugar Cane Technol.* 24:560–562.

McLeod RS, McMahon GG, Allsopp PG (1999). Costs of major pests and disease to the Australian sugar industry. *Plant Protect. Quart.* 14:42–46.

McQualter RB, Burn P, Smith GR, Dale JL, Harding RM (2003a). Molecular analysis of *Fiji disease fijivirus* genome segments 5,6,8 and 10. *Arch. Virol.* 149:713–721.

McQualter RB, Dale JL, Harding RH, McMahon JA, Smith GR (2004a). Production and evaluation of transgenic sugarcane containing a Fiji disease virus (FDV) genome segment S9-derived synthetic resistance gene. *Aust. J. Agric. Res.* 55:139–145.

McQualter RB, Fong Chong B, O'Shea M, Meyer K, Van Dyk DE, Viitanen PV, Brumbley SM (2004b). Initial evaluation of sugarcane as a production platform for a p-hydroxybenzoic acid. *Plant Biotechnol. J.* 2:1–13.

McQualter RB, Harding RM, Dale JL, Smith GR (2001). Virus derived resistance to Fiji disease in transgenic plants. *Proc. Int. Soc. Sugar Cane Technol.* 24:584–585.

McQualter RB, Harding RM, Dale JL, Smith GR (2003b). A transgene derived from segment 9 ORF I of the genome of *Fiji disease fijivirus* confers resistance in transgenic plants. *Proc. Int. Soc. Sugar Cane Technol. Molecular Biology Workshop,* 4:18.

McQualter RB, Smith GR, Dale JL, Harding RM (2003c). Molecular analysis of *Fiji disease fijivirus* genome segments 1 and 3. *Virus Genes* 26:283–289.

Ming R, Monte TA, Hernandez E, Moore PH, Irvine JE, Paterson AH (2002a). Comparative analysis of QTLs affecting plant height and flowering among closely-related diploid and polyploidy genomes. *Genome* 45:794–803.

Ming R, Liu SC, Bowers JE, Moore PH, Irvine JE, Paterson AH (2002b). Construction of a *Saccharum* consensus genetic map from two interspecific crosses. *Crop Sci.* 42:570–583.

Ming R, Liu SC, Lin YR, Da Silva J, Wilson W, Braga D et al. (1998). Detailed alignment of *Saccharum* and Sorghum chromosomes: Comparative organization of closely related diploid and polyploidy genomes. *Genetics* 150:1663–1682.

Ming R, Liu SC, Moore PH, Irvine JE, Paterson AH (2001). QTL analysis in a complex autopolyploid: Genetic control of sugar content in sugarcane. *Genome Res.* 11:2075–2084.

Ming R, Wang YW, Draye X, Moore PH, Irvine JE, Paterson AH (2002c). Molecular dissection of complex traits in autopolyploids: Mapping QTLs affecting sugar yield and related traits in sugarcane. *Theor. Appl. Gen.* 105:332–345.

Moonan F, Molina J, Mirkov TE (2000). Sugarcane yellow leaf virus: An emerging virus that has evolved by recombination between luteoviral and poleroviral ancestors. *Virology* 269:156–171.

Moore PH (1999). Progress and development in sugarcane biotechnology. *Proc. Int. Soc. Sugar Cane Technol.* 23:241–258.

Monacco LC et al. (1977). In: J Heinrt, YPS Bajaj, eds, *Applied and Fundamental Aspects of Plant Cell Tissue and Organ Culture,* 109–129. Berlin: Springer Verlag.

Morel G (1963). Botanique-La Culture in Vitro Du Meristeme Apical De Certaines Orchidees 123456. *C.R. Acad. Sci.* 256:4955–4957.

Mourgues F, Briset MN, Chevreau E (1998). Strategies to improve plant resistance to bacterial diseases through genetic engineering. *Trends Biotechnol.* 16:203–210.

Mudge J, Andersen WR, Kehrer RL, Fairbanks D (1996). A RAPD genetic map of *Saccharum officinarum. Crop Sci.* 36:1362–1366.

Murashige T (1974). Plant propagation through tissue cultures. *Ann. Rev. Plant Physiol.* 25:135–166.

Murashige T, Skoog F (1962). A revised medium for rapid growth and bioassays with tobacco tissue cultures. *Physiol. Plant.* 15:473–497.

Murray C, Christeller JT (1994). Genomic nucleotide sequence of a proteinase inhibitor II gene. *Plant Physiol.* 106:1681.

Nadar HM, Heinz DJ (1977). Root and shoot development from sugarcane callus tissue. *Crop. Sci.* 17:814–816.

Nair NV, Nair S, Sreenivasan TV, Mohan M (1999). Analysis of genetic diversity and phylogeny in *Saccharum* and related genera using RAPD markers. *Genet. Res. Crop. Evol.* 46:73–79.

Napoli C, Lemieux C, Jorgensen R (1990). Introduction of a Chimeric Chalcone Synthase Gene into Petunia Results in Reversible Co-Suppression of Homologous Genes in trans. *Plant Cell* 1990 Apr;2(4):279–289.

Nawrath C, Poirier Y, Somerville C (1994). Targeting of the polyhydroxybutyrate biosynthetic pathway to the plastids of *Arabidopsis thaliana* result in high levels of polymer accumulation. *Proc. Natl Acad. Sci. USA* 91:12760–12764.

Nickell LC (1964). Tissue and cell cultures of sugarcane: Another research tool. *Hawaii Plant Rec.* 57:223–229.

Nobécourt P (1939). Sur la pérennité et l'augmentation de volume des cultures de tissues végétaux. *Compt. Rendus Soc. Biol. Lyon.* 130:1270–1271.

Nutt KA, Allsopp PG, Geijskes RJ, McKeon MC, Smith GR (2001). Canegrub resistant sugarcane. *Proc. Int. Soc. Sugar Cane Technol.* 24:584–585.

Nutt KA, Allsopp PG, McGhie TK, Shepherd KM, Joyee PA, Taylor GO, McQualter RB, Smith GR (1999). Transgenic sugarcane with increased resistance to canegrubs. *Proc. Aust. Soc. Sugarcane Technol.* 21:171–176.

Oropeza M, de Garcia E (1997). Use of molecular markers for the identification of varieties of sugarcane (*Saccharum sp.*). *Phyton* 61:81–85.

Oropeza M, Guevara P, de Garcia E, Ramirez JL (1995). Identification of somaclonal variants of sugarcane (*Saccharum spp.*) resistant to sugarcane mosaic virus via RAPD markers. *Plant Mol. Biol. Rep.* 13:182–189.

Osusky M, Osuska L, Hancock RE, Kay WW, Misra S (2004). Transgenic plants expressing cationic peptide chimeras exhibit broad-spectrum resistance to phytopathogens. *Transgenic Res.* 13:181–190.

Osusky M, Zhou G, Osuska L, Hancock RE, Kay WW, Misra S (2000). Transgenic plants expressing cationic peptide chimeras exhibit broad-spectrum resistance to phytopathogens. *Nature Biotechnol.* 18:1162–1166.

Pan YB, Grisham MP, Burner DM, Damann KE (1998). A polymerase chain reaction protocol for the detection of *Clavibacter xyli* subsp. *Xyli*, the causal bacterium of sugarcane ratoon stunting disease. *Plant Dis.* 82:285–290.

Parmessur Y, Aljanabi S, Saumtally S, Dookun-Saumtally A (2002). Sugarcane yellow leaf virus and sugarcane yellows phytoplasma: Elimination by tissue culture. *Plant Pathol.* 51:561–566.

Patel SR, Patel CL, Patel AA, Prajapati SB (1999). The effect of media composition on establishment and growth of in vitro sugarcane meristem var. CoLk 8001. *Indian Sugar* 49:25–29.

Pawar SV, Patil SC, Jambhale VM, Mehetre SS (2002a). Effect of growth regulators on *in vitro* multiplication of sugarcane varieties. *Indian Sugar* 109–111.

Pawar SV, Patil SC, Jambhale VM, Naik RM, Mehetre SS (2002b). Rapid multiplication of commercial sugarcane varieties through tissue culture. *Indian Sugar* 11:183–186.

Pierik RLM (1979). *In Vitro Culture of Higher Plants.* Springer, Wageningen, Netherlands.

Piperidis G, Christopher MJ, Carroll BJ, Berding N, D'Hont A (2000). Molecular contribution to selection of intergeneric hybrids between sugarcane and the wild species *Erianthus arundinaceus. Genome* 43:1033–1037.

Piperidis G, Rattey AR, Taylor GO, Cox MC (2004). DNA markers: A tool for identifying sugarcane varieties. *Proc. Aust. Soc. Sugarcane Technol.* 26.

Poirier Y, Dennis DE, Klomparens K, Somerville C (1992). Polyhydroxybutyrate, a biodegradable thermoplastic, produced in transgenic plants. *Science* 256:520–523.

Preece JE, Compton ME (1991). Biotechnology in agriculture and Forestry. *High Tech. and Micropropagation,* 17. Berlin: Springer-Verlag.

Price S, Warner JN (1959). The possible use of induced mutations for sugarcane improvement. *Proc. Int. Soc. Sugar Cane Technol.* 10:782–794.

Quark F (1977). Meristem culture and virus free plants. In: *Applied and Fundamental Aspects of Plant Cell, Tissue and Organ Culture,* 598–615. New York: Springer-Verlag.

Que L, Hancock JF (2001). Detecting and mad mapping repulsion-phase linkage in polyploids with polysomic inheritance. *Theor. Appl. Genet.* 103:136–143.

Rae AL, Grof CPL, Casu RE, Bonnett GD (2005a). Sucrose accumulation in the sugarcane stem: Pathways and control points for transport and compartmentation. *Field Crop Res.* 92:159–163.

Rae AL, Perroux J, Grof CPL (2005b). Sucrose partitioning between vascular bundles and storage parenchyma in the sugarcane stem: A potential role for the ShSUTI sucrose transporter. *Planta* 220:817–825.

Rangel P, Gomes L, Victoria JL, Angel F (2003). Transgenic plants of CC 84-75 resistant to the virus associated with the sugarcane yellow leaf syndrome. *Proc. Int. Sugar Cane Technol. Mol. Biol. Workshop* 4:30.

Rathus C, Birch RG (1992). Stable transformation of callus from electroporated sugarcane protoplasts. *Plant Sci.* 82:81–89.

Rathus C, Bower R, Birch R (1993). Effects of promoter, intron and enhancer elements on transient gene expression in sugarcane and carrot protoplasts. *Plant Mol. Biol.* 23: 613–618.

Roach, BT (1989). Origin and improvement of the genetic base of sugarcane. *Proc. Aust. Soc. Sugarcane Technol.* 11:34–47.

Roberts S, Grof CPL, Bucheli CS, Robinson SP, Wilson JR (1996). Genetic engineering of sugarcane for low colour raw sugar. In: JR Wilson, DM Hogarth, JA Campbell, AL Garside, eds, *Sugarcane: Research towards efficient sustainable production,* 130–132. Brisbane: CSIRO Division of Tropical Crops and Pastures.

Rohwer JM, Botha F (2001). Analysis of sucrose accumulation in the sugarcane culm on the basis of *in vitro* Kinetic data. *Biochem. J.* 358:437–445.

Rossi M, Araujo PG, Paulet F, Carsmeur O, Dias VM, Chen H, Van Sluys MA, D'Hont A (2003). Genomic distribution and characterization of EST-derived resistance gene analogs (RGAs) in sugarcane. *Mol. Gen.* 269:406–409.

Rott P, Bailey RA, Comstock JC, Croft BJ, Saumtally AS (2000). *A guide to sugarcane disease.* Montpellier, France: CIRAD Publication Service.

Salter B, Bonnett GD (2000). High soil nitrate concentration during autumn and winter increase suckering. *Proc. Aust. Soc. Sugarcane Technol.* 22:322–327.

Savangikar VA, Savangikar CV, Joshi MS, Ravtkar DM (1991). Agricultural papers. Annu. Conv. DSTA, 41:239–240.

Sauvaire D, Glozy R (1978). Multiplication vegetative de la canne a sucre (*Saccharum* sp.) par bouturage in vitro. *C. R. Acad. Sci Paris, Ser. D.* 467–470.

Schenk PM, Remans T, Sagi L, Elliott AR, Dietzgen RG, Swennen R, Ebert PR, Grof CPL, Manners JM (2001). Promoters for pregenomic RNA of banana streak badnavirus are active for transgene expression in monocot and dicot plants. *Plant Mol. Biol.* 47:399–412.

Selvi A, Nair NV, Balasundaram N, Mohapatra T (2003). Evaluation of maize microsatellite markers for genetic diversity analysis for fingerprinting in sugarcane. *Genome* 46:394–403.

Setamou M, Bernal JS, Legaspi JC, Mirkov TE, Legaspi BC (2002). Evaluation of lectin-expressing transgenic sugarcane against stalkborers (Lepidoptera: pyralidae): Effects on life history parameters. *J. Econ. Entomol.* 95:469–477.

Shimizu-Sato S, Mori H (2001). Control of outgrowth and dormancy in axillary buds. *Plant Physiol.* 127:1405–1413.

Shukla R, Garg GK, Khan AQ (1994). *In vitro* clonal propagation of sugarcane: Optimization of media and hardening of plants. *Sugarcane* 4:21–23.

Sills GR, Bridges W, Al-Janabi SM, Sobral BWS (1995). Genetic analysis of agronomic traits in a cross between sugarcane (*Saccharum officinarum* L.) and its presumed progenitor (*S. robustum Brandes & Jesw. Ex Grassl*). *Mol. Breed.* 1:355–363.

Singh G, Chapman SC, Jackson PA, Lawn RJ (2000). Lodging a major constraint to high yield and CCS in the wet and dry tropics. *Proc. Aust. Soc. Sugarcane Technol.* 22:315–321.

Singh JP, Marwaha RS, Srivastava OP (1995). Processing and nutritive qualities of potato tubers as affected by fertilizer nutrients and sulphur application. *J. Indian Potato Assoc.* 22:32–37.

Singh V, Lal RJ, Lal S, Sinha OK, Singh AP, Srivastava SN (2004). Bioefficient strains of Trichoderma harzianum against red rot pathogen of sugarcane. In: Proceedings of Symposium on Recent Advances in Fungal Bioagents and Their Social Benefit, Zonal Chapter, Ann. Meet., Mycol. & Pl. Pathol, NBRI, Lucknow, September 10, 2004, pp. 47–48.

Skirvin RM (1979). Natural and induced variation in Tissue culture. *Euphytica.* 27:241–266.

Skoog F, Miller CO (1957). Chemical regulation of growth and organ formation in plant tissues cultured in vitro. *Symp. Soc. Exp. Biol.* 11:118–131.

Smith GR, Clarke ML, Van de Velde R, Dale JL (1994). Chemiluminescent detection of Fiji disease virus in sugarcane with biotinylated DNA probes. *Arch. Virol.* 136:325–334.

Smith GR, Ford R, Frenkel MJ, Shukla DD, Dale JL (1992). Transient expression of the coat protein of sugarcane mosaic virus in sugarcane protoplasts and expression in *Escherichia coli*. *Arch. Virol.* 125:15–23.

Smith GR, Joyee PA, Handley JA, Sithisarn P, Maugeri MM, Bernad MJ, Berding N, Dale JL, Harding RM (1996). Genetically engineering resistance to sugarcane mosaic and Fiji disease viruses in sugarcane. In: JR Wilson, DM Hogarth DM, JA Campbell, AL Garside, eds, *Sugarcane: Research towards efficient and sustainable production*, 138–140. Brisbane: CSIRO Division of Tropical Crops and Pastures.

Smith, G, van de Velde R (1994). Detection of sugarcane mosaic virus and Fiji disease virus in diseased sugarcane using the polymerase chain reaction. *Plant Dis.* 78:557–561.

Snell KD, Peoples OP (2002). Polyhydroxyalkanoate polymers and their production in transgenic plants. *Metabol. Eng.* 4:29–40.

Somers DA, Makarevich I (2004). Transgene integration in plants: Poking or patching holes in promiscuous genomes? *Curr. Opin. Biotechnol.* 15:126–131.

Sommer PW, Van Sambeck W, Preece E, Graffney G, Mayers O (1986). In vitro micropropagation of black walnut (Juglans *nigra* L.). Proc. of 8th Ntl. Amer. for Bio. Workshop, pp. 26–28.

Soo HM, Handley JA, Mauger MM, Burns P, Smith GR, Dale JL, Harding RM (1998). Molecular characterization of Fiji disease reovirus genomes segment 9. *J. Gen. Virol.* 79:3155–3161.

Sreenivasan J, Sreenivasan TV, Alexander C (1987). Somaclonal variation for rust resistance in sugarcane. *Indian J Genet.* 47:109–114.

Sreenivasan TV, Jalaja NC (1982). Production of sub-clones from the callus culture of Saccharum–Zea hybrids. *Plant Sci. Lett.* 24:255–259.

Sreenivasan TV, Jalaja NC (1985). Utilisation of tissue culture technique in sugarcane improvement. Sugarcane Breeding Institute Annual Report, p. 33.

Sreenivasan TV, Sreenivasan J (1992). Micropropagation of sugarcane varieties for increasing cane yield. *SISSTA Sugar J.* 18:61–64.

Sreenivasan TV, Jalaja MC (1995). Utility of tissue culture technology in sugarcane improvement. *Proc. Annual Conv.* DSTA; Pune. 1–9.

Staskawiez BJ, Ausubel FM Baker BJ, Ellis JC, Jones JDC (1995). Molecular genetics of plant disease resistance. *Science* 268:661.

Steward FC (1958). Growth and organized development of cultured cells. III. Interpretations of the growth from free cell to carrot plant. *Am. J. Bot.* 45:709–713.

Street HE (1973). *Plant Tissue and Cell Culture*. Oxford: Blackwell.

Taylor PWJ, Dukic S (1993). Development of an *in vitro* culture technique for conservation of *Saccharum* spp. hybrid germplasm. *Plant Cell Tiss. Organ Cult.* 34:217–222.

Taylor PWJ, Geijskes JR, Ko HL, Fraser TA, Henry RJ, Bich RG (1995). Sensitivity of random amplified polymorphic DNA analysis to detect genetic change in sugarcane during tissue culture. *Theor. Appl. Genet.* 90:1169–1173.

Terras FRG, Eggermont K, Kovaleva V, Raikhel NV, Osborn R, Vanderleyden J, Cammue BPA Broekaert WF (1995). Small cysteine-rich antifungal proteins from radish: Their role in host defense. *Plant Cell* 7:573–588.

Thompson N, Choi Y, Randles JW (1998). Sugarcane striate mosaic disease: Development of a diagnostic test. 7th International Congress of Plant Pathology, Edinburgh, August 9–16. (Abstract 3.3.21).

Tlaskal J (1975). A note on disease resistant subclones derived from tissue culture techniques. *Sugarcane Breed Newsl.* 36:71–72

Twyman RM, Stoger E, Schiliberg S, Christou P, Fischer R (2003). Molecular farming in plants: Host systems and expression technology. *Trend. Biotechnol.* 21:570–578.

Ulian E (2000). Functional genomics for sugar accumulation gene discovery in sugarcane. Sugarcane Genomics Workshop, Brisbane, Australia.

Upadhyay SK (1994). M.Sc. Thesis, RAU, Bihar, Pusa.

Van Damme EJM, Allen AK, Peumans WJ (1987). Isolation and characterization of a lectin with exclusive specificity towards mannose from snowdrop (*Galanthus nivialis*) bulbs. *FEBS Lett.* 215:140–144.

Van der Merwe MJ, Groenewald JH, Botha FC (2003). Isolation and evaluation of a developmentally regulated sugarcane promoter. *Proc. S. Afri. Sugar Cane Technol.* 77:146–169.

Vettore AL, da Silva FR, Kemper EL, Souza GM., da Silva AM, Ferro MIT et al. (2003). Analysis and functional annotation of an expressed sequence tag collection for tropical crop sugarcane. *Genome Res.* 13:2725–2735.

Vickers JE, Grof CPL, Bonnett GD, Jackson PA, Knight DP, Roberts SE, Robinson SP (2005c). Directed over-expression and general elevation of polyphenol oxidase activity in transgenic sugarcane is related to juice and raw sugar colour. *Crop Sci.* 45:354–362.

Vickers JE, Grof CPL, Bonnett GD, Jackson PA, Knight DP, Roberts SE, Robinson SP (2005a). Overexpression of polyphenol oxidase in transgenic sugarcane results in darker juice and raw sugar. *Crop Sci.* 45:354–362.

Vickers JE, Grof CPL, Bonnet GD, Jackson PA, Morgan TE (2005b). Effects of tissue culture, biolistic transformation, and introduction of PPO and SPS gene constructs a performance of sugarcane clones in the field. *Aust. J. Agric. Res.* 56:57–68.

Victoria JI, Guzman ML (1993). Rapid in vitro multiplication of sugarcane varieties. Multiplication rapida in vitro de variedades de cane de azucar. Serie Tecnica- Centro de investigation de la cane de Azucar de Colombia. No. 12, 28.

Vijaya M (1997). Varietal resistance to smut in sugarcane. *J. Mycol. Plant Pathol.* 27:74–75.

Viswanathan R (2005). Impact of mosaic infection on growth and yield of sugarcane. *Sugar Tech.* 7(1):61–65.

Wagih M, Gordon GH, Ryan CC, Adkins SW (1995). Development of an axillary bud culture technique for Fiji disease virus elimination in sugarcane. *Aust. J. Bot.* 43:135–143.

Waithaka K (1992). In: G Thottapilly et al., eds, *Biotechnology: Enhancing Research on Tropical Crops in Africa*, 183–188. United Kingdom: Sayee Pub Exeter.

Walkey DCA (1980). In: Ingram DS and Helgeson, JP, eds, *Tissue Culture Methods for Plant Pathologists*, 109–117. Oxford, UK: Blackwell Sci. Pub.

Wang ML, Goldstein C,Su W, Moore PH, Albert HH (2005). Production of biologically active GM-CSF in sugarcane: A secure biofactory. *Transgenic Res.* 14:167–178.

Watt D, McCormics A, Govender C, Cramer M., Huckett B, Botha FC (2005). Increasing the utility of genomics in unraveling sucrose accumulation. *Field Crop Res.* 92:149–158.

Wei, H (2001). Isolation and characterization of two sugarcane polyubiquitin gene promoters and matrix attachment regions. University of Hawaii, Manoa.

Wei H, Albert HH, Moore PH (1999). Differential expression of sugarcane polyubiquitin genes and isolation of promoters from two highly expressed members of the gene family. *J. Plant Physiol.* 155:513–519

Wei H, Moore PH, Albert HH (2003). Comparative expression analysis of two sugarcane polyubiquitin promoters and flanking sequences in transgenic plants. *J. Plant Physiol.* 160:1241–1251.

Wenzl P, Carling J, Kudrna D, Jaccoud D, Huttner E, Kleinhofs A, Kilian A (2004). Diversity Arrays Technology (DarT) for whole genome profiling of barley. *Proc. Natl. Acad. Sci. USA* 101:9915–9920.

White PR (1939). Potentially unlimited growth of excised plant callus in an artificial nutrient. *Am. J. Bot.* 26:59–64.

Wong KP and Fletcher J (2004). Sugarcane white leaf phytoplasma in tissue culture: Long term maintenance, transmission, and oxytetracycline remission. *Plant Cell Tissue and Organ Cult.* 1:149–164.

Worrell AC, Bruneau JM, Summerfelt K, Boersig M, Voelker TA (1991). Expression of a maize sucrose phosphate synthase in tomato alters leaf carbohydrate partitioning. *Plant Cell* 3:1121–1130.

Wu KK, Burnquist W, Sorrels ME, Tew TL, Moore PH, Tanksley SD (1992). The detection and estimation of linkage in polyploids using single-dose restriction fragments. *Theor. Appl. Genet.* 83:294–300.

Yang MZ, Bower R, Burow MD, Paterson AH, Mirkov TE (2003). A rapid and direct approach to identify promoters that confer high levels of gene expression in monocots. *Crops Sci.* 43:1805–1814.

Yang ZN, Mirkov TE (1997). Sequence and relationships of sugarcane mosaic and sorghum mosaic strains and development of RT-PCR-based RFLPs for strain discrimination. *Phytopathology* 87:932–939.

Zhang L, Xu J, Birch RG (1999). Engineered detoxification confers resistance against a pathogenic bacterium. *Nature Biotechnol.* 17:1021–1024.

chapter twelve

Proline metabolism as sensors of abiotic stress in sugarcane

*Ashu Singh, Kalpana Sengar, Manoj Kumar Sharma,
R.S. Sengar and Sanjay Kumar Garg*

Contents

Introduction

Plants are exposed to different types of environmental stress. Among these stresses, osmolyte production stress, in particular that due to drought and salinity, is the most serious problem that limits plant growth and crop productivity in agriculture (Boyer, 1982). Plant growth and productivity is greatly affected by environmental stresses such as drought, high salinity and low temperature. Expression of a variety of genes is induced by these stresses in various plants. The products of these genes function not only in stress tolerance but also in stress response. In the signal transduction network from perception of stress signals to stress-responsive gene expression, various transcription factors and *cis*-acting elements in the stress-responsive promoters function for plant adaptation to environmental stresses. Recent progress has been made in analysing the complex cascades of gene expression in drought and cold stress responses, especially in identifying specificity and crosstalk in stress signaling. In this chapter, we highlight transcriptional regulation of gene expression in response to drought and cold stresses, with particular emphasis on the role of transcription factors and *cis*-acting elements in stress-inducible promoters. Genes produced during stress conditions function not only in protecting cells from stress by producing important metabolic proteins, but also in regulating genes for signal transduction in the stress response. Thus, these gene products are classified into two groups (Fowler et al., 2002; Kreps et al., 2002; Seki et al., 2002). The first group includes proteins that probably function in stress tolerance, such as chaperones, LEA (late embryogenesis abundant) proteins, osmotin, antifreeze proteins, mRNA-binding proteins, key enzymes for osmolyte biosynthesis such as proline, water channel proteins, sugar and proline transporters. The second group contains protein factors involved in further regulation of signal transduction and gene expression that probably function in stress response. They included various transcription factors, suggesting that various transcriptional regulatory mechanisms function in the drought-, cold-, or high salinity-stress signal transduction pathways (Seki et al., 2003). Many workers suggest that there are more than 300 genes that have been identified as being stress-inducible (Kazuko and Shinozaki, 2006). Among these genes, more than half of the drought-inducible genes are also induced by high salinity, indicating the existence of significant crosstalk between the drought and high-salinity responses. The class of small molecules known as 'compatible osmolytes' includes certain amino acids (notably proline), quaternary ammonium compounds (e.g. glycinebetaine, proline-betaine, β-alaninebetaine and choline-O-sulfate), and the tertiary sulfonium compound 3-dimethylsulfoniopropionate (DMSP). The quaternary ammonium compounds and DMSP are derived from amino acid precursors. These compounds share the property of being uncharged at neutral pH and are of high solubility in water (Ballantyne and Chamberlin, 1994). Moreover, at high concentrations they have little or no perturbing effect on macromolecule-solvent interactions (Low, 1985; Somero, 1986; Timasheff, 1993; Yancey, 1994; Yancey et al., 1982).

Unlike perturbing solutes (such as inorganic ions) which readily enter the hydration sphere of proteins, favouring unfolding, compatible osmolytes tend to be excluded from the hydration sphere of proteins and stabilize folded protein structures (Low, 1985). These compounds are thought to play a pivotal role in plant cytoplasmic osmotic adjustment in response to osmotic stresses (Wyn Jones et al., 1977).

Sensors of abiotic stress: Osmolytes

It has been well-documented during stress responses in plants an accumulation of osmolytes occurs. Although their actual roles in plant-stress tolerance remain obscure, these bio-molecules are thought to have positive effects on enzyme and membrane integrity, along with adaptive roles in providing osmotic adjustment in plants grown under salinity/drought conditions. Recent research has demonstrated that the manipulation of genes involved in the biosynthesis of these osmolytes have improved plant tolerance to drought and salinity in a number of crops. There is hope of understanding the molecular basis of osmolyte accumulation under stress and manipulating these processes via genetic engineering. Future research involves generating transgenic plants with still higher levels of tolerance.

There are many cellular mechanisms by which organisms ameliorate the effects of environmental stresses. Accumulation of compatible solutes such as proline is one such phenomenon. Many plants, including halophytes, accumulate compatible osmolytes, such as proline (Pro), glycine betaine and sugar alcohols, when they are exposed to drought or salinity stress (Csonka, 1989; Delauney and Verma, 1993; Hellebust, 1976; Yancey et al., 1982). (See Figure 12.1.) It has been suggested that compatible osmolytes do not interfere with normal biochemical reactions and act as osmoprotectants during osmotic stress. Among known compatible solutes, proline is probably the most widely distributed osmolyte. The accumulation of proline has been observed not only in plants but also in eubacteria, marine invertebrates, protozoa and algae (Delauney and Verma, 1993; McCue and Hanson, 1990). Genes for enzymes involved in the biosynthesis and metabolism of proline have been isolated from various plants, and their expression and the functions of their gene products have been characterized. Results of investigations of the relationship between the expression of these genes and the accumulation of proline under water stress indicate that the level of proline in plants is mainly regulated at the transcriptional level during water stress. Moreover, the overproduction of proline results in the increased tolerance of transgenic tobacco plants to osmotic stress. Thus tolerance to abiotic stress, especially to salt and improved plant growth, was observed in a variety of transgenics that were engineered for overproduction of proline (Kishor et al., 1995; Roosens et al., 2002). Proline seems to have diverse roles under osmotic stress conditions, such as stabilization of proteins, membranes and subcellular structures, and protecting cellular functions by scavenging reactive oxygen species (Bohnert et al., 1999).

Proline and its function in osmoregulation

Proline plays diverse functions in plants. As an amino acid it is a structural component of proteins, but it also plays a role of stress osmolyte solute under environmental stress conditions. Proline metabolism involves several subcellular compartments and contributes to the redox balance of

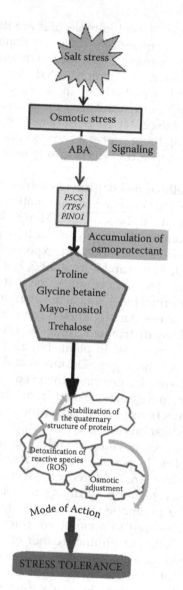

Figure 12.1 Overview of salinity stress.

the cell. Proline synthesis has been associated with tissues undergoing rapid cell divisions, such as shoot apical meristems, and appears to be involved in floral transition and embryo development. High levels of proline can be found in pollen and seeds, where it serves as compatible solute, protecting cellular structures during dehydration. The proline concentrations of cells, tissues and plant organs are regulated by the interplay of

biosynthesis, degradation and intracellular as well as intercellular transport processes. Among the proline transport proteins characterized so far,
both general amino acid permeases and selective compatible solute transporters were identified, reflecting the versatile role of proline under stress
and non-stress situations.

Proline accumulation is a common metabolic response of higher
plants to water deficits and salinity stress (Figure 12.2), and has been the
subject of numerous reviews over the last 20 years (see e.g. Delauney and
Verma, 1993; Hanson and Hitz, 1982; Rhodes, 1987; Rhodes et al., 1999;
Samaras et al., 1995; Stewart, 1981; Stewart and Larher, 1980; Taylor, 1996;
Thompson, 1980). This highly water soluble amino acid is accumulated
by the leaves of many halophytic higher plant species grown in saline
environments (Briens and Larher, 1982; Stewart and Lee, 1974; Treichel,
1975), in leaf tissues and shoot apical meristems of plants experiencing
water stress (Barnett and Naylor, 1966; Boggess et al., 1976; Jones et al.,
1980), in desiccating pollen grains (Hong-qi et al., 1982), in root apical
regions growing at low water potentials (Voetberg and Sharp, 1991), and
in suspension cultured plant cells adapted to water stress (Rhodes et al.,
1986; Tal and Katz, 1980) or NaCl stress (Katz and Tal, 1980; Rhodes and
Handa, 1989; Thomas et al., 1992). Proline protects membranes and proteins against the adverse effects of high concentrations of inorganic ions
and temperature extremes (Pollard and Wyn Jones, 1979; Paleg et al.,
1981; Nash et al., 1982). Proline may also function as a protein-compatible

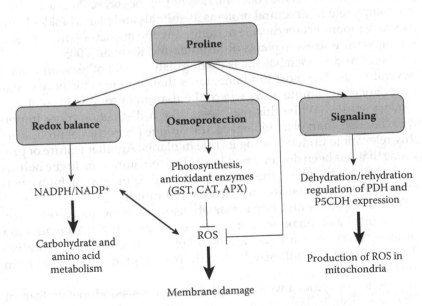

Figure 12.2 Potential role of proline during abiotic stress.

hydrotrope (Srinivas and Balasubramanian, 1995), and as a hydroxyl radical scavenger (Smirnoff and Cumbes, 1989). The proline accumulated in response to water stress or salinity stress in plants is primarily localized in the cytosol (Ketchum et al., 1991; Leigh et al., 1981; Pahlich et al., 1983). In addition, the transient accumulation of proline, might serve as a safety valve to adjust cellular redox state during stress (Kuznetsov and Shevyakova, 1999; Shen et al., 1999).

Proline metabolism and its implications for plant–environment interaction

In plants, under abiotic stress, low molecular weight osmoprotectants and osmolytes such as proline, glycine betaine and sugar alcohols are synthesized and accumulated, and their synthesis and degradation have been studied well (Ishitani et al., 1993; Kiyosue et al., 1996). Very high accumulation of cellular proline (up to 80% of the amino acid pool under stress and 5% under normal conditions) has been documented in many plant species (Choudhary et al., 2005; Widodo et al., 2009). The increase in osmoprotectants is achieved either by altering metabolism (increasing biosynthesis and/or decreasing degradation) or by transport (increase uptake and/or decrease export), which also depends on the species and the extent of stress. In plants proline is synthesized from glutamate as well from arginine/ornithine. Unlike other amino acids, proline has cyclized amino nitrogen that has significant influence on the conformation of polypeptides. Proline is also a major component of structural proteins in animals and plants besides being a known osmoprotectant capable of mitigating the impacts of drought, salt, and temperature stress in plants (Rodriguez and Redman, 2005).

Proline, and its metabolism, is distinguished from other amino acids in several ways. The most fundamental is that proline is the only one of the proteinogenic amino acids where the α-amino group is present as a secondary amine. While this may seem like a distinction more important to chemists than plant biologists, the unique properties of proline are highly relevant to understanding its role in plants. Another feature of proline, one that has been documented in studies too numerous to cite here, is that several types of plant stress cause proline to accumulate to high levels in many plant species (as well as bacteria and fungi).

The role of proline and sulphur metabolism during osmotic stress tolerance in plants has been recently emphasized (Verma, 1999). Accumulation of proline could be due to *de novo* synthesis or decreased degradation (or both). Proline is synthesized not only from glutamate but also from arginine/ornithine.

Genes for enzymes involved in the biosynthesis and metabolism of proline have been isolated from various plants, and their expression and the functions of their gene products have been characterized. Results of investigations of the relationship between the expression of these genes and the accumulation of proline under water stress indicate that the level of proline in plants is mainly regulated at the transcriptional level during water stress. Moreover, the overproduction of proline results in the increased tolerance of transgenic tobacco plants to osmotic stress.

The core of proline metabolism involves two enzymes catalysing proline synthesis from glutamate in the cytoplasm or chloroplast, two enzymes catalysing proline catabolism back to glutamate in the mitochondria, as well as an alternative pathway of proline synthesis via ornithine (Figure 12.1). The interconversion of proline and glutamate is sometimes referred to as the 'proline cycle'. The transcriptional upregulation of proline synthesis from glutamate and downregulation of proline catabolism during stress are thought to control proline levels, although exceptions to this pattern have been observed (Stines et al., 1999). Much data are consistent with this straightforward model of proline metabolism being mainly regulated by transcription of genes encoding the key enzymes. This is likely to not be the whole story, however, as posttranslational regulation of these enzymes has been little explored and the role of ornithine as a proline precursor remains unclear. Likewise, the proline cycle may at first seem to be a futile cycle; however, understanding the coordinate regulation of this cycle and metabolic flux through it is key to overall understanding of proline metabolism and function.

Thus in addition to being a major constituent of proteins, proline also acts as an osmotic protectant in bacteria, plants and animals that are under osmotic stress. In *Arabidopsis*, proline can account for up to 20% of the free amino acid pool after salt stress (Verbruggen et al., 1993). There are two alternative routes in proline biosynthesis in higher plants: the ornithine and the glutamate pathways. It is also known that, as in plants, both ornithine and glutamate are precursors of proline biosynthesis in microorganisms and mammals. The plant glutamate pathway differs from that in bacteria and human. In bacteria and humans, the conversion of glutamate to glutamate-5-semialdehyde (GSA) is catalysed by two enzymes via two consecutive reactions, whereas, in higher plants the conversion is catalysed by a bi-functional enzyme in a single reaction (Hu et al., 1992). Many research activities have been devoted to understand the relative contributions of the two alternative pathways to the increased proline accumulations under stress.

Pathways for the biosynthesis and metabolism of proline in plants

The pathway for the biosynthesis of proline in plants was elucidated by reference to the pathway in *Escherichia coli* (Leisinger, 1987). Figure 12.3

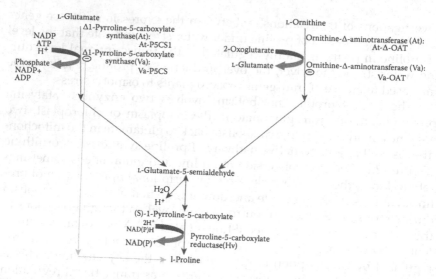

Figure 12.3 Proline biosynthesis pathway.

shows this pathway in bacteria, and the pathways for the biosynthesis and metabolism of proline in plants. The pathway in bacteria begins with the ATP-dependent phosphorylation of the y-carboxy group of L-glutamic acid (L-G1U) by y-glutamyl kinase (y-GK). The product of y-GK is reduced to glutamic-y-semialdehyde (GSA) by GSA dehydrogenase (GSADH), with which y-glutamyl kinase forms an obligatory enzyme complex. GSA cyclizes spontaneously to form A'-pyrroline-5- carboxylate (P5C), which is finally reduced to proline by P5C reductase (P5CR). It has been suggested that, in plants, proline is synthesized either from Glu or from ornithine and that the pathway from Glu is the primary route for the synthesis of Pro under conditions of osmotic stress and nitrogen limitation, while the pathway from ornithine predominates at high levels of available nitrogen (Delauney et al., 1993).

The second important factor that controls levels of proline in plants is the degradation or metabolism of proline. LPro is oxidized to P5C in plant mitochondria by proline dehydrogenase (oxidase) (ProDH; EC 1.5.99.8), and P5C is converted to L-G1U by P5C dehydrogenase (P5CDH) (Boggess et al., 1977; Elthon and Stewart, 1981). Such oxidation of proline is inhibited during the accumulation of proline under water stress and is activated in rehydrated plants (Rayapati and Stewart, 1991). ProDH and P5CDH catalyse reactions that are the reverse of those catalysed by P5CS and P5CR, respectively, in the biosynthesis of proline.

Although stress-induced proline accumulation is evolutionarily conserved in a wide range of plants, its regulatory mechanism is subject to considerable variation (Kishor et al., 1995). In most plant species studied, proline accumulation during stress is the result of reciprocal action of increased biosynthesis and inhibited degradation (Kishor et al.).

Degradation of proline

Proline degradation in eukaryotes takes place in mitochondria and thus in plants is spatially separated from the biosynthetic pathway. Proline catabolism starts with the oxidation of proline to P5C by proline dehydrogenase (PDH), using FAD as cofactor. P5C is subsequently converted to glutamate by pyrroline-5-carboxylate dehydrogenase (P5CDH) using NAD+ (Figure 12.3). Whereas in eukaryotes two enzymes catalyse these subsequent steps in proline degradation, in bacteria both mono- and bifunctional enzymes exist (Tanner, 2008). Two homologous genes have been identified to encode proline dehydrogenases in *Arabidopsis* and tobacco (Funck, unpublished; Mani et al., 2002; Ribarits et al., 2007; Verbruggen and Hermans, 2008), while data from other plant species are scarce. In contrast, the enzyme catalysing the second step of proline degradation (P5CDH) is encoded by a single copy gene in all monocot and dicot species analysed so far (Ayliffe et al., 2005; Deuschle et al., 2001; Mitchell et al., 2006). Biochemical analysis revealed the presence of two P5CDH activities with slightly different characteristics in *Nicotiana plumbaginifolia* and *Zea mays* (Elthon and Stewart, 1982; Forlani et al., 1997). At present it is not clear if both activities arise from a single gene or if a second P5CDH gene is present in these species.

Genetic engineering for abiotic stress tolerance

Metabolic engineering is the directed improvement of cellular properties through the modification of species biochemical reactions or the introduction of new ones, with the use of recombinant DNA technology (Stephanopoulos, 1999). Osmoprotectant accumulation is only one facet of a myriad of stress-tolerant traits found in nature. Since oxidative stress is a component of drought and salinity, manipulations aimed at improving oxidative stress tolerance have also resulted in salinity tolerance (Roxas et al., 1997). This could be done either via repeatedly engineering the gene or by crossing and selecting transgenic plants engineered for different traits (Table 12.1). For example, manipulation of genes involved in ion transport together with osmoprotectant synthesis can be expected to increase a cell's ability to withstand salinity stress. The gene products involved in ion homeostasis have been identified by the use of yeast model systems (Serrano et al., 1999) and by analysing mutants altered for salt sensitivity (Liu et al., 2000; Wu et al., 1996). Osmoprotectant synthesis in naturally stress-tolerant species is highly regulated by stress. In addition to the use of stress inducible promoters for engineering osmoprotectant synthesis

Table 12.1 Genetic engineering of crop plants for abiotic stress tolerance

Gene	Targeted trait	Species	References
Pyrroline-5-carboxylate synthetase (*P5CS*)	Salinity stress tolerance	Tobacco	Kishor et al., 1995
	Increased proline accumulation		Zhang et al., 1995
	Drought and salinity tolerance	Rice	Zhu et al., 1998
	Increased proline accumulation and osmotic stress tolerance		Hong et al., 2000
	Salinity stress tolerance		Anoop and Gupta, 2003
	Faster plant growth under drought and salt stress conditions		Su and Wu, 2004
	Enhanced salt and cold stress tolerance	*Arabidopsis*	Hur et al., 2004
	Hypersensitivity to osmotic stress	Wheat	Nanjo et al., 1999
	Salinity stress tolerance	*Chlamydomonas*	Sawahel and Hassan, 2002
	Tolerance to toxic heavy metals	Carrot	Siripornadulsil et al., 2002
	Salinity stress tolerance	Citrus	Han and Hwang, 2003
	Drought stress tolerance	Potato	Molinaria et al., 2004
	Proline production and salt tolerance	*Larix leptoeuropaea*	Hmida-Sayari et al., 2005
	Resistant to cold, salt and freezing stresses	*Saccharum officinarum*	Gleeson et al., 2005
	Water-deficit stress		Molinari et al., 2007
	Salt stress tolerance	*Oryza sativa*	Kumar et al., 2010
	Tolerance against drought and salinity	*Oryza sativa*	Priya et al., 2015
	Water stress tolerance through proline accumulation	Wheat	Vendruscoloa et al., 2007

(Continued)

Table 12.1 (Continued) Genetic engineering of crop plants for abiotic stress tolerance

Gene	Targeted trait	Species	References
Pyrroline-5-carboxylate reductase (*P5CR*)	Salinity stress tolerance	Tobacco	LaRosa et al., 1991
	Enhanced heat and drought stress tolerance	Soybean	De Ronde et al., 2000
	Osmotic and drought stress tolerance		De Ronde et al., 2004
	Salt and freezing stress tolerance	*Arabidopsis*	Nanjo et al., 1999
Proline dehydrogenase (*ProDH*)	Hypersensitivity to proline		Mani et al., 2002
	Elevated salt tolerance	Tobacco	Kolodyazhnaya et al., 2006
	Increased proline content and drought stress tolerance	Tobacco	Kochetov et al., 2004
	Enhanced water stress tolerance	*Arabidopsis*	Ueda et al., 2008
	Overexpression for osmotic stress tolerance	*Arabidopsis* and tobacco	Miller et al., 2009
	Salinity and water stress tolerance through elevated proline content	Tobacco	Ibragimova et al., 2012
	Enhanced oxidative stress tolerance	*Arabidopsis*	Monteoliva et al., 2014
Ornithine-δ-aminotransferase (δ-OAT)	Increased proline biosynthesis and osmo-tolerance	Tobacco	Roosens et al., 2002
	Drought and salinity tolerance	Rice	Wu et al., 2003
GmDREB3	Enhances drought tolerance through proline accumulation	Soybean	Qi, 2012

pathways, genes involved in stress signal sensing are additionally useful for engineering stress-tolerant plants.

Stress tolerance via proline

Kishor et al. (2005) reported overproduction of proline in transgenic tobacco, and the transgenic plants produced enhanced root biomass under water stress. In one recent study, the *OsP5CS1* and *OsP5CS2* genes were co-expressed in tobacco that conferred transgenic plants increased proline accumulation and reduced oxidative damage to cells under abiotic stress conditions (Zhang et al., 2005). Similar proline production was also reported in P5CS-transgenic rice, wheat and carrot plants that showed tolerance to salt stress (Han et al., 2003; Sawahel et al., 2002; Zhu et al., 1998). Transgenic *Arabidopsis* plants that expressed the P5CS antisense gene were found with morphological abnormalities, and the plants were hypersensitive to osmotic stress, which was observed by Nanjo et al. (1999). In addition to its role in protecting vital proteins, it was proposed that proline would play a possible role in ROS (reactive oxygen species) scavenging; this was conceptualized by Smirnoff and Cumbes (1989). In transgenic *Arabidopsis p5cs* mutant lines, it was reported that the ROS scavenging enzymes showed significantly lower activities. This evidence suggested that proline either protects enzymes of the glutathione–ascorbate cycle or enhances their activities during osmotic stress. In order to enhance salt stress tolerance (Hmida-Sayari et al., 2005), the *Arabidopsis P5CS* gene was transferred to potato under a stress inducible promoter. The effect of its expression was observed in plant growth, tuber morphology and yield. Transgenic potato plants accumulated high proline content compared to control under high salt stress (100 mM NaCl) and, in turn, showed improved salt tolerance by reduced tuber yield and weight compared to that of non-transgenic control. In addition, some other studies were conducted on transgenic petunias and pigeon pea (*Cajanus cajan*) with the *P5CS* gene that conferred these plants drought and salt tolerance, respectively (Surekha et al., 2013). Petunia was transformed with pyrroline-5-carboxylate synthetase genes (*AtP5CS* from *A. thaliana* L. or *OsP5CS* from *Oryza sativa* L.). Transgenic plants accumulated more proline that resulted in drought tolerance for a period of 14 days and transformed pigeon pea with the mutagenized version (*P5CSF129A*) of wild *P5CS* gene from *Vigna aconitifolia*. The resultant transgenic plants accumulated more proline content than their non-transgenic plants. About four times higher proline content was observed in the T1 transgenic plants compared to that of non-transgenic under 200 mM NaCl stress (Surekha et al., 2013). As a result of comparatively high proline accumulation, the transgenic plants exhibited better growth, more chlorophyll and relative water content and lower levels of lipid peroxidation under salt stress. These findings suggest the important role of proline biosynthesis in transgenic plants against osmotic stress induced by salt and drought stresses.

Huang et al. (2013) examined HtP5CS, HtOAT and HtPDH enzyme activities, and gene expression patterns of putative *HtP5CS1, HtP5CS2, HtOAT, HtPDH1* and *HtPDH2* genes. The objective of our study was to characterize the proline regulation mechanisms of Jerusalem artichoke, a moderately salt-tolerant species, under NaCl stress. Jerusalem artichoke plantlets were observed to accumulate proline in roots, stems and leaves during salt stress. HtP5CS enzyme activities were increased under NaCl stress, while HtOAT and HtPDH activities generally repressed. Transcript levels of *HtP5CS2* increased, while transcript levels of *HtOAT, HtPDH1* and *HtPDH2* generally decreased in response to NaCl stress. Our results support that for Jerusalem artichoke, proline synthesis under salt stress is mainly through the Glu pathway, and *HtP5CS2* is predominant in this process, while *HtOAT* plays a less important role. Both *HtPDH* genes may function in proline degradation. Ashfaque et al. (2014) treated plants with H_2O_2, which significantly influenced the parameters both under non-saline and salt stress. The application of both 50 and 100 nM H_2O_2 reduced the severity of salt stress through the reduction in Na+ and Cl− content; and the increase in proline content and N assimilation. This resulted in increased water relations, photosynthetic pigments and growth under salt stress. However, maximum alleviation of salt stress was noted with 100 nM H_2O_2 and 50 nM H_2O_2 proved less effective. Under non-saline condition also application of H_2O_2 increased all the studied parameters. Szymon et al. (2015) showed the role of priming-induced modulation of activities of particular genes and enzymes of proline turnover, and its relationship with higher content of hydrogen peroxide, in improving seed germination under salinity stress. The accumulation of proline during priming and post-priming germination was associated with strong upregulation of the *P5CSA* gene, downregulation of the *PDH* gene and accumulation of hydrogen peroxide. The upregulated transcript level of *P5CSA* was consistent with the increase in P5CS activity, and the genes involving pyrroline-5-carboxylate synthetase (P5CS), ornithine-δ-aminotransferase (OAT) and proline dehydrogenase (PDH) were determined.

Conclusion

Plants exposed to salt stress undergo changes in their environment. The ability of plants to tolerate salt is determined by multiple biochemical pathways that facilitate retention and/or acquisition of water, to protect chloroplast functions and to maintain ion homeostasis. Essential pathways include those that lead to synthesis of osmotically active metabolites, which are collectively known as osmoprotectants. An important feature of osmoprotectants is that their beneficial effects are generally not species-specific, so that alien osmoprotectants can be engineered into plants and protect their new host, specific proteins, and certain free radical scavenging enzymes that control ion and water flux and support scavenging of oxygen radicals or chaperones. The ability of plants to detoxify radicals under conditions of salt stress is probably the most critical requirement. Many salt-tolerant species accumulate methylated metabolites, which play crucial dual roles as osmoprotectants and as radical scavengers. Their synthesis is correlated with stress-induced enhancement of photorespiration. In this chapter, plant responses to salinity stress are reviewed with emphasis on physiological, biochemical and molecular mechanisms of salt tolerance. This review may help in interdisciplinary studies to assess the ecological significance of salt stress. In naturally stress-tolerant plants there is a wide variety of adaptations to stress, many of which have not yet been identified at the molecular level. Understanding the function of such genes, determining these factors will improve our understanding of the complexity of plant metabolism and may provide unique opportunities for the metabolic engineer. Progress made in understanding the salt stress mechanism and improvement in development of salinity tolerant sugarcane crop.

Acknowledgements

The authors express their acknowledgments to the Woman Scientist Scheme-A, Department of Science and Technology, Ministry of Science and Technology, and SVP University of Agriculture and Technology (Meerut, India) for the financial and technical support to conduct activities on research program.

References

Anoop N, Gupta AK (2003) Transgenic indica rice cv IR-50 over-expressing *Vigna aconitifolia* Δ^1-pyrroline-5-carboxylate synthetase cDNA shows tolerance to high salt. *J Plant Biochem Biotechnol* 12: 109–116.

Ashfaque F, Khan MIR, Khan NA (2014) Exogenously applied H2O2 promotes proline accumulation, water relations, photosynthetic efficiency and growth of wheat (*Triticum aestivum* L.) under salt stress. *Annu Res Rev Biol* 4: 105–120.

Ayliffe MA, Mitchell HJ, Deuschle K, Pryor AJ (2005) Comparative analysis in cereals of a key proline catabolism gene. *Mol Genet Genomics* 274: 494–505.

Bajaj S, Mohanty A (2005) Recent advances in rice biotechnology towards genetically superior transgenic rice. *Plant Biotech J* 3: 275–307.

Ballantyne JS, Chamberlin ME (1994) Regulation of cellular amino acid levels. In *Cellular and molecular physiology of cell volume regulation*, K Strange (ed), 111–122. Boca Raton, FL: CRC Press.

Barnett NM, Naylor AW (1966) Amino acid and protein metabolism in Bermuda grass during water stress. *Plant Physiol* 41: 1222–1230.

Boggess SF, Aspinall D, Paleg LG (1976) Stress metabolism. IX. The significance of end-product inhibition of proline biosynthesis and of compartmentation in relation to stress-induced proline accumulation. *Aust J Plant Physiol* 3: 513–525.

Boggess SF, Paleg LG, Aspinall D (1975) Δ^1-Pyrroline-5-carboxylic acid dehydrogenase in barley, a proline-accumulating species. *Plant Physiol* 56: 259–262.

Bohnert HJ, Shen B (1999) Transformation and compatible solutes. *Sci Hortic* 78: 237–260.

Briens M, Larher F (1982) Osmoregulation in halophytic higher plants: A comparative study of soluble carbohydrates, polyols, betaines and free proline. *Plant, Cell Environ* 5: 287–292.

Choudhary NL, Sairam RK, Tyagi A (2005) Expression of delta1-pyrroline-5-carboxylate synthetase gene during drought in rice (*Oryza sativa* L.). *Ind J Biochem Biophys* 42: 366–370.

De Ronde JA, Cress WA, Kruger GHJ, Strasser RJ, Van Staden, J (2004) Photosynthetic response of transgenic soybean plants, containing an *Arabidopsis P5VR* gene, during heat and drought stress. *J Plant Physiol* 161: 1211–1224.

De Ronde JA, Spreeth MH, Cress WA (2000) Effect of antisense-Δ^1-pyrroline-5-carboxylate reductase transgenic soybean plants subjected to osmotic and drought stress. *Plant Growth Regul* 32: 13–26.

Delauney AJ, Verma DPS (1993) Proline biosynthesis and osmoregulation in plants. *Plant J* 4: 215–223.

Deuschle K, Funck D, Hellmann H, Diischner K, Binder S, Frommer WB (2001) A nuclear gene encoding mitochondrial Δ^1-pyrroline-5-carboxylate dehydrogenase and its potential role in protection from proline toxicity. *Plant J* 27: 345–355.

Elthon TE, Stewart CR (1981) Submitochondrial location and electron transport characteristics of enzymes involved in proline oxidation. *Plant Physiol* 67: 780–784.

Elthon TE, Stewart CR (1982) Proline oxidation in corn mitochondria: Involvement of NAD, relationship to ornithine metabolism, and sidedness on the inner membrane. *Plant Physiol* 70: 567–572.

Flowers TJ, AR Yeo (1995) Breeding for salinity resistance in crop plants. Where next? *Aust J Plant Physiol* 22: 875–884.

Forlani G, Scainelli D, Nielsen E (1997) Two Ll,'-pyrroline-5-carboxylate dehydrogenase isoforms are expressed in cultured *Nicotiana plumbaginifolia* cells and are differentially modulated during the culture growth cycle. *Planta* 202: 242.

Funck D, Stadelhofer B, Koch W (2008) Ornithine-a-aminotransferase is essential for arginine catabolism but not for proline biosynthesis. *BMC Plant Bio* 8: 40.

Glesson D, Lelu-Walter MA, Parteinson M (2005) Overproduction of proline in transgenic hybrid larch cultures renders them tolerant to cold, salt and frost. *Mol Breed* 15: 21–29.

Gsonka LN (1989) Physiological and genetic responses of bacteria to osmotic stress. *Micorbiol Rev* 53: 121–147.

Ha KH, Hwang CH (2003) Salt tolerance enhanced by transformation of a *P5CS* gene in carrot. *J Plant Biotechnol* 5: 149–153.

Hanson AD, Hitz WD (1982) Metabolic responses of mesophytes to plant water deficits. *Annu Rev Plant Physiol* 33: 163–203.

Hare PD, Cress WA, Van Staden J (1998). Dissecting the roles of osmolyte accumulation in plants. *Plant Cell Environ* 21: 535–553.

Hellebust JA (1976). Osmoregulation. *Annu Rev Plant Physiol* 27: 485–505.

Hmida-Sayari A, Gargouri-Bouzid R, Bidani A, Jaoua L, Savoure A, Jaoua S (2005) Overexpression of D1-pyrroline-5-carboxylate synthetase increases proline production and confers salt tolerance in transgenic potato plants. *Plant Sci* 169: 746–752.

Hong Z, Lakkineni K, Zhang Z, Verma DPS (2000) Removal of feedback inhibition of pyrroline-5-carboxylate synthetase results in increased proline accumulation and protection of plants from osmotic stress. *Plant Physiol* 122: 1129–1136.

Hong-Qi Z, Croes AF, Linskens HF (1982) Protein synthesis in germinating pollen of *Petunia*: Role of proline. *Planta* 154: 199–203.

Hu CA, Delauney AJ, Verma DP (1992) A bifunctional enzyme (delta 1-pyrroline-5-carboxylate synthetase) catalyzes the first two steps in proline biosynthesis in plants. *Proc Natl Acad Sci USA* 89(19): 9354–9358.

Huang Z, Zhao L, Chen D, Liang M, Liu Z, Shao H, Long X (2013) Salt stress encourages proline accumulation by regulating proline biosynthesis and degradation in Jerusalem artichoke plantlets. *PLoS One* 8(4): e62085.

Hur J, Hong Jong K, Lee C-H, An G (2004) Stress-inducible *OsP5CS2* gene is essential for salt and cold tolerance in rice. *Plant Sci* 167: 417–426.

Ibragimova SS, Kolodyazhnaya YS, Gerasimova SV, Kochetov AV (2012) Partial suppression of gene encoding proline dehydrogenase enhances plant tolerance to various abiotic stresses. *Russ J Plant Physiol* 59(1): 88–96.

Ishitani M, Arakawa K, Mizuno K, Kishitani S, Takabe T (1993) Betaine aldehyde dehydrogenase in the Gramineae: Levels in leaves of both betaine-accumulating and nonaccumulating cereal plants. *Plant Cell Physiol* 34: 493–495.

Jones MM, Osmond CB, Turner NC (1980). Accumulation of solutes in leaves of sorghum and sunflower in response to water deficits. *Aust J Plant Physiol* 7: 193–205.

Kishor PBK, Hong Z, Miao G, Hu C-AA, Verma, DPS (1995) Over expression of Δ^1-pyrroline-5-carboxylate synthetase increases proline overproduction and confers osmtolerance in transgenic plants. *Plant Physiol* 108, 1387–1394.

Kishor PBK, Sangam S, Amrutha RN, Laxmi PS, Naidu KR, Rao KRSS, Rao S, Reddy KJ, Theriappan P, Sreenivasulu N (2005) Regulation of proline biosynthesis, degradation, uptake and transport in higher plants: Its implications in plant growth and abiotic stress tolerance. *Curr Sci* 88: 424–438.

Ketchum REB, Warren RC, Klima LJ, Lopez-Gutierrez F, Nabors MW (1991) The mechanism and regulation of proline accumulation in suspension cultures of the halophytic grass *Distichlis spicata* L. *J Plant Physiol* 137: 368–374.

Kiyosue T, Yoshiba Y, Yamaguchi-Shinozaki K, Shinozaki K (1996) A nuclear gene encoding mitochondrial proline dehydrogenase, an enzyme involved in proline metabolism, is upregulated by proline but downregulated by dehydration in *Arabidopsis*. *Plant Cell* 8: 1323–1335.

Kochetov AV, Kolodyazhnaya YS, Komarova ML, Koval VS, Makarova NN, Lysinkyi YY, Trifonova EA, Shumny VK (2004) Tobacco transformants bearing antisense suppressor of proline dehydrogenase gene are characterized by higher proline content and cytoplasm osmotic pressure. *Genetika* 40: 282–285.

Kolodyazhnaya YS, Titov SE, Kochetov AV, Komarova ML, Romanova AV, Koval VS, Shumny VK (2006) Evaluation of salt tolerance in *Nicotiana tabacum* plants bearing an antisense suppressor of the proline dehydrogenase gene. *Genetika* 42: 278–281.

Kumar V, Shriram V, Nikam TD, Jawali N, Shitole MG (2008) Sodium chloride induced changes in mineral elements in indica rice cultivars differing in salt tolerance. *J Plant Nutr* 31(11): 1999–2017.

Kushiro T, Okamoto M, Nakabayashi K, Yamagishi K, Kitamura S et al. (2004) The *Arabidopsis* cytochrome P450 CYP707A encodes ABA 8'-hydroxylases: Key enzymes in ABA catabolism. *EMBO J* 23: 1647–1656.

Kuznetsov VV, Shevyakova NI (1999). Proline under stress: Biological role, metabolism, and regulation. *Russ J Plant Physiol* 46: 274–287.

LaRosa PC, Rhodes D, Rhodes JC, Bressan RA, Csonka LN (1991) Elevated accumulation of proline in NaCl-adapted tobacco cells is not due to altered Δ^1-pyrroline-5-carboxylate reductase. *Plant Physiol* 96: 245–250.

Leigh RA, Ahmad N, Wyn Jones RG (1981) Assessment of glycine betaine and proline compartmentation by analysis of isolated beet vacuoles. *Planta* 153: 34–41.

Leisinger T (1987) Biosynthesis of proline. In: *Escherichia coli* and *Salmonella typhimurium: Cellular and molecular biology*, FC Neidhardt, JL Ingraham, KB Low, B Magasanik, M Schaechter, HE Umbarger (eds), 346–351. Washington, DC: American Society for Microbiology.

Liu J, Ishitani M, Halfter U, Kim CS, Zhu JK (2000) The *Arabidopsis thaliana* SOS2 gene encodes a protein kinase that is required for salt tolerance. *Proc Nat Acad Sci USA* 97: 3730–3734.

Low PS (1985) Molecular basis of the biological compatibility of nature's osmolytes. In: *Transport processes, iono- and osmoregulation*, R Gilles, M Gilles-Baillien (eds), 469–477. Berlin: Springer-Verlag.

Mani S, Van de Cotte B, Van Montagu M, Verbruggen N (2002) Altered levels of proline dehydrogenase cause hypersensitivity to proline and its analogs in *Arabidopsis*. *Plant Physiol* 128: 73–83.

McCue KF, Hanson AD (1990) Drought and salt tolerance: Towards understanding and application. *Trends Biotechnol* 8: 358–362.

Miller G, Honig A, Stein H, Suzuki N, Mittler R, Zilberstein A (2009) Unraveling Δ^1-pyrroline-5-carboxylate-proline cycle in plants by uncoupled expression of proline oxidation enzymes. *J Biol Chem* 284(39): 26482–26492.

Mitchell HJ, Ayliffe MA, Rashid KY, Pryor AJ (2006) A rust-inducible gene from flax (fis 1) is involved in proline catabolism. *Planta* 223: 213–222.

Molinaria HBC, Marura CJ, Filhoa Joao CB, Kobayashib AK, Pileggic M, Juniora RPL, Pereirad LFP, Vieira LGE (2004) Osmotic adjustment in transgenic citrus rootstock Carrizo citrange (*Citrus sinensis* Osb. × *Poncirus trifoliate* L. Raf.) overproducing proline. *Plant Sci* 167: 1375–1381.

Monteoliva MI, Rizzi YS, Cecchini NM, Hajirezaei MR, Alvarez ME (2014) Context of action of proline dehydrogenase (*Pro*DH) in the hypersensitive response of *Arabidopsis*. *BMC Plant Biol* 14: 21–32.

Munns R, Tester M (2008) Mechanisms of salinity tolerance. *Annu Rev Plant Biol* 59: 651–681.

Nanjo T, Kobayashi M, Yoshiba Y, Sanada Y, Wada K, Tsukaya H, Kakubari Y, Yamaguchi-Shinozaki K, Shinozaki K (1999) Biological functions of proline in morphogenesis and osmotolerance revealed in antisense transgenic *Arabidopsis thaliana*. *Plant J* 18: 185–193.

Nanjo T, Kobayashi M, Yoshiba Y, Kakubari Y, Yamaguchi-Shinozaki K, Shinozaki K (1999) Antisense suppression of proline degradation improves tolerance to freezing and salinity in *Arabidopsis thaliana*. *FEBS Lett* 461: 205–210.

Nash D, Paleg LG, Wiskich JT (1982) Effect of proline, betaine and some other solutes on the heat stability of mitochondrial enzymes. *Aust J Plant Physiol* 9: 47–57.

Pahlich E, Kerres R, Jager HJ (1983) Influence of water stress on the vacuole/extravacuole distribution of proline in protoplasts of *Nicotiana rustica*. *Plant Physiol* 72: 590–591.

Paleg LG, Douglas TJ, van Daal A, Keech DB (1981) Proline, betaine and other organic solutes protect enzymes against heat inactivation. *Aust J Plant Physiol* 8: 107–114.

Peng Z, Lu Q, Verma DP (1996) Reciprocal regulation of delta^1-pyrroline-5-carboxylate synthetase and proline dehydrogenase genes control proline levels during and after osmotic stress in plants. *Mol Gen Genet* 253: 334–341.

Pollard A, Wyn Jones RG (1979) Enzyme activities in concentrated solutions of glycinebetaine and other solutes. *Planta* 144: 291–298.

Qi Q (2012) Effect of transgenic DREB3 drought resistant soybean on soil enzyme activity and soil functional microorganism. Master's thesis, Northeast Agricultural University, Harbin, China.

Rayapati PJ, Stewart CR (1991) Solubilization of a proline dehydrogenase from maize (*Zea Mays* L.) mitochondria. *Plant Physiol* 95: 787–791.

Rhodes D (1987) Metabolic responses to stress. In: *The biochemistry of plants*, vol 12, DD Davies (ed), 201–241. New York: Academic Press.

Rhodes D, Handa S (1989) Amino acid metabolism in relation to osmotic adjustment in plant cells. In: *Environmental stress in plants: Biochemical and physiological mechanisms*, NATO ASI Series, vol G19, JH Cherry (ed), 41–62. Berlin: Springer.

Rhodes D, Verslues PE, Sharp RE (1999). Role of amino acids in abiotic stress resistance. In: *Plant amino acids: Biochemistry and biotechnology*, BK Singh (ed), 319–356. New York: Marcel Dekker.

Ribarits A, Abdullaev A, Tashpulatov A, Richter A, Heberle-Bors E, Touraev A (2007) Two tobacco proline dehydrogenases are differentially regulated and play a role in early plant development. *Planta* 225: 1313–1324.

Rodriguez R, Redman R (2005) Balancing the generation and elimination of reactive oxygen species. *PNAS* 102: 3175–3176.

Roosens NH, Bitar FA, Loenders K, Angenon G, Jacobs M (2002) Over expression of ornthine-*d*-aminotransferase increases proline biosynthesis and confers osmo-tolerance in transgenic plants. *Mol Breed* 9: 73–80.

Roxas VP, Smith RK Jr, Allen ER, Allen RD (1997) Overexpression of glutathione S-transferase/glutathione peroxidase enhances the growth of transgenic tobacco seedlings during stress. *Nat Biotech* 15: 988–991.

Sairam RK, Srivastava GC, Agarwal S, Meena RC (2005) Differences in response to salinity stress in tolerant and susceptible wheat genotypes. *Biol Plant* 49(1): 85–91.

Sairam RK, Tyagi A (2004) Physiology and molecular biology of salinity stress tolerance in plants. *Curr Sci* 86(3): 407–421.

Sawahel WA, Hassan AH (2002) Generation of transgenic wheat plants producing high levels of the osmoprotectant proline. *Biotechnol Lett* 24: 721–725.

Serrano R, Mulet JM, Rios G, Marquez JA, Larrinoa IF, Leube MP et al. (1999) A glimpse of the mechanisms of ion homeostasis during salt stress. *J Exper Bot* 50: 1023–1036.

Shen B, Hohmann S, Jensen RG, Bohnert HJ (1999) Roles of sugar alcohols in osmotic stress adaptation. Replacement of glycerol by mannitol and sorbitol in yeast. *Plant Physiol* 121: 45–52.

Siripornadulsil S, Traina S, Verma DPS, Sayre RT (2002) Molecular mechanisms of proline-mediated tolerance to toxic heavy metals in transgenic microalgae. *Plant Cell* 14: 2837–2847.

Smirnoff N, Cumbes QJ (1989) Hydroxyl radical scavenging activity of compatible solutes. *Phytochemistry* 28: 1057–1060.

Somero GN (1986). Protons, osmolytes, and fitness of internal milieu for protein function. *Am J Physiol* 251: R197–R213.

Srinivas V, Balasubramanian D (1995) Proline is a protein-compatible hydrotrope. *Langmuir* 11: 2830–2833.

Stephanopoulos G (1999) Metabolic fluxes and metabolic engineering. *Metabol Eng* 1: 1–11.

Stewart CR (1981) Proline accumulation: Biochemical aspects. In: *Physiology and biochemistry of drought resistance in plants,* LG Paleg, D Aspinall (eds), 243–259. Sydney: Academic Press.

Stewart GR, Larher F (1980). Accumulation of amino acids and related compounds in relation to environmental stress. In: *The biochemistry of plants,* vol 5, BJ Miflin (ed), 609–635. New York: Academic Press.

Stewart GR, Lee JA (1974) The role of proline accumulation in halophytes. *Planta* 120: 279–289.

Stines AP, Naylor DJ, Hoj PB, van Heeswijck R (1999) Proline accumulation in developing grapevine fruit occurs independently of changes in the levels of Δ1-pyrroline-5-carboxylate synthetase mRNA or protein. *Plant Physiol* 120: 923–931.

Su J, Wu R (2004) Stress-inducible synthesis of proline in transgenic rice confers faster growth under stress conditions than that with constitutive synthesis. *Plant Sci* 166: 941–948.

Surekha CH, Nirmala Kumari K, Aruna LV, Suneetha G, Arundhati A, Kavi Kishor PB (2013) Expression of the Vigna aconitifolia P5CSF129A gene in transgenic pigeon pea enhances proline accumulation and salt tolerance. *Plant Cell Tissue Organ Cult* 116: 27–36.

Szymon K, Lukasz Q, Katarzyna L, Stanley L, Malgorzata G (2015) Enhanced expression of the proline synthesis gene *P5CSA* in relation to seed osmopriming improvement of *Brassica napus* germination under salinity stress. *J Plant Physiol* 183: 1–12.

Tal M, Katz A (1980) Salt tolerance in the wild relatives of the cultivated tomato: The effect of proline on the growth of callus tissue of *Lycopersicon esculentum* and *L. peruvianum* under salt and water stress. *Z Pflanzenphysiol Bd* 98: 283–288.

Tanner J (2008) Structural biology of proline catabolism. *Amino Acids* 35: 719–730

Taylor CB (1996) Proline and water deficit: Ups and downs. *Plant Cell* 8: 1221–1224.

Thompson JF (1980) Arginine synthesis, proline synthesis, and related processes. In: *The biochemistry of plants,* vol 5, BJ Miflin (ed), 375–403. New York: Academic Press.

Timasheff SN (1993) The control of protein stability and association by weak interactions with water: How do solvents affect these processes? *Annu Rev Biophys Biomol Struct* 22: 67–97.

Treichel S (1986). The influence of NaCl on delta^1-pyrroline-5-carboxylate reductase in proline-accumulating cell suspension cultures of *Mesembryanthemum nodiflorum* and other halophytes. *Plant Physiol* 67: 173–181.

Ueda A, Shi W, Shimada T, Miyake H, Takabe T (2008). Altered expression of barley proline transporter causes different growth responses in *Arabidopsis*. *Planta* 227(2): 277–286.

Vendruscoloa ECG, Schusterb I, Pileggic M, Scapimd CA, Molinarie HBC, Marure CJ, Vieirae LGE (2007). Stress-induced synthesis of proline confers tolerance to water deficit in transgenic wheat. *J Plant Physiol* 164: 1367–1376.

Verbruggen N, Hua XJ, May M, Van Montagu M (1996) Environmental and developmental signals modulate proline homeostasis: Evidence for a negative transcriptional regulator. *Proc Natl Acad Sci* 93: 8787–8791.

Verbruggen N, Villarroel R, Van Montagu M (1993) Osmoregulation of a pyrroline-5-carboxylate reductase gene in *Arabidopsis thaliana*. *Plant Physiol* 103(3): 771–781.

Verma DPS (1999) Osmotic stress tolerance in plants: Role of proline and sulfur metabolisms. In: *Molecular responses to cold, drought, heat and salt stress in higher plants*, K Shinozaki, K Yamaguchi-Shinozaki (eds), 153–168. Austin, Texas: Landes Company.

Voetberg GS, Sharp RE (1991) Growth of the maize primary root tip at low water potentials. III. Role of increased proline deposition in osmotic adjustment. *Plant Physiol* 96: 1125–1130.

Widodo, Patterson JH, Newbigin E, Tester M, Bacic A, Roessner U (2009) Metabolic responses to salt stress of barley (*Hordeum vulgare* L.) cultivars, Sahara and Clipper, which differ in salinity tolerance. *J Exp Bot* 60: 4089–4103.

Wu LQ, Fan ZM, Guo L, Li YQ, Zhang WJ, Qu LJ, Chen ZL (2003) Over-expression of an *Arabidopsis delta-OAT* gene enhances salt and drought tolerance in transgenic rice. *Chin Sci Bull* 48: 2594–2600.

Wu SJ, Ding L, Zhu JK (1996) SOS1 a genetic locus essential for salt tolerance and potassium acquisition. *Plant Cell* 8: 617–627.

Wyn Jones RG, Storey R, Leigh RA, Ahmad N, Pollard A (1977) A hypothesis on cytoplasmic osmoregulation. In *Regulation of cell membrane activities in plants*, E Marre, O Cifferi (eds), 121–136. Amsterdam: Elsevier.

Yancey PH (1994) Compatible and counteracting solutes. In *Cellular and molecular physiology of cell volume regulation*, K Strange (ed), 81–109. Boca Raton, FL: CRC Press.

Yancey PH, Clark ME, Hand SC, Bowlus RD, Somero GN (1982) Living with water stress: Evolution of osmolyte systems. *Science* 217: 1214–1222.

Yamaguchi-Shinozaki K, Shinozaki K (2006) Transcriptional regulatory networks in cellular responses and tolerance to dehydration and cold stresses. *Annu Rev Plant Biol* 57: 781–803.

Zhang C-S, Lu Q, Verma DPS (1995) Removal of feedback inhibition of Δ^1-pyrroline-5-carboxylate synthetase, a bifunctional enzyme catalyzing the first two steps of proline biosynthesis in plants. *J Biol Chem* 270: 20491–20496.

Zhang X, Tang W, Liu J, Liu Y (2014) Co-expression of rice OsP5CS1 and OsP5CS2 genes in transgenic tobacco resulted in elevated proline biosynthesis and enhanced abiotic stress tolerance. *Chin J Appl Environ Biol* 717–722.

Zhu B, Su J, Chang M, Verma DPS, Fan YL, Wu R (1998) Overexpression of a Δ^1-pyrroline-5-carboxylate synthetase gene and analysis of tolerance to water-and salt-stress in transgenic rice. *Plant Sci* 139: 41–48.

Index

Page numbers followed by f and t indicate figures and tables, respectively.

Index

Index

Printed in the United States
by Baker & Taylor Publisher Services